The
EVERGLADES HANDBOOK

Understanding the Ecosystem

Second Edition

Thomas E. Lodge

CRC PRESS

Boca Raton   London   New York   Washington, D.C.

Library of Congress Cataloging-in-Publication Data

Lodge, Thomas E., 1943–
  The Everglades handbook : understanding the ecosystem / Thomas E. Lodge. -- 2nd ed.
    p. cm.
  Includes bibliographical references (p.).
  ISBN 1-56670-614-9 (alk. paper)
    1. Swamp ecology--Florida--Everglades. 2. Swamp animals--Florida--Everglades. 3.
  Swamp plants--Florida--Everglades. 4. Everglades (Fla.) 5. Ecosystem
  management--Florida--Everglades. I. Title.

  QH105.F6L63 2004
  577.68′09759′39--dc22   2004045726

Visit the CRC Web site at www.crcpress.com

© 2005 by CRC Press LLC

No claim to original U.S. Government works
International Standard Book Number 1-56670-614-9
Library of Congress Card Number 2004045726
Printed in the United States of America 1 2 3 4 5 6 7 8 9 0
Printed on acid-free paper

Dedication

To the memory of:

Thomas Sebree Baskett
(January 23, 1916 – December 6, 1999)
Sustaining my family's "Missoura" roots in his USFWS/University
of Missouri career,
my mentor of high standards – and uncle.

William B. Robertson, Jr.
(August 22, 1924 – January 28, 2000)
A man of birds and fire, in his guidance to countless
Everglades scientists,
"Dr Bill" lives on in our memory and respect.

Peter C. Rosendahl
(March 6, 1944 – September 30, 2001)
Friend, fellow student at UM, recommendation that finally
got me employed,
and inspiration to publish the first edition of this book –
many thanks, Pete.

Catherine Hauberg Sweeney
(April 11, 1914 – January 25, 1995)
My introduction to the world of David Fairchild,
and nurturing as if I were family – thank you, Kay.

My guiding light:

In the end, we conserve only what we love. We love only what we understand. We understand only what we are taught.

Baba Dioum
Senegalese poet

COVER PHOTOS

Front cover:

Left center: A great white heron, a tropical species in the Everglades, takes a golden shiner, a fish of temperate North America. The scene depicts two unifying themes of this book: *ecosystem functions* represented by a wading bird eating a fish — an important indicator of ecosystem health; and *biogeography*, showing the juxtaposition of tropical and temperate influences that characterize the plants and animals of the Everglades (Photo by R. Hamer).

Clockwise from upper left — Four other aspects of the Everglades emphasized in this book: (1) mangrove swamps and tidal rivers in Everglades National Park looking southwest along Roberts River (at right) toward Whitewater Bay. This estuarine habitat was historically the breeding grounds of Everglades wading birds, and is still critical to fisheries of the Gulf of Mexico; (2) the flower of the cardinal airplant near Flamingo, Everglades National Park, one of the myriad of tropically related plants that characterize the Everglades region; (3) ridge-and-slough landscape with a tree island, the flow-related signature of the heart of the original Everglades, here preserved in Water Conservation Area 3A; and (4) an alligator feeding in rushing water at the Anhinga Trail, Everglades National Park. The alligator is a keystone species based on its role in enhancing Everglades food chains that support the ecosystem's suite of top predators — the wading birds. (Photos by T. Lodge)

Back cover:

Tom Lodge, Taylor Alexander, and Fred Dayhoff, relaxing after discussing Fred's experiences with the '47 hurricane, gator poachers, and other adventures.

Contents

PART I Background

PART II Environments of the Everglades Region

PART III   The Flora and Fauna of Southern Florida

PART IV   ENVIRONMENTAL IMPACTS

Contents

List of Maps, Diagrams, and Graphs

List of Color Figures

(Color insert follows page 94)

List of Tables

Acknowledgments

Upon the outset of revising the first edition, I thought the job would be easy, even in the face of the greatly expanded literature. I have been wrong before, but this assumption was one of my worst. Not only was the work difficult far beyond my estimation, but I needed considerable outside technical and editorial advice as well as encouragement.

The two new landscape chapters, The Big Cypress Swamp (Chapter 7) and Lake Okeechobee and the Everglades Headwaters (Chapter 11), required the help of several experts. Taylor Alexander, Fred and Sandy Dayhoff, and Allyn Golub provided assistance in supplying information and in reviewing the final Big Cypress chapter. Clyde Butcher contributed the photo for Figure 7.3, facilitated by Allyn Golub, and Jeff Ripple provided the photo for Figure 7.5. For Chapter 11, Paul Gray of the National Audubon Society and Karl Havens of the South Florida Water Management District critiqued text and other technical material, and Susan Gray and Tom James of the District helped supply data. Larry Harris of Friends of Lake Okeechobee provided data display ideas.

Numerous people provided information or reviews for Chapter 1 (geology), including Kevin Cunningham of USGS and Brad Waller of Hydrologic Associates USA. Pattie Gertenbach of E Sciences and Michael Grunwald of *The Washington Post* provided reviews of the entire chapter manuscript.

I would also like to thank Wes Jurgens, who modified the art work for Figures 3.1 and 3.9, Hal Wanless for coastal processes and sea-level-rise data, David Black for botany, Tiffany Gann for southeastern Everglades tree island characteristics, Scott Zona for critique of text on palms, Frank Mazzotti and Walter Meshaka for comments on reptiles, Christina Ugarte for information on amphibian food chains, Bob Doren for guidance on regulations regarding introduced species, Steve Carney for his critique on the invertebrate and synthesis chapters, Joe Zigler and Grenville Barnes for guidance on NGVD/NAVD, Bob Carr and Sally Channon for critique of the text on Native Americans, Peter Frederick for mercury and wading bird publications, Michelle Diffenderfer for critique of text regarding the Seminole Tribe, Don DeAngelis for supplying the ATLSS graphic and caption for Color Figure 7, Alice Clarke for critique of house cat impacts, and Jay Slack of the U.S. Fish and Wildlife Service and Beth Carlson Lewis for their critiques of the text on the Endangered Species Act.

The South Florida Water Management District supplied two color figures — Chris McVoy and Winnie Said facilitated obtaining the historic Everglades graphic, Color Figure 2; and Gary Goforth supplied the EAA map, Color Figure 6.

I appreciate Bob Hamer's effort providing the center cover photograph and four other photographs from the first edition. I am still indebted to Bob for his idea to do a book on the Everglades in the first place, dating from our joint photographic trips of the early 1970's.

I would like to express gratitude to my previous employer, EA Engineering, Science, and Technology. Nearly the entire second edition was written while I worked for EA. Numerous coworkers assisted. Dan Levy and Danny Spector critiqued Chapter 1. Suzie Boltz coordinated data reduction with Anita Corum for Figure 11.8. Tina Roberts critiqued text on the Endangered Species Act and Ted Couillard edited the several plant lists and compiled the data for Figure 21.7. Loren Jensen, CEO, generously offered company graphics support, provided by Jennifer Zoukis, who often worked from my crude, hand-drawn sketches, but also improved detailed maps and diagrams to meet publication standards.

Finally, I am indebted to Marilyn Paris, companion and long-time friend. She compiled the hurricane tracks for Figure 9.1, and provided tireless referencing, editing, indexing and encouragement for the final and toughest year of the effort to complete this second edition.

Authors

Thomas E. Lodge, Ph.D., is a self-employed ecologist. During most of the preparation of this book, he was a Senior Ecologist with the Miami office of Baltimore-based EA Engineering, Science, and Technology, Inc., an environmental consulting firm. Born in Cleveland, Ohio, he has a B.A. with a major and Departmental Honors in Zoology from Ohio Wesleyan University (1966) and a Ph.D. in biology from the University of Miami in Florida (1974). Nurturing his childhood interest in aquatic biology, he worked part-time for the Cleveland Museum of Natural History during high school and college. There, he became an expert on the fishes of northern Ohio and contributed extensively to the museum's fish collection and to its summer educational programs. In graduate school, Dr. Lodge became fascinated with the Everglades, both academically and personally. In addition to publishing magazine articles on the Everglades, he wrote and directed an educational film ("The Everglades Region, An Ecological Study," John Wiley & Sons, 1973), and published on the fishes of the region. After receiving his Ph.D., he became an environmental consultant, specializing in wetlands and aquatic environments. From 1996 to 1998, he was Director of The Kampong, a botanical garden and former home of Dr. David Fairchild. He has led numerous environmental projects directly relating to the Everglades, including the development of methodology for evaluating the ecological function of historic Everglades wetlands. He serves on the Board of Directors of the Tropical Audubon Society and is an appointed member of the U.S. Fish and Wildlife Service's "Multi-Species Ecosystem Restoration Team," which assists in Everglades restoration strategies dealing with listed species. He has also been an Invited Faculty, teaching South Florida Ecology at Florida International University, where the first edition of *The Everglades Handbook* has been used regularly as a course text. His personal interest in the region has outweighed his professional activities. For more than 30 years, he has been a regular observer and photographer of Everglades wildlife.

Marjory Stoneman Douglas (April 7, 1890 — May 14, 1998) is the single name that has become permanently linked to the Everglades. She held a B.A. degree from Wellesley College (1912), and seven honorary doctorates, including Litt. D. from the University of Miami, and LL.D. from Barry University. She arrived in Miami in 1915 to work with her father, founder and editor of *The Miami Herald*. In the 1920s, she joined the movement to create Everglades National Park, working with Ernest Coe, David Fairchild, and many others (see Introduction). Subsequently, her authority in conservation — wilderness, wildlife, and water alike — grew continuously stronger. In 1970, she formed the Friends of the Everglades, an organization in which she remained active well into the 1990's. Her petite, five-foot-one-inch frame abruptly contrasted with her stature in the environmental arena where she was internationally recognized as a giant by friends and adversaries alike. In November 1993, President Clinton awarded her the highest honor given to civilians by the United States, the Presidential Metal of Freedom; and on October 7, 2000, she was posthumously inducted into the National Women's Hall of Fame. But in spite of her early frustrations and confrontations with a male-dominated world, she ultimately said, "I'd like to hear less talk about men and women and more talk about citizens."

A writer for all of her professional life, her best-known book is *The Everglades: River of Grass*, originally published in 1947 and now in its "50th Anniversary Edition" (1997). Her other books include *Florida: The Long Frontier, Alligator Crossing, Hurricane, Freedom River, Adventures in a Green World—the Story of David Fairchild and Barbour Lathrop, Nine Florida Stories by Marjory Stoneman Douglas*, and her autobiography, *Marjory Stoneman Douglas: Voice of The River*.

Marjory Stoneman Douglas' ashes are scattered in the Everglades, her "river of grass."

Taylor R. Alexander was born in Hope, Arkansas, on May 27, 1915. He spent his early life working on a small ornamental plant farm until he graduated from high school. He received his A.B. degree with a major in biology from Ouachita College, Arkadelphia, Arkansas (1936), his M.S. degree in plant physiology from the University of Chicago (1938), and his Ph.D. in plant physiology and ecology from the University of Chicago (1941).

Dr. Alexander was associated with the University of Miami from September 1940 until his retirement in June 1977, as Botanist (1940–47), Professor (1947), and Chairman (1948–1965) of the Department of Botany, and as Professor of Botany (1965–1977) following merger of the botany and zoology departments into the Department of Biology. During World War II, he took temporary leave from the university to train soldiers headed for battle in various areas including health and procedures for poisonous-gas protection. He authored an identification manual for distinguishing friendly aircraft and ships from those of the enemies for soldiers headed for battle. Following retirement from the University of Miami, he worked for the consulting company, Tropical Bioindustries, for several years.

Some of Dr. Alexander's professional associations have been:

* Member of the Governor's Committee on Natural Resources (1967–70)
* President of Florida Academy of Sciences (1970–71)
* Advisory Committee to Florida's Endangered Lands Purchasing Program under Governor Kirk
* Florida Big Cypress Oil Well Site Advisory Committee (1971–1984)
* Metropolitan Dade County Environmental Quality Control Board (1981–1985)

Dr. Alexander's most important contribution to our understanding of the Everglades has centered around the effects of various environmental factors, notably fire and hydrology, on plant community succession. He authored two books, *Botany* (Golden Press, 1970) and *Ecology* (Golden Press, 1973). *Botany* gained acclaim when Harvard botany graduate students used it in preparing for comprehensive examinations All of his periodical publications and special reports pertaining to South Florida are listed in this book (see references 8–28, 163, 176, and 586). In 2002 he received Florida Native Plant Society's Mentor Award. He published some of the first detailed descriptions of Florida's plant communities. Dr. Alexander is one of Florida's outstanding and earliest environmentalists. He still resides in South Miami in the house he and his late wife, Edith, built in 1947. The house was designed by Marian Manley, who lived with Marjory Stoneman Douglas.

Preface

THE FIRST EDITION

The first edition of this book was published by St. Lucie Press in May 1994. The book had been conceived in 1982 as a large-format color photographic book on the Everglades. The intent of the text was not only to accompany the excellent Everglades wildlife and habitat photographs by Robert Hamer, with whom I had conspired since 1971, but was also to provide a more comprehensive coverage of the Everglades than available in other "coffee-table" publications. He and I both enjoyed "escaping" to the Everglades for photography and wildlife observation, so the project initially was much more play than work. The effort to find a publisher for the large-format book failed, but through the several years in that effort, I continued revising the text. Then, at the Everglades Coalition meetings in Tallahassee in early 1993, I took a suggestion from friend, Pete Rosendahl, to move ahead and publish the text (requiring extensive revision), as a service to the environmental community for its understanding of the ecosystem — needed for the growing momentum of Everglades restoration. My approach came from my then 27 years of experience as a biologist in southern Florida, first as a graduate student and then as a professional environmental consultant in projects that often dealt directly with Everglades ecology. Documenting how to minimize project impacts to ecological functions, and how to provide appropriate mitigation for unavoidable impacts, requires an understanding of how an ecosystem works. This kind of work served as the basis of my evolving understanding.

THE SECOND EDITION

Overall, the second edition, as the first, is centered around the question, "What would one need to know about the Everglades to have a reasonable understanding of what it is and how it works?" Perhaps an objective would be to stand before a county commission, the state legislature, or even Congress, and not be discredited for lack of knowledge. Relevant issues may include water management decisions, permits and mitigation for projects located in wetlands, water quality problems in urban and agricultural planning, wildlife management, and foremost, Everglades restoration.

It was a surprise to me when the first edition was widely used as a college-level ecology text. Initially, I kept track and identified about a dozen university orders that indicated its use as a text. The list included several northern universities that conducted spring break field trips to southern Florida. By chance encounters, I signed many copies of the book for such students at Mrazek Pond and other locations in Everglades National Park. In 1998, I was invited to teach South Florida ecology at Florida International University, where *The Everglades Handbook* was in regular use (and still is as of this writing). I agreed to teach on a limited weekend or evening schedule, mostly out of the desire to find how the handbook could be improved for course use. An immediate revelation was the need for many more pictures of plants and animals than could be accommodated in the book's scope. To fulfill that need, I initially carried slides and a projector to class. But I soon found the *National Audubon Society Field Guide to Florida* (see reference 6), and I used it as a supplemental text, which the students obviously appreciated. I recommend it, and the availability of works of this kind guided my decision not to include significantly more pictures in the second edition. Also worthy of mention in this regard, for those more botanically inclined, is Roger Hammer's excellent field guide, *Everglades Wildflowers* (see reference 254).

The second edition has been expanded for use as a college text and for the environmental community's interest in ecosystem restoration. Added chapters are on Big Cypress Swamp, on Lake Okeechobee and its headwaters, and on succession and food chains. The last of this list is Chapter 20, called *Synthesis*, which contains diagrams aimed at aiding in the understanding of ecosystem processes.

Numerous text improvements are directed toward the environmental community for understanding ecosystem restoration. For example, there is more geologic background needed for underground "aquifer storage and recovery" (ASR) of surface water, which has become an important and controversial topic. Much of the consternation arises out of an inadequate understanding of the physical aspects of the process — where does the water go underground, what does it interact with, and what happens when we withdraw it? The revised first chapter sets the stage for this understanding. The final chapter contains updated information about other ongoing and planned restoration efforts. When the first edition came out, Everglades restoration was in its conceptual infancy, but it has now become an insecure teenager.

REFERENCES AND CITATIONS

As an aid to readers wishing to research specific topics more deeply, citations (indicated by superscript numbers) are provided throughout the text and listed in the references at the end of the book. While such referencing helps support facts and theories, the larger intent is to act as a guide to more detailed sources. References that apply to entire topics are indicated with the chapter or section headings. More specific citations are indicated in the text. It should be noted that the reference section is more than twice the size of that in the first edition, even though I have endeavored to omit nonrelevant "gray" literature (e.g., popular magazines and newspaper accounts) unless there is specific relevance. The literature "explosion" has resulted from the enormous attention and research money related to Everglades restoration. For me, collecting and understanding the new literature was a formidable task. Had I understood the hours and agony required, I might have been intimidated at the outset.

ABOUT THE INTRODUCTION TO THE FIRST EDITION

Out of professional honesty, I have to admit that I wrote the introduction to the first edition — that is, my hand put the words on paper, words dictated by Marjory Stoneman Douglas (see Color Figure 11). Our agreement included my assisting in order to accommodate her failed eyesight, and necessitated reading the entire text aloud to her prior to embarking on the introduction. It was the summer of 1989 when she was "only" 99 years old. We spent numerous weekends working in her combination living room and office. During one particularly long session of my reading, I noticed her tap her "speaking" clock. It announced the time, and upon my next pause she interrupted, "I usually have a drink at 5:00. It's 4:30 now, and that's close enough." So we put papers aside and drank some Scotch whiskey.

Her attentiveness during my reading was extraordinary. Some sessions lasted nearly 3 hours, with short breaks only for suggested changes or incidental comments. She appeared to show nothing but fascination with an admittedly lengthy text, and initially I had wondered how much she retained. Then, upon beginning the third reading session she said, "I don't believe you mentioned the harlequin snake (her vernacular for either of the Florida scarlet snake or the scarlet kingsnake) in your reptile chapter last week. I once saw one in my woodpile. I want you to include it." And so it was done, and my question of her comprehension was rested. Actually she suggested several additions to the text, always wanting more details to satisfy her own curiosity. For whatever apprehension I had at the outset — whether we would have unpleasant disagreements, difficulty in communicating, and so forth — I now look back at those weekends with confidence: there was nothing less than friendship and lots of fun.

In the years following the initial writing, I visited Marjory numerous times. Among other things, I asked her about the statement of her age in the first sentence, as it was getting out of date. With her concurrence, the first change was merely from 99 to 100, with a sense of pride to both of us. But as time advanced without publication, she joked that I should skip directly to 102 — which I did. Our final meeting to discuss the book was on March 21, 1993, just a few weeks prior to her 103rd birthday. Following that meeting, however, I realized that there was really no need to obscure those enjoyable days when we actually worked over the details of the text, and I elected to return to the original "99 years."

The last time I visited Marjory was March 6, 1997, when she signed my personal copy of the first edition at her house. Her ability to hear and to recall details of her marvelous life had faded considerably, but she still affectionately recognized my voice. Marjory died in May 1998 at the age of 108.

Thomas E. Lodge, Ph.D.
May 2004

Introduction to the First Edition*

by Marjory Stoneman Douglas

As many of you may know, I have devoted the greater portion of my 99 years to the Everglades and its related issues, particularly those having to do with that vital ingredient: *water*. Although I have been almost totally blind for some years now, I can still see clearly that the Everglades continues to need help — probably now more than ever before.

The story of my love for Florida and my concern for the Everglades begins with my father, Frank Bryant Stoneman, who had lived in Florida since 1896. First settling in Orlando, he studied for the bar and became an attorney there. But soon he became interested in the new city of Miami where he moved in 1906 — with a flatbed press he had taken for a bad debt — and started the first morning paper in Miami, which he called *The News-Record*. Without adequate money, the paper was about to fail, when Frank B. Shutts of the law firm of Shutts and Bowen of Miami, bought control. In 1910, it was renamed *The Miami Herald*, which became the most important newspaper between North and South America.

My father started the paper at the time when Napoleon Bonaparte Broward had run for governor, under the then-popular slogan of draining the Everglades, primarily to provide new lands for agriculture. Having become aware early of water problems in the western United States, my father wrote vehement editorials protesting the idea of drainage, of the consequences of which, he said, people were entirely ignorant. This attitude so enraged Governor Broward that later, when my father won an election for circuit judge on the east coast, Broward refused to confirm his election. My father was always very humorous about that. He said it saved him from a life of shame — of having to run for election every 2 years thereafter.

I became interested in the Everglades immediately following my arrival in Miami in 1915. My parents had been separated in the North when my mother, who was ill, had taken me home to my grandparents in Massachusetts where I grew up and where I graduated from public schools and from Wellesley College. After my mother's death and my own brief unworkable marriage, I came to Florida to be with my father and to get a divorce. He had recently remarried a lovely woman who became my first friend as I settled in to establish residence with a job at the *Herald*. I found not only that the new relationship with my father would be one of the most important in my life, but that newspaper writing was the thing that I most wanted to do in the world, and that Florida, with its wonderful climate and new associations, was the place I wanted to live the rest of my life.

The city of Miami itself then contained 5000 people and was not impressive, but it was the country — the flat tropic land, the sea, the great sky, and all the excitement of a new world that stirred my enthusiastic loyalty. I can remember that in those days, the Tamiami Trail went from Miami westward only to the Dade County line, some 40 miles. It was great fun to go out on the Trail to its end for picnics and to see all beyond the wonderful untouched expanse of Everglades.

The idea that the end of the Florida peninsula should be established as a great national park had already been conceived by the man who gave the rest of his life to that pursuit. He was Ernest F. Coe, who had been a nationally known landscape architect but had been stranded penniless in Miami after the devastating 1926 hurricane interrupted the development of those early days. My father, as editor of the *Herald*, was a strong promoter of Mr. Coe's idea for a park so that — together with Mr. Coe and others — I was immediately made a member of the first committee that

---

* Published in May, 1994.

worked for it. The committee was chaired by Dr. David Fairchild, and we received great support and assistance from Horace Albright, then director of the National Park Service.

The policy of the Federal Government was that it did not buy land for national parks. All parks had to be reserved from the public domain or donated to the Federal Government by states or private interests. We were 20 years working to acquire the land, originally by donation of the tract owned by the Florida Federation of Women's Clubs, but also of State lands and privately owned lands. The park was finally dedicated in 1947 by President Truman in the town of Everglades on the southwest coast. It has become the second most visited national park in the country. I am, at present, the only living member of the original committee.

We know now that the basin of the Everglades begins with the once-meandering flow of the Kissimmee River down into Lake Okeechobee from the north. Prior to man-made changes of the lake's drainage, Lake Okeechobee overflowed its southern rim during the wet season into the broad expanse of the Everglades. It all worked as a unit: the Kissimmee–Lake Okeechobee–Everglades watershed. But beginning in 1882, this unit was changed. First, the Caloosahatchee River was "canalized" and connected to the lake, making an open outlet for the lake's waters toward the west to the Gulf of Mexico. Then, great canals were excavated right through the Everglades, further diverting southeastward to the Atlantic coast would-be Everglades waters from the lake. These were the Miami, the North New River, the Hillsboro, and the West Palm Beach Canals. In 1916, a final canal — the St. Lucie — was begun, adding a more direct, eastward outlet from Okeechobee to the Atlantic by way of the Indian River at the town of Stuart. The lake's entire southern rim was diked by a huge levee, so that the only outlets were the canals, all fitted with gates to control the waters in an effort to put man — not nature — in charge of the Everglades. All this provided an enormous area of reclaimed land upon which agriculture, mostly sugarcane, developed to the south and southeast of the lake.

Through the years, the canals have been enlarged and interconnected into a vast network, and even the once-meandering beauty of the Kissimmee and Caloosahatchee has been lost to dredges and straight-line engineering. Today, the original, natural order of flow has been so interrupted by canals, levees, and water-control structures (dams, gates, locks, and even pumps) that the movement of water through the watershed towards Everglades National Park has become entirely unnatural. Nevertheless, the condition of the Park still depends upon the proper maintenance of the Kissimmee–Lake Okeechobee–Everglades watershed. It is ironic that we have now started to undo past mistakes on the Kissimmee River by reconnecting some of its original meanders and allowing its waters again to spread out over parts of its once natural, pollutant-cleansing flood plain.

With a growing number of others, my voice has been heard through many decades in the efforts to preserve the historic flows of water to the Park. Today, the powerful Washington-based Everglades Coalition — comprised of over 20 national, state, and local organizations, including my own Friends of the Everglades — continues this work to strengthen the control of water for the Everglades and the Park. But the Coalition's work must not be taken for granted. Its continuing success requires a never-ending fight against the forces of agricultural and commercial development. The future of this wonderful Park depends on the public support of these conservation efforts. In that regard, I hope this book will prove to be an important contribution to the understanding of the nature and beauty of the Everglades as well as to the many-faceted challenges of keeping the Everglades alive and well.

Recollections — an Introduction to the Second Edition

by Taylor R. Alexander, Ph.D.

I arrived in Miami in the fall of 1940, and despite my education as a botanist, South Florida's plants bewildered me. I had grown up in Arkansas and worked in the nursery business through college, graduating in 1936 from Ouachita College in Arkadelphia, Arkansas, where, incidentally, evolution was not allowed to be discussed. Good fortune included a fellowship, and I went on to graduate school in botany at the University of Chicago, completing my Ph.D. in 1941. Some of the first courses in the pioneering science of ecology were given during my tenure at Chicago. The new concept of ecological succession — the natural replacement of one plant community by another in response to environmental conditions — was then being explored. Succession was to become the unifying concept in my South Florida work.

At Chicago I came to know science students in various disciplines, as we were all housed in one location. Suddenly and without explanation, all of the physics students disappeared. Years later we learned that they had been sequestered in their involvement in the A-bomb development. Those of us in the biological sciences received far less attention — employment prospects were bleak. It was pure luck that University of Miami's new president, Jay Pearson, a Ph.D. graduate of Chicago, asked the head of my department about a botanist to head the Botany Department at the fledgling University of Miami. I was recommended, and, sight unseen and my degree unfinished, I accepted. My contract was verbal and pay was about $200 per month for the academic year provided that enough students enrolled. The promise was fulfilled.

My knowledge of South Florida's flora grew rapidly due largely to my associations with the Botany Department's Walter Buswell and a recent botany graduate, Roy Woodbury, and with John Henry Davis, then of the Florida Geological Survey. Roy later became renowned for his work on Puerto Rican flora. Walter had been Thomas Edison's botanist in the search for alternative rubber sources during World War I. He had developed a large South Florida plant collection that was helpful to me. The collection became the hub of the university's noted "Buswell Herbarium" (now at Fairchild Tropical Garden). John Davis, also a Ph.D. graduate of Chicago, had just completed a monumental work on Florida's mangroves[144] and was beginning the first overall study and mapping of Everglades vegetation, which he completed in 1943.[145] I accompanied him on several field trips, including one to Loop Road where a woman, disgruntled with her husband's drafting into the war the previous day, denied us passage through her property — at gunpoint. John's contract provided official access to tires and gas, both of which were limited by the war effort. We became good friends, and I followed John's continuing work on peat deposits of Florida.[146] In the 1950's and later, I benefited from friendships with Bill Robertson, Everglades National Park's first resident scientist, and Fred Dayhoff. "Dr. Bill" pioneered the ecological role of fire in the Everglades, work that shaped my views on plant succession.

I met Fred Dayhoff about 1960. He had grown up hunting and fishing in the open glades and cypress country and became an Everglades park ranger, later assigned to inventorying conditions in the Big Cypress when the National Park Service was preparing to establish the Big Cypress National Preserve. I found his detailed knowledge of the backcountry, wildlife, and vegetation exceptionally helpful. One unique adventure Fred described to me concerned the 1947 hurricane, when flood waters as deep as a foot were still running over the Tamiami Trail. Not yet 8 years old, he and his older brother sat on opposite fenders of their 1937 Packard, their legs around the fender-mounted headlights, as their father drove slowly. Fred recalled that giant whirlpools marked the submerged north ends of the culverts that passed under the road, and that the floodwater over the road was swift. With their rifle, they took turns shooting bass that were running over the road against the current. Upon each hit, the other brother would jump off and attempt to grab the fish. They had to be quick because the current was fast enough that the fish would be quickly carried into deep water over the sawgrass south of the road. He said the bass were most abundant about where the L-67 levee hits the trail today, in the middle of the Shark River Slough. They ended that day's adventure on Loop Road about eight and a half miles westward from the Trail's "Fortymile Bend." The flow there was too swift and deep — about 2 feet, he recalled — for them to go farther. Fred is lucky to be alive today, but not from his many daring recreational adventures. In his work for the park, he became a key enforcement agent against alligator poaching. Bullet holes in the park's airboats attested to the risk.

Through the mid-1940's, there were not only wartime shortages of gas and tires, but there were also few roads. The main ones were the road to Paradise Key — now Royal Palm in Everglades National Park — and its very rough and often impassible extension all the way to the fishing settlement of Flamingo. And there was the Tamiami Trail. Gravel roads followed most of the drainage canals. Scant road access then led to the further problem of interior access. There were only Indian canoe trails to follow, either in hand-propelled skiffs or on foot. Airboats and swamp buggies had not yet been developed, so my initial explorations were limited. I recall when, in 1946, the exploratory oil well was drilled seven miles south of Tamiami Trail to the west of Shark River Slough. The rock road to the well site opened a new area for our exploration. The well site later became the Shark River Tower location in Everglades National Park.

An exception to restricted access was the Fakahatchee Strand. Logging of the huge cypress stands was in progress through the 1940s, and I was invited to ride the steam locomotives over the tramways of the operation. It was a bitter reward to see it happen, but an easy opportunity to collect plants in an otherwise nearly inaccessible wilderness. Many of my photographs of that Fakahatchee work were donated to the Collier County Museum in Naples.

The abundance of wildlife stands out in my earliest memories of the Everglades. Along the oil well road embankment, I recall applesnail shells by the thousands. Sometimes they were piled several deep, attesting to the abundance of the living snails. Their predators, limpkins and snail kites, were common. Limpkins were probably responsible for most of the shell piles. Crayfish were common everywhere, and during droughts, their "chimneys" (entrances to their burrows) littered the dry marsh. Locals would pick up crayfish on the roads in rainy weather — it was easy to get enough for a meal. I remember how deep dry-season cracks in the muck soil along the Tamiami Trail were once filled with crayfish and hoards of young catfish in the early rainy season. Great flocks of wading birds and grackles came to feast on them.

Seeing birds in the Everglades required little effort. In February 1942, a colleague and I took a short morning trip from the University to see the Everglades in what is now southwestern Hialeah. We went north around the Miami Airport, crossed the Miami Canal on the iron swing bridge at Miami Springs, and proceeded only a short distance, maybe a mile, northwest on to Okeechobee Road. There we saw wading birds in numbers that defied our imagination. They were flying and feeding everywhere, and I took the picture shown on the next page. Ducks and especially coots were also abundant in the Everglades, and the head of the university's Zoology Department, Henry Leigh, would occasionally go duck hunting in the early morning prior to classes.

After airboats came into regular use in the 1950's, I recall that abundance of treefrogs in the Everglades was a problem. The inch-long frogs would be clinging to vegetation and would jump just as the airboat approached. Traveling at airboat speeds, the passengers were constantly slapped by these small amphibians. Airboaters fitted their rigs with screens to prevent the problem, but by the 1970's and 1980's the screens were seldom needed. Gar were also more abundant then. The Indians along the Tamiami Trail would spear them and cook them in 55-gallon drums. The gar's

peculiar scales were used in Indian jewelry, and the gar flesh was eaten. Gar are still seen today, but their numbers are only a fraction of the former profusion. Other prominent memories included the abundance of marsh hawks gliding low over the sawgrass in the Everglades. The variously colored tree snails were common in the hammocks. Otters were also regularly seen, and marsh rabbits were everywhere along the roadsides. My young daughters would make a contest of counting marsh rabbits on trips across the Everglades — so many that they held a child's short attention span.

My recollection of road kills is vivid, especially on the Tamiami Trail. They were a gruesome measure of wildlife abundance. Huge numbers of crayfish, lubber grasshoppers, frogs, turtles, snakes, marsh rabbits, otters, and wading birds were hit — sometimes making the road slippery and the stench obnoxious. Owls were often hit at night as they swooped low over the road in pursuit of prey, and by day vultures would feed on road kills, making daytime driving hazardous.

The abundance of large frogs in the Everglades supported many airboat "froggers" in the frog-legs business, even into the 1960's. Fred Dayhoff said that an experienced frogger could harvest over 150 pounds of frog legs in a good night. I recall watching from atop a levee one night when at least 50 different airboat lights could be seen at one time — I was so amazed that I counted them. Snakes were hunted for their skins, which were sold for snake-leather wallets, belts, boots and so forth. These Everglades livelihoods have long since subsided or ceased, mostly for lack of profitable numbers of frogs and snakes. Despite my intense work throughout the Everglades in the 1970's, I never saw wildlife resembling the abundance I recall from my early days. Recent comers to South Florida cannot imagine what existed then. Large-scale changes in the Everglades were obviously the main cause, but the carnage on the Tamiami Trail and other roads must have contributed to the demise of many species — roads in the Everglades were formidable wildlife barriers.

A notable exception to the abundance of wildlife in the early days was deer — they were rare. Because of a tick-borne disease that endangered cattle, there was a government program in the late 1930s and early 1940s to reduce deer numbers.[145, 376] Hunters, mostly helping to feed their families in the war years, got a dollar in exchange for a pair of ears as evidence of a kill. The effect on deer populations was dramatic. I do not recall seeing many deer until years later.

I saw progressive changes in the Everglades' vegetation. I remember that tall sawgrass was common in the early 1940's, even in the vicinity of the Tamiami Trail. At my short height of only 5' 6," attempting to hike through tall, dense sawgrass was foolhardy. Sawgrass up to 9 feet tall continued to be found farther north, such as in Water Conservation Area No. 1, but much taller than 5 feet became rare in the southern Everglades, so that I could see over the top well enough to find my way most places. I suspect that altered water and fire regimes and the loss of peat soils were responsible for these changes. I watched as successive fires of the drought-stricken mid-1940's altered the over-drained 'glades, and created dense smoke all over South Florida. I witnessed the area south of the Tamiami Trail in the vicinity of Krome Avenue change from sawgrass and tree islands eventually to willow with the consumption of the former muck soil — much of it burned down to the rock surface. Other vegetation changes were obviously the doing of collectors. In the hammocks and tree islands, I recall that many kinds of orchids and ferns that I saw regularly in the 1940s became very rare. Butterfly orchids were everywhere, and it was not unusual to find clamshell and night-blooming orchids.

In the early 1970's, I was contracted by the Department of the Interior to study changes in Everglades vegetation.[23, 24, 25, 26] With my associate, Alan Crook, we interpreted new and old aerial photographs and ground-truthed selected locations, reaching remote areas by helicopter. In all, we evaluated 100 one-square-mile quadrats, comparing their current vegetation with what we could see in the 1940 aerial photographs, the earliest available and, incidentally, the same photography that John Davis had used as the basis of his vegetation mapping some 30 years earlier. Among extensive changes we found were obvious influences of rising sea level: mangroves were documented to be moving inland, and coastal cypress and sabal palms, variously intolerant of salt, were found dead and dying. In the freshwater Everglades, the effects of man-made changes in hydrology and fire regimes were apparent. The already known conditions in the water conservation areas were verified, namely the effects of too little water in the northern portions and too much flooding in the southern ends — a simple outcome of impoundment in a sloping landscape. The new influence of Alligator Alley — then only eight years old — was found to be drastic with its induced invasion of wax myrtle, dahoon holly, and saltbush into former sawgrass prairie. Related to that effect, the most surprising of our findings was that a widespread reduction in fire frequency was evident. By the time of our study, extensive fire suppression had been both intentional and the passive result of man-made fire breaks, namely the roads and canals that had been constructed through the ecosystem. People think of South Florida environments as having too much fire, which was obviously the case in the early years of uncontrolled drainage, as I had seen. But the more recent effect of too little fire on the many natural, fire-dependent communities was very apparent. Extensive shrub invasions in many areas of the Big Cypress Swamp and the Everglades were examples. The importance of scientific prescribed-burn programs was obvious.

I have always associated the final closure of the water conservation areas with the major decline of wildlife and many of the changes in vegetation. It was apparent to me that the food base provided by the Everglades in those early years was far larger, and I am convinced that the original continuity and water flow was the basis of it. After the 1940's, I watched wildlife dwindle to fractions of its original abundance as the massive Everglades flood-control project was constructed, with its diversion of waters and impoundment of the former free-flowing ecosystem. By the late 1970's, I had become so pessimistic about the deteriorating conditions that it was difficult for me to impart any enthusiasm about the future of the Everglades. My standards, based on early observations, were far too high. It is unfortunate to hold those images merely as distant recollections, and I can only hope that the efforts now underway can bring back some semblance of the functioning habitats and wildlife abundance of those earlier days.

Part I

Background

1 The Everglades in Space and Time[*]

The underlying geology and the overlying climate are at the core of understanding the Everglades. With the modern focus on the ecosystem's restoration, knowing the region's geology has gained added importance. To set the stage for exploring the Everglades, this chapter probes beneath the surface into the geologic setting of the Everglades region.

FLORIDA, GEOLOGIC TIME, AND PLATE TECTONICS [267, 391, 417, 426, 556]

It could be called an accident of geography that the Florida peninsula extends from North America's temperate climate to the edge of the Caribbean tropics (Figure 1.1). South (or southern) Florida — the portion of the peninsula from Lake Okeechobee southward — is the home of the Everglades and is especially appreciated because of its proximity to the tropics. But in delight for its warm weather and its diverse mix of temperate and Caribbean tropical vegetation, few people would question the geologic origin of the region. Among the few are geologists studying the region's "basement" rocks, lying as deep as three miles beneath the surface. Those ancient rocks have an intriguing story to tell about continental drift, or more properly, plate tectonics.

When the earth's first fishes were evolving, the predecessors of the modern continents of North America and Africa were in relative positions comparable to those of today. But Africa and South America were then connected as parts of a larger continent called Gondwana by geologists, joined below what is now Africa's west bulge. (Gondwana also included Antarctica, Australia, and India, which adjoined the east side of Africa.) Between North America and Gondwana was an ocean, known to geologists as the Iapetus. (In Greek mythology Iapetus was the father of Atlantis.) This ocean was in a position similar to that of the modern North Atlantic. A map of ancient North America would have lacked a prominent modern feature: Florida. The terrain that was to become Florida was then part of the west bulge of the African portion of Gondwana.

In the millions of years that followed, North America slowly approached Gondwana, narrowing the Iapetus Ocean until the landmasses collided about 300 million years ago. The collision created the Appalachian Mountains as it fused North America's east coast with Africa's west bulge,[†] just north of South America's position. It locked Florida's terrain deep in the interior of the giant continent that was formed. Geologists call that ancient supercontinent Pangaea.

After a long term of quiet, about 200 million years ago while early reptiles were evolving (the Triassic period, see Table 1.1), Pangaea began to tremble with volcanic activity centering near Florida's east side. A "hot spot" had developed deep in the earth's crust, initially perhaps resembling the geologic environment of today's Yellowstone National Park. Volcanic eruptions added massive igneous rock, transforming Florida's earlier sedimentary terrain. The upheaval continued, and about 180 million years ago (the Jurassic period), it began forcing North America to rift from Pangaea,

---

[*] The main sources for this chapter are references 195, 316, 417, and 486.

[†] The location of the actual collision, which occurred during the Paleozoic Era, is termed the "Alleghanian suture" and is thought to be represented by an area of magnetic deviation called the "Brunswick magnetic anomaly." This feature runs from offshore along the southeastern coast of the United States, through southern Georgia and southeastern Alabama to the western panhandle of Florida.[588]

3

FIGURE 1.1 The Everglades in the Gulf of Mexico and Caribbean regions.

giving slow but violent birth to the North Atlantic Ocean, the Caribbean Sea, and the Gulf of Mexico. In the new configuration, however, the terrain of Florida, the Bahamas, and parts of southern Georgia and Alabama, was welded to the North American continent.[313]

Upheaval in the remainder of Pangaea continued, and about 125 million years ago, South America and Africa began to separate, opening the South Atlantic, thus completing the birth of the Atlantic Ocean, which was initially a long, thin north-south rift in the earth's surface. Opposing motions of the earth's crust between North and South America created volcanic conditions that produced the West Indies in the evolving Caribbean, as the two continents continued their departure from Africa with the widening Atlantic. Unlike its configuration today, the Caribbean Sea was open to the Pacific as well as the Atlantic Ocean. Only about 2.5 million years ago (during the Pliocene epoch) did geologic movements finally uplift the area of Panama and Costa Rica, completing the land bridge through Central America that now connects North and South America and separates the Pacific from the Caribbean. Even today, the earth's crust continues to move. Volcanos in Central America evidence continental uplift, and the Atlantic Ocean grows an inch or so wider every year, spreading from the mid-Atlantic rift that runs the north-south length of the entire ocean.[183, 373, 461, 642]

Looking back to Florida's initial rifting from Africa, the peninsular terrain (the Florida Platform) was land at first. But by late Jurassic time, the underlying earth's crust had cooled enough that the southern part of the platform had subsided to become a shallow, tropical marine environment. By early Cretaceous time, the entire platform had submerged, with the southern part falling even deeper. Sediments accumulated rapidly over the now volcanically quiet and stable underlying crust, keeping pace with the platform's slow subsidence as the crust continued cooling. Because of sedimentation, most of the platform remained shallow — enough so that minor changes in sea level occasionally exposed the surface, but only for geologically brief intervals (until much more recently).[484]

TABLE 1.1
Geologic time table pertaining to South Florida: Paleozoic to present[a]

Era	Period	Epoch	Years before present*	South Florida relevance
Cenozoic	Quaternary	Holocene	Present to 17,000	Warming climate and rising sea level to its present stand
		Pleistocene	17,000 to 2M	Four glaciations with attendant low sea levels, and interglacial high levels with deposits
	Tertiary	Pliocene	2-5M	Final deposition of the Hawthorn Formation followed by high sea level and deposition of the Tamiami Formation
		Miocene	5-25M	Main deposition of the Hawthorn Formation
		Oligocene	25-37M	Limestone strata of the Floridan aquifer deposited
		Eocene	37-55M	
		Paleocene	55-65M	
Mesozoic	Cretaceous	Upper	65-100M	Continued limestone sedimentation and aquiclude formation
		Lower	100-135M	Oil-bearing Sunniland Formation laid down
	Jurassic	Upper	135-150M	Peninsula submerged — limestone sedimentation began over the igneous basement rocks
		Middle	150-175M	Florida peninsula not yet submerged
		Lower	175-190M	North America began to rift from Pangaea
	Triassic		190-245M	Hot spots with volcanos develop in Pangaea
Paleozoic	7 periods: Cambrian through Permian		245-600M	Collision of Proto-North America with Gondwana to form Pangaea during the Devonian period about 300M years ago

[a] The main sources for this table are references 176 and 486

* M = million (years before present)

Various processes converted the sediments into limestone and related rocks,* thicker over the southern platform area. One early sequence, the Sunniland Formation, is of particular interest. It is about 135 million years old (lower Cretaceous time) and now lies 11,000 feet (about 2 miles) beneath the Big Cypress Swamp and western Everglades (Figure 1.2). It is the modest oil bearing formation of the region. Production comes from wells located along the northwest-southeast, reef-like Sunniland Trend, a series of limestone mounds, primarily the fossilized remains of long-extinct oyster-like mollusks called rudists.[192, 346]

LIMESTONES AND AQUIFERS

It is important that the sediments, up to Miocene time about 25 million years ago, formed mostly limestones and related rocks on the Florida Platform. Dissolved calcium carbonate is abundant in sea water, and, in a warm tropical environment, it can be precipitated by physical-chemical or biological processes to form sediments that solidify into various forms of limestone (described in Chapter 10). However, as easily as limestones form, they can also dissolve. Partial dissolving forms water-conducting channels and pockets, especially if sea level falls and exposes the rock to air and rainfall. Many layers of Florida's limestones resemble the character of an imaginary rigid sponge

---

* Limestones are calcium carbonate ($CaCO_3$). Under certain conditions within subsurface limestone layers, magnesium from seawater seeping through passageways slowly replaces the calcium, forming dolomite ($MgCO_3$). Other processes make additional minerals such as anhydrite ($CaSO_4$) and related gypsum. The older rock layers of the Florida Platform contain these related minerals, but limestone predominates. [486]

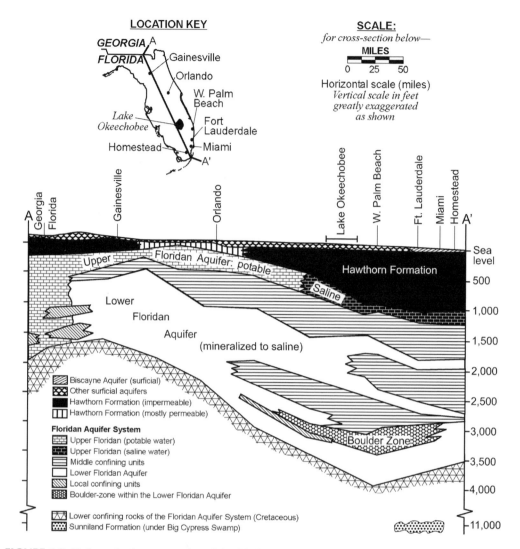

FIGURE 1.2 Hydrogeologic cross section of the Florida peninsula's centerline. Surface locations are shown as if they are directly on the centerline, but most, such as West Palm Beach, are offset. The upper four units (see key) lie above the Floridan aquifer. The only two units shown below the Floridan are the Cretaceous "floor" of the Floridan and the much deeper oil-bearing Sunniland Formation (see text). (From Fernald and Purdum,[195] modified by permission from the Institute of Science and Public Affairs, Florida State University.)

— highly permeable, with spaces that hold and transmit water. When such a layer can be used as a water source, as by drilling a well for water supply, the layer is called an *aquifer* (pronounced *ä-kwa-fer*). Other layers, however, may remain nearly impervious, thus preventing water exchange between layers above and below. An impermeable layer confines water movement and is called an *aquiclude* or *confining unit*.

The oldest sedimentary rocks that formed over the original basement rocks and until about 70 million years ago (the end of the Cretaceous period) are thousands of feet thick. Despite their abundance, they lie too deep to be of much economic interest, except for the oil-bearing formation mentioned above. Furthermore, the upper portion of the Cretaceous layers acts as an aquiclude, so there is no water exchanged between it and the younger overlying rocks.

The limestone layers that were deposited on the Florida Platform between 70 million and 25 million years ago are of considerable interest because they contain a mass of water — the Floridan aquifer — that is slowly moving generally southward and toward the coasts from mid-peninsula. Water enters the Floridan aquifer from rainfall percolating through soils and permeable rock in northern and central Florida, and it flows very slowly southward through the rock. From central Florida and southward, there are intermediate confining layers (aquicludes) within the Floridan aquifer, dividing it into two major levels (see Figure 1.2). The lower level is salty, but the upper level, called the Upper Floridan aquifer, is widely used as a drinking water source in northern and central Florida where it lies close to the surface. However, from the north side of Lake Okeechobee and continuing southward, the Upper Floridan aquifer lies deeper, buried under newer sedimentary impermeable layers, and its water is too salty (brackish) to be a drinking water source.[195, 316, 416]

The importance of the Floridan aquifer system in southern Florida comes from two very different uses. The most interesting is called aquifer storage and recovery (ASR), where excess freshwater is injected through a well into the Upper Floridan aquifer and later retrieved. Recovery of injected water is surprisingly high. There is little mixing with adjacent salty water, and no loss to evaporation, which is excessive in a lake or reservoir. A massive application of ASR is planned in Everglades restoration, as detailed in the final chapter, but smaller scale uses for drinking water supply have been operational in Florida since 1985.[38, 489, 612]

The second use of the Floridan aquifer involves a much deeper layer that exists in the Lower Floridan aquifer in South Florida. Separated from the Upper Floridan by at least two aquicludes is a unique layer of the Lower Floridan aquifer called the "bolder zone," so named because it has cavernous spaces, as if it were made of large boulders. Some of these spaces, filled with salt water, are large enough to contain a house. Because of the large volume of void spaces — and its hydrologic isolation around 3000 feet below the surface — the boulder zone has been used for oil-field brine waste since 1943.[411] More recently, deep disposal wells mostly from West Palm Beach to Miami pump semi-treated wastes into the boulder zone[159, 416] where it is thought that it will remain for many decades before possibly leaking out into deep water off the coast of Florida.[*]

EMERGENCE OF LAND ON THE FLORIDA PLATFORM: THE PENINSULA

The sediments that formed the rocks of the Floridan aquifer included very little eroded material, such as clay or silica beach sand, generically called *clastic* sediments, from the adjacent North American landmass. Clays, in particular, block the flow of water. Sediments containing abundant clay-sized particles nearly always form an aquiclude. The Floridan aquifer characteristics — few clastics and abundant limestone — tells geologists that the Florida Platform was isolated from sediments eroded from the mainland during its formation. Evidence shows that there was a deeper channel with swift currents across northern Florida and southern Georgia, connecting between the Gulf of Mexico and the Atlantic. By the Miocene epoch, that ancient seaway, called the Suwannee Channel, Suwannee Strait, or Gulf Trough (see Figure 1.1), had filled, an event that transformed the Florida peninsula.[134, 642]

The Suwannee Channel apparently filled in because the current through it slowed, allowing clastic sediments from the mainland to encroach without being swept away. Also, there was uplift of the Appalachian Mountains that increased sediment loads carried down rivers to the adjacent coastal waters. These clastic sediments ranged from coarse silica sand to fine clay particles. So prolific were these sediments that they blanketed the entire platform, even a few miles beyond the modern location of the Florida Keys into the Straits of Florida. In north and central Florida, where

---

[*] Controversy also centers around possible upward leakage of wastewaters that are lighter (less saline) than the ambient saltwater of the boulder zone. Imperfect plugging of test wells, faulty seals around operational wells, and incomplete aquicludes are concerns.[411, 610]

the platform was very shallow, they formed the first permanent land on the Florida Platform: dunes, created by winds that piled beach sand into mounds and ridges. On the southern platform, the new sediments created shallows where the water had become too deep for carbonate sediment formation, thus extending the shallows of the Florida Platform. In southern Florida, these Miocene and early Pliocene sediments are now between 500 and 1000 feet thick.[134]

Geologists call the sedimentary sequences that were produced — some of which consolidated into rock while others remained unconsolidated — the Hawthorn Formation, named for a small town (spelled *Hawthorne*) near Gainesville, Florida, where it was first described in 1892. It is important to remember that clastic sediments, initially carried by rivers, graded from coarse to fine. In many locations of northern and central Florida, the Hawthorn Formation is composed of coarse sediments that transmit water so that parts of it are used as a groundwater supply. But from north of Lake Okeechobee and southward, the Hawthorn is composed of fine, impermeable clays. Thus, the Hawthorn Formation in southern Florida effectively isolates the Floridan aquifer beneath from the surface. If a well is drilled through the Hawthorn in Miami, for example, saline water from the Floridan aquifer comes up through the hole. Such abandoned artesian wells (nearly 6500 in South Florida) have contaminated surface waters, and there have been expensive programs to find and plug them with concrete in an effort to stop salty contamination of surface waters.[195]

The Hawthorn Formation is also significant as the source of Florida's most valuable mineral: phosphate. Phosphate is mined for fertilizer in a large region east of Tampa, mainly in the headwaters of the Peace River around Lakeland and Bartow. Here is the richest such deposit in the world, and it is extracted by strip mining a layer of gravel, sand, and clay that lies about 60 feet underground. The mining, processing, and subsequent mining-area reclamation are highly controversial for reasons of environmental impact. The particular Miocene and early Pliocene sequences are called the Bone Valley Formation, so named because of their abundance of marine and terrestrial mammalian and other fossil bones. The geologic basis of how phosphate became concentrated there is complex, involving particularly rich, shallow marine waters and fluctuating sea level.[113, 373, 392]

By the conclusion of the Miocene and Pliocene epochs, the Florida Platform closely resembled its modern form. To the east, it drops off abruptly into the Atlantic. Southward, it slopes gently to a rim just seaward of the Florida Keys. It then also recedes quickly into the deep trough called the Straits of Florida, which is located between Cuba and Florida and carries the Florida Current.[*] To the west, the Florida Platform slopes gradually far out into the Gulf of Mexico before receding into deep water.[274, 523]

SEA LEVEL, CLIMATE, AND THE BIRTH OF THE EVERGLADES
185, 237, 365, 468, 642

The emergence of a large landmass on the Florida Platform was facilitated by the thick accumulations of Miocene and early Pliocene clastics, but the most impressive effects were the world's glacial cycles. During the last two million years (the glacial age or Pleistocene epoch of the Quaternary Period), four major (and some minor) episodes of glaciation occurred,[†] when the Florida peninsula stood much higher above sea level than it does today, and the climate was cool and dry.

---

[*] Most people, including Florida fishermen, refer to this current as the Gulf Stream, or just "The Stream." To be technically correct, we should refer to the Gulf Stream as the great northward flowing current of the North Atlantic, formed by the confluence of the Florida Current and a westwardly current that arrives well off the central Florida coast just north of the Bahamas.

[†] Continental drift is thought to have been responsible for the rather recent configuration of continents and oceans that has restricted circulation of currents to and from the Arctic Ocean and has positioned Antarctica squarely over the south pole. This arrangement has probably caused the glacial age that began about two million years ago. The glaciations (actual accumulations and advances of polar ice) during this age are thought to result from cyclic interactions ("Milankovitch cycles") of oscillations in the earth's distance from the sun, the shape of its elliptical orbit, and tilt of its axis. Together, they affect the amount of heat the earth receives from the sun.[465]

Between glaciations, marine waters inundated the region, sometimes to depths exceeding a hundred feet above today's sea level, and the climate was warm and humid. During these "interglacial" times of high sea level, new sediments were deposited. Wind, waves, and currents moved clastic sediments, mostly silica beach sands that reshaped the hilly, interior beach ridges of northern and central Florida as well as forming flat, sandy coastal areas there and in southern Florida. In shallow marine and freshwater environments of southern Florida, additional new carbonates were also deposited, often mixed with sand and shells. These interglacial deposits formed relatively thin layers over most, but not all, of the older sediments, resulting in the land surfaces that we encounter today (Figure 1.3). While they were relatively minor amounts compared to Hawthorn and earlier sequences, these recent sediments are critically important to modern vegetation and wildlife, and to human uses of southern Florida.

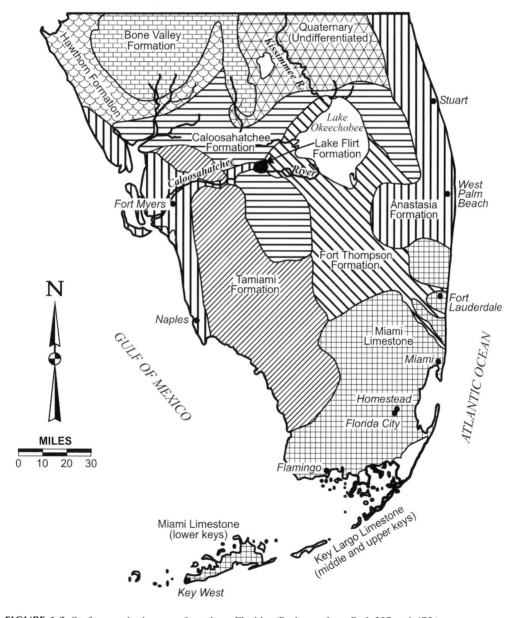

FIGURE 1.3 Surface geologic map of southern Florida. (Redrawn from Ref. 237 and 475.)

After the Hawthorn Formation was laid down, there was an extended high stand of sea level into the early Pliocene Epoch, preceding the glaciations. New sand and carbonate sediments then accumulated throughout South Florida, making an assemblage called the Tamiami Formation (named for the Tamiami Trail). It is up to 150 feet thick on top of the older Hawthorn. The limestone portion of the Tamiami Formation is now exposed at the surface in extreme southwest Florida, where it forms the upper bedrock in most of the Big Cypress Swamp,[176] with newer rocks covering it only in patches (see Figure 7.2). Elsewhere in South Florida, it is covered completely.

Following the deposition of the Tamiami Formation, a period of alternating shallow inundation and exposure spread a thin sheet of sand over much of the Tamiami. Then sea level rose to 140 feet above today's level and remained high for nearly a half million years, to the end of the Pliocene Epoch. Deposits of sand and seashells called the Caloosahatchee Formation were laid down around the area of Lake Okeechobee. The Pleistocene epoch then followed with its four major glaciations and intervening high sea levels followed. Interglacial sea levels of the Pleistocene, however, never reached the earlier high levels of the Pliocene. Interglacial deposits produced the Fort Thompson Formation and, most recently, the Anastasia Formation, the Miami Limestone, and the Key Largo Limestone (see Figure 1.3).

These most recent deposits were laid down during the interglacial period, called the Sangamon interglacial, when sea level was 25 feet higher than today. Waves then shaped the well-known Pamlico Terrace, a modern topographic feature representing the shoreline of that time. This terrace is recognizable around much of the peninsula at the +25-foot contour (Figure 1.4). The southernmost land was near the town of Immokalee, more than 100 miles north of Key West. The relatively short Sangamon interglacial lasted from about 180,000 to 100,000 years ago.[417]

The Anastasia Formation, Miami Limestone, and Key Largo Limestone are the products of the Sangamon interglacial period. The Anastasia Formation is primarily silica sand and seashells deposited along the east coast from Fort Lauderdale and northward, as well as the west coast in the vicinity of Ft. Myers. Along the east coast, it merges southward into the contemporary Miami Limestone, which forms the floor of the southern Everglades and Florida Bay, the coastal ridge in the Miami area, and the rock surface of the lower Florida Keys. In the Miami-to-Fort Lauderdale area, the granular, non-coralline Miami Limestone (a maximum of only 40 feet thick), and the underlying Fort Thompson Formation carry a plentiful supply of freshwater — the famous Biscayne aquifer. Because it is located at the surface — a *surficial aquifer* — rainfall and waters of the interior Everglades enter it directly. In the middle and upper keys, the contemporary deposit is the Key Largo Limestone, made of fossil coral reefs. U.S. Highway 1 through the keys lies atop this ancient coral reef from Key Largo to Bahia Honda. At Big Pine Key, the road switches to the Miami Limestone, which continues to Key West (see Figure 1.3). The difference between the lower and middle/upper keys is obvious from the air, but it is also apparent even on crude road maps.

The onset of the most recent glaciation drastically reversed Florida's character. Surface deposits were worn or even erased by weathering, exposing older sedimentary layers in many places. Importantly, the Miami Limestone was partially dissolved by percolating rainwater, producing its labyrinth of finger-sized holes and channels that now carry the Biscayne aquifer. At the glacial maximum only 17,000 years ago, the north polar ice cap reached as far south as New York City, the location of Chicago lay under mile-thick ice, and the basins of the Great Lakes were gouged-out by the southward motion of mountainous ice. So much water was removed from the world's oceans — locked up in ice and snow piled above sea level in the northern and southern hemispheres' ice caps — that sea level was between 300 and 400 feet lower than it is today. The lower sea level exposed much more of the Florida Platform, so that land extended far into the Gulf of Mexico (see Figure 1.4). The peninsula was about twice as wide as it is today, but it did not extend significantly eastward or much farther south than the Florida Keys, because it was constrained by the deep Straits of Florida. Based on several lines of evidence, southern Florida was never connected to islands of the Caribbean region.[461, 631]

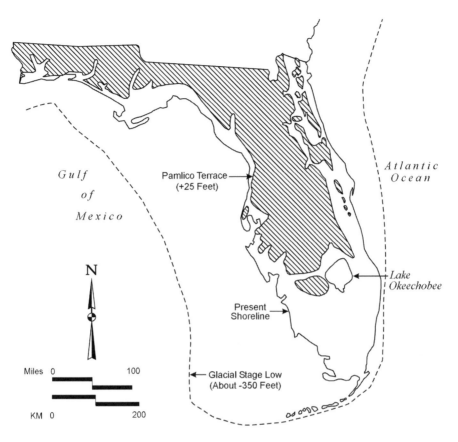

FIGURE 1.4 The present shoreline of Florida, and two former shorelines: the Pamlico Terrace (last interglacial shoreline about 125,000 years ago), and the most recent glacial stage low (about 20,000 years ago). (Adapted from Refs. 280 and 442.)

The glacial periods also had important climatic effects in Florida. The peninsula was substantially cooler and drier than it is today, and surface waters (lakes and streams) were probably rare or absent because of the leaky characteristic of limestone. The climate was not tropical, and plants and animals requiring tropical conditions were unable to persist, even in the southernmost parts. Alternatively, fossil evidence shows that animals, such as the porcupine, and numerous plants that are now found only in areas far north of Florida, occurred in the Florida peninsula at that time.[373, 436, 507]

The Holocene epoch,* the warming trend that followed the last glacial cycle, again changed Florida. Sea level rose by as much as six feet per century, rapidly shrinking the area of land, and the first humans arrived in South Florida — between 12,000 and 13,000 years ago.[248] About 6000 years ago tropical plants again began to flourish in extreme southern Florida, invading from the West Indies as they had in past interglacial times. Rainfall increased, and the Everglades finally began forming — at about the same time that the Great Pyramid of Cheops in Egypt was constructed. The story of the Everglades is but a glimpse — the last 5000 years — in the vast realm of time elapsed in the origin of Florida.

---

* It is now widely believed that the Holocene is merely another cyclic warming phase of the Pleistocene, i.e. we are still in the Pleistocene. Thus, the Holocene is merely a convenient label for the most recent post-glacial warming period, and a time when mankind's influence over the earth has grown dramatically.

2 An Ecosystem Overview

Marjory Stoneman Douglas described the Everglades as the "River of Grass," and, before her, the Miccosukee Indians called it "Pa hay okee," meaning "grassy water." Why have people given special names to and written whole books about the Everglades?

The Everglades is unique to Florida, and the term itself requires some comment. Technically it refers to the expanses of freshwater marsh originally extending from Lake Okeechobee to nearly the southern tip of the Florida mainland. The word "Everglades" has an obscure and apparently accidental origin, with the first part, "ever," originally "river." The second part, "glade," is probably the English word meaning an opening in a forest where grasses cover the ground. Marjory Stoneman Douglas traced these origins in her famous book, *The Everglades: River of Grass*, published in 1947 (its 1997 updated edition still in print).[170] She identified the term's first published appearance to an 1823 map. Since our introduction at The Kampong[*] in Coconut Grove in 1974, I have appreciated Marjory's intolerance of improperly researched work, so I have had to consider carefully my use of the term "Everglades." Marjory's book begins, "There are no other Everglades in the world." She treated the word as plural, so I have had to choose between singular and plural, as it is found both ways in the Everglades literature. My decision is to treat Everglades as singular because it is the name of one physiographic region, and there is no subdivision of the Everglades that could be called an Everglade. However, I can agree with Marjory that the Everglades is unique. (Actually, we discussed the issue, and she came to agree that singular is appropriate although both are grammatically correct.)

Discovery of the nature of the Everglades was slow in coming. Virtually the entire United States was mapped prior to the Everglades. The inaccessibility of its watery wilderness — with unpleasant expanses of sawgrass, having stiff, toothed leaves useless for cattle — inhibited exploration. No reasonably inviting use of the land was apparent that would spur exploration. However, one feature of the region's interior intrigued people — the unexplained outpouring of freshwater from Florida's southern end. Early literature is full of observations of huge flows from innumerable rivers into the mangrove forests at the southwest edge of the peninsula and of springs gushing water into Biscayne Bay. There was considerable speculation regarding the source of this abundant freshwater. Some imagined a huge lake in the interior. While there was some credible information of a large lake, finally resolved to be Lake Okeechobee (see Chapter 11), the mystery of the area south of the lake remained. In 1840, a military expedition led by Colonel William S. Harney (origin of the name Harney River in Everglades National Park) pursued Indians in the region. The group crossed the southern Everglades from Miami, and an account published in early 1841 provided a reasonable description of the Everglades. But there was little to excite further exploration until the 1880's when drainage and navigation activity in Lake Okeechobee excited many of the possibility of using the Everglades.[591]

---

[*] The former estate of Dr. David Fairchild (1869-1954), botanist and plant explorer for whom the famous Fairchild Tropical Garden, located near Miami, was named. In 1963, after the death of Mrs. Fairchild, Catherine H. Sweeney purchased and maintained The Kampong in the spirit of Dr. Fairchild's interests. In 1984, she donated it to the Pacific Tropical Botanical Garden, which was then renamed the National Tropical Botanical Garden by act of Congress specifically for the acquisition of The Kampong, to be maintained as a tropical garden in perpetuity. Over the years, a great many people involved in botanical and related work have met at the Kampong, and some of the earliest meetings to lay the framework for having a national park in the Everglades were held at The Kampong — Dr. Fairchild was the first president of the citizens' group that formed in 1928 to promote the idea.[171, 670]

This book focuses on the Everglades, its relatively recent origin, and its probable future. Considerable attention is devoted to the types of environments that surround it, which are inextricably linked biologically, geologically, and hydrologically. The Everglades must be understood within the context of a larger region, encompassing more than just southern Florida. The watershed of the Everglades comes from as far north as the Orlando area of central Florida.

A UNIQUE AND VALUABLE ECOSYSTEM

The state of Florida contains enormous acreages of marshlands, ranging in size from tiny isolated depressions within pinelands to the vastness of the Everglades, originally about 4000 square miles in size. But the Everglades is not special just to Florida — it is unique in the world. Of the world's larger wetlands, most receive their water and nutrients from associated rivers that overflow their banks. Examples that superficially resemble the Everglades (some even much larger in size) are the Pantanal of Brazil, the Llanos of Venezuela, and especially the delta region of the Usumacinta and Grijalva Rivers in southern Mexico[115, 451] (see Figure 1.1). Except for the historic overflows from Lake Okeechobee, which principally affected the northern Everglades, the bulk of the Everglades ecosystem received nutrients only from the atmosphere, primarily in the form of rainfall. Thus, configuration of the Everglades as a "sheet flow" ecosystem independent of river or stream channels is one of a kind.[*][252]

The Everglades would probably arouse little interest but for values associated with it. These values include the enormous populations of wading birds (Figure 2.1) and the peculiar diversity of tropical plants, which are the main reasons an earlier generation sought to protect the area as a national park,[190] an abundant, high quality water supply for the human population of Florida's southeastern coast, and highly valued agricultural resources. How the Everglades functions as an ecosystem and provides these values will be described in the following chapters. In the final chapter, in which it is made obvious that these values have not been in harmony, the features planned for restoration of the damaged ecosystem and for continued use by the human population are explained.

TERMS AND DEFINITIONS

Some important definitions are appropriate at this point. First is the term *wetland*, generally describing areas of the earth's surface that are regularly or periodically covered by shallow water. Legal definitions may include lands where little or no surface water occurs but where ground water periodically rises close enough to the surface to saturate the soil, making a *hydric soil*, which favors a prevalence of wetland plants and restricts the growth of upland plants. [186, 230]

Most wetlands alternate between flooded and "dry" (i.e., lacking surface water) on a regular basis, as with tidal cycles or seasonal cycles of changing rainfall. For non-tidal wetlands, the average annual duration of continuous flooding is called the *hydroperiod*, which is based only on the presence of surface water and not its depth. Relative to the Everglades, *long hydroperiod* is in excess of ten months (often with continuous flooding for a few years), and *short hydroperiod* is about seven or fewer months. Large annual variations are typical of individual locations because of year-to-year differences in rainfall.

The term that refers to depth as well as hydroperiod is *hydropattern*.[250] Hydropatterns are best understood by a graphic depiction, a *hydrograph*, showing the water level (above as well as below the ground) through annual cycles. An average hydrograph for a sawgrass community in the Everglades would show water one to one-and-a-half feet deep for several months during the wet season and falling to a foot or so below the ground at the end of the dry season. The various areas differ greatly however, and annual and multi-year cycles are highly variable.

---

[*] The historic overland flow-way from Lake Istokpoga to Lake Okeechobee's northwest side resembled a miniature Everglades: see Chapter 11.

FIGURE 2.1 The essence of the Everglades: a mixed group of wading birds feeding just north of the Tamiami Trail in Water Conservation Area 3A. Seasonal high water produces the abundant prey, and falling water concentrates the prey — mostly small fishes and aquatic invertebrates — for such a feeding frenzy. (Photo by T. Lodge.)

Wetlands come in many varieties, but the two main types are *swamps* and *marshes*. Although the definitions are not universal, swamps are usually considered to be wetlands in which the dominant plants are trees. Examples of swamps in the Everglades region are mangrove swamps in the tidal brackish and saltwater areas and cypress swamps in freshwater areas. If trees are so scattered that their shade is insignificant, the wetland is better called a marsh. Marshes are dominated by much lower, herbaceous (non-woody) vegetation and often look like prairies.

Strictly speaking, the term Everglades applies only to southern Florida's huge, interior freshwater marsh variously dotted with "islands" of trees. The term is often shortened to just *the glades*. Everglades National Park is situated at the southern end of the Everglades, and it encompasses far more kinds of habitats than just "Everglades." In fact, the true Everglades occupies only about a quarter of the park. Much larger portions are the marine waters of Florida Bay and the mangrove-dominated estuaries of the southwest tip of the Florida mainland. The principal kinds of environments of the Everglades region south of Lake Okeechobee are:[*]

- Freshwater marshes
- Wetland tree islands (broad-leaved types)
- Cypress heads, domes, and dwarf cypress forests
- Tropical hardwood hammocks
- Pinelands
- Mangrove swamps and mangrove islands
- Coastal saline flats, prairies, and forests

---

[*] The historic Everglades also included a narrow band of cypress swamp along the eastern edge primarily in Palm Beach and Broward Counties, and a pond-apple swamp around the south and southeast shore of Lake Okeechobee.[145, 150, 650] These communities are not listed here because they no longer exist excpt for a fragment of cypress swamp in Loxahatchee National Wildlife Refuge (see the following section on the historic Everglades, Chapter 4, and Chapter 11).

- Tidal creeks and bays
- Shallow, coastal marine waters

Other authors may use various different names for these environments or may even use other kinds of divisions.[128, 252, 369, 452] It is important to bear in mind that any system of categorizing the features of our environment is for our convenience. We too often expect nature to conform to our notions of its organization, only to become frustrated with nature's lack of cooperation. Thus, while these divisions are useful for organizing our thoughts, there will be some exceptions: places that do not conform to these divisions and divisions that might better be revised or subdivided for particular purposes.

SOUTH FLORIDA CLIMATE AND WEATHER [106, 177, 271, 375, 657]

Together with geology (see Chapter 1), climate and weather are the most important controlling factors that regulate any ecosystem. Climate describes the general, long-term characteristics of the atmosphere. It includes both averages and the statistics of extremes, as for rainfall, temperature, days of sunny conditions, wind, and the frequency of severe weather systems such as hurricanes etc. Weather is the actual day-to-day description of these atmospheric conditions.

Of major importance to plants and animals as well as the general comfort for people are the climatic temperature and rainfall. Whether southern Florida is tropical is often debated. Climatologists use various classification systems, and the most common (the Köppen classification) defines much of the region south of Lake Okeechobee (and extending slightly northward close to each coast) as *Tropical Savannah* because the average temperature of every month is above 64 degrees Fahrenheit and because there are pronounced wet and dry seasons. The "savannah" designation is because this climate would normally characterize a grassland with only scattered trees. In South Florida, that norm is offset by the occurrence of high rainfall during the warmest part of the year and by the limitation of most lightning to the wet weather. The remainder of Florida is classified as "humid subtropical" in that system.

In contrast to climatologists, ecologists define "tropical" by the lack of freezing temperatures. By that definition, only the Florida Keys and highly protected environments on the mainland of extreme southern Florida are tropical (see Chapter 12) because freezing temperatures occur often enough in the rest of South Florida to limit sensitive tropical plants. Whatever definition is used, the important fact is that the abundance of tropical vegetation in southern Florida is impressive, which is an abrupt contrast to central Florida and lands north. In mapping the distribution of tropical vegetation, a correlation has been found with the January 54-degree (F) isotherm (Figure 2.2), with the more cold-tolerant tropical species being the only ones nearing the northern edge of this line. Significant numbers of tropical tree species were historically found growing together only as far north as Pompano Beach[13] (just north of Fort Lauderdale), where they occurred near the protection of warm coastal waters (see Chapter 5).

The average annual rainfall in South Florida varies between 50 and 60 inches, depending upon location, with significantly more falling close to the east coast. As a long-term average, three quarters of the annual total comes in the six-month period from May through October. However, May and October are pivotal months with inconsistent rainfall. May is particularly enigmatic with summer temperatures approaching, water levels still low, and rainfall often sparse. The situation sometimes incites water-use restrictions and tricky situations for fire fighters, namely abundant lightning and dry environments. Thus, the most impressive statistic omits May and October. In an average year, about 60% of the rainfall falls in the four-month summer period of June through September (Figure 2.3), and only 25% falls in the six-month dry season of November through April.[177]

The onset of the wet season is often abrupt, and, as a result, water levels in the Everglades typically rise rapidly but do not reach their maximum until late in the season. Trade winds influence the weather during the warm summer months, bringing a continuous supply of moist, warm air

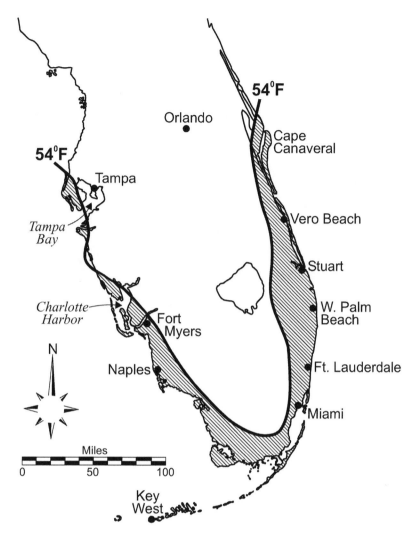

FIGURE 2.2 The line (isotherm) where the average January temperature is 54° Fahrenheit. The January 54-degree isotherm approximates the northern limit of the least sensitive tropical plants. (Redrawn and modified from Tomlinson.[599])

westward over the Atlantic. Localized afternoon thunderstorms develop daily as a result of heating of the air as it passes over land, and the frequency of lightning is high. And, as most long-term residents are painfully aware, the rainfall is sometimes increased catastrophically by tropical weather events, namely tropical depressions, tropical storms, and hurricanes.

Now consider the six-month, winter-spring dry season when only one quarter of the annual rainfall occurs. Mid-October or early November nearly always brings the first abrupt change in weather after a hot, humid, and wet summer. The arrival of the first cold front is a pleasant relief, at least for the human residents. In mid-winter, some of the cold fronts are unpleasantly cold, but the mild first arrivals do little more than signal the change in dominant weather from the tropical trade-wind system to the temperate, prevailing westerly system. The latter is characterized by "frontal" weather, pulsing a new cold front through South Florida every week or two from December through February, then less frequently through March and April. As these high-pressure air masses move down the peninsula, they warm, reducing the atmosphere's relative humidity. In short,

FIGURE 2.3 Summer thunderstorm seen from the Anhinga Trail, Everglades National Park, September 1978. (Photo by T. Lodge.)

southern Florida "dries out," and water levels in the Everglades recede (Figure 2.4) but with minor increases, due to sporadic dry-season rains in which lightning is rare. It is fortunate that South Florida's dry season coincides with cooler weather. If the seasonality were reversed, the hot summer temperatures would greatly accentuate drying conditions. It is also noteworthy that central and northern Florida do not have the pronounced winter dry season but get more rainfall from cold fronts than does southern Florida.

Substantial deviations from the normal dry season regime occur as a result of the El Niño phenomenon, which occurs sporadically every three to seven years and lasts one to two or more years. This phenomenon is based on unusually warm sea-surface temperatures in the eastern Pacific Ocean, which brings abnormally wet weather to the southern United States, including Florida, during the winter (early and middle dry season). The winter of 1982-3 and the three consecutive winters of 1991-2 through 1993-4 (the longest El Niño on record) were exceptionally wet because of El Niño events, with only a short, late dry season occurring in those years. El Niño is also associated with reduced hurricanes during the summer wet season.[195, 258, 445]

THE HISTORIC EVERGLADES[80, 145, 237, 456, 591, 656]

As described in the introduction, before changes began in 1882, the entire ecosystem that included the Everglades was a watershed beginning near the present location of Orlando: the Kissimmee-Lake Okeechobee-Everglades watershed (Figure 2.5), often given the acronym KLOE. Surface water flowed from the Kissimmee River, and several smaller streams ended in Lake Okeechobee, from which there was no outlet during the dry season. But in normal summer rainy seasons, the water level of Lake Okeechobee rose, and upon reaching an elevation of about 18 feet (above sea level), the lake began overflowing its southern rim through a forest of pond apple trees, also called custard apple.* Early explorers tried to find navigable outlets from the lake through streams leading

---

* Not to be confused with a related tropical American fruit tree known also as "custard apple." See Table 4.1.

FIGURE 2.4 Dry algal mat, Shark River Slough, Everglades National Park, April 1975. (Photo by T. Lodge.)

into this swamp, which fringed the south and southeastern rim of the lake like a necklace up to three miles thick. All of these streams ended in the profusion of swamp forest or in dense sawgrass. Beyond the ends of the short streams lay tall, nearly impenetrable sawgrass of the northern Everglades.*

---

* It is noteworthy that passage through the swamp and into the Everglades marshes was possible in canoes during very high water, and that the Everglades itself was regularly traversed by the Native Americans (and a few early explorers) in such craft except in very low water.[590, 591, 656]

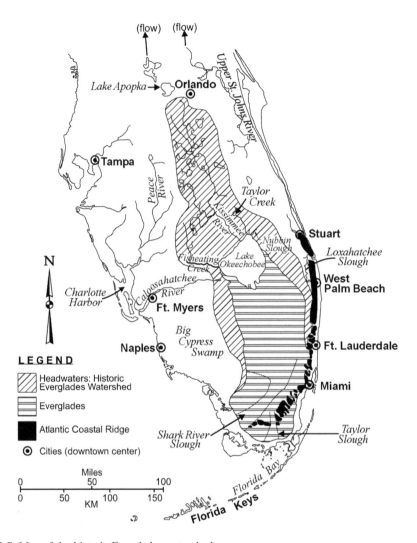

FIGURE 2.5 Map of the historic Everglades watershed.

Sawgrass dominated the northern Everglades interior but was far from the only plant community of this vast ecosystem, which was 40 miles wide and extended almost 100 miles to the tidal waters of Florida Bay and the Gulf of Mexico. The northern Everglades "sawgrass plain" extended about 15 miles south and southeast from the pond apple swamp before giving way to a more varied landscape (see Color Figure 2). Along the edges of the Everglades were more open wet prairies that harbored a rich variety of marsh plants. Through the interior there were deeper open-water sloughs (pronounced "slews") where lily pads adorned the surface. Sloughs were bordered by slightly raised "ridges" of sawgrass, and the landscape was dotted with elevated tree islands. This mosaic covered the major portion of the Everglades where the elongated shapes of sloughs, sawgrass ridges, and tree islands were conspicuously aligned with the southward flow of water (see Chapter 3).

Although sloughs were easy to traverse in canoes, there were no open channels through the Everglades. The water moved in a vast "sheet," more rapidly through relatively open sloughs and more slowly through sawgrass. The slope from Lake Okeechobee to tidal waters averaged about two inches per mile. With slightly higher terrain in the Big Cypress Swamp to the west and the Atlantic Coastal Ridge to the east (its highest elevations scarcely over twenty feet), the Everglades waters were guided slowly southward by the imperceptible slope of the terrain.[128, 456] The rate of flow, slowed by the marsh vegetation, ranged from negligible to about two feet per minute.[512, 628] The depth varied considerably with seasons and years, but rarely exceeded four feet in the deepest marshes. Average wet-season maximum depths in the various marsh plant communities now range between one and two feet.[*]

Throughout the entire ecosystem, the flat terrain and seasonally abundant water made small variations in topography important. Differences of a few inches in elevation cause large responses in vegetation.[128] Ranging from a mosaic of freshwater marsh and tree island communities to cypress, pinelands, and upland forests and finally to saline environments of the coast, these variations added intriguing dimensions to the landscapes of the region. The interesting details of the various plant communities of the Everglades region are discussed in Part II.

ORIGIN OF THE EVERGLADES [581]

The environmental factors that control the ecology of the Everglades — the abundance and seasonality of rainfall, water quality, fire, and unique geographic/geologic features — were also responsible for its evolution from uplands that existed when the sea level was much lower. This evolution began very recently in geologic time, with the accumulation of peat soil in the Everglades starting only about 5000 years ago.[233, 237] From its much lower stand during the last glacial period, sea level had risen enough so that the area that was to become the Everglades could no longer drain rapidly. Florida's rainfall had increased to the point where it resembled today's, a much wetter climate than had existed earlier. An increasing occurrence of flooding began to stress then-existing upland plants and to favor the growth of wetland plants. Under the protection of water, wetland soils began to accumulate, and the Everglades was "born."

A region previously dominated by upland plant communities (including pinelands like those along the edges of the Everglades today) began to disappear.[97, 365, 452] First in the areas of lowest elevation and then progressively higher, the Everglades vegetation advanced, its developing soils covering the original scanty soil and widely exposed limestone, the bedrock "floor" of today's Everglades. Because the accumulation of wetland soils depended upon water level, the developing terrain was very flat, far more level than the original rock floor beneath.[145, 237]

This "growth" of the Everglades proceeded to the level where wet-season waters began to spill over the lower areas of the Atlantic Coastal Ridge. Some water passed from the northern Everglades eastward through the Loxahatchee Slough and Loxahatchee River (in Palm Beach County) to the Atlantic at Jupiter. Farther south, overflow produced some short rivers, such as the New River in Fort Lauderdale and the Miami River. From the southern Everglades there were numerous "transverse glades" — wet season overflows resembling Everglades habitat (see Color Figure 2). A transverse glade existed, for example, near the present location of the South Miami Hospital where high water drained from the Everglades to Biscayne Bay.[259]

Thus, given the elevation of the Atlantic Coastal Ridge, the Everglades evolved to its maximum possible extent, the giant wetland that existed when the first Europeans visited Florida. Native Americans, however, saw it all. They had already inhabited southern Florida for more than 5000 years before the first soils of the Everglades were formed (see Chapter 21: Native Americans and the Everglades).[98, 248]

---

[*] Flooding in 1947 rendered much of the Everglades six to eight feet deep.[456] Water depths in the original Everglades will never be known, but based on descriptions of the vegetation, such excessive depths must have been unusual and of short duration. The depths stated above are from an evaluation of 32 years (1954-1985) of data obtained in Everglades National Park. Because these data were collected after the Everglades was subject to water management control, they are undoubtedly biased to some extent by control of flooding extremes, and by water retention north (upstream) of the park during dry conditions. These recorded levels have correlated with the general integrity of most wetland plant communities but not the animal communities as discussed in the final chapter. Recent evaluation of historic documents indicates that the original, pre-drainage Everglades was significantly deeper, with wet-season slough depths probably averaging three feet. [402]

Part II

Environments of the Everglades Region

3 Freshwater Marshes: Water, Weather, and Fire[*]

Expanses of freshwater marsh are the essence of the Everglades. The productivity and habitat of these marshlands provide the value upon which Everglades wildlife thrives. Seasonally changing water levels alternately produce and then concentrate aquatic organisms, making them easy targets for the ecosystem's top predators. While the expanses of marshes often give us the impression of tranquility, they have been formed and maintained through powerful stresses imposed by water, weather, and fire.

MARSH VEGETATION AND PLANT COMMUNITIES

If a single word had to be used to describe the Everglades, it would be *sawgrass* (specifically called Jamaica swamp sawgrass). However, many other marsh plants occur in the Everglades (Table 3.1), with well over 100 species present. From a biogeographic perspective, most of these species, including sawgrass, have such extensive ranges in temperate and tropical regions that the distinction is unimportant. Overall, however, the marsh flora is more related to temperate North America, with but a few examples (such as alligatorflag) of tropical origin.[366]

Variations in marsh vegetation over the expanses of the Everglades have prompted numerous efforts to define plant communities. Regional variations result from geologic and hydrologic conditions, but many localized variations do not have obvious causes. Because of the region's pronounced wet and dry seasons, conditions promoting and restricting growth of various kinds of marsh plants are always at work, including the rigors of excessive flooding and the opposite, concurrent drought and fire. These stresses have caused the mosaic pattern of growth seen through most of the Everglades, a pattern most easily appreciated from an aerial view. The following categories are generally recognizable and widely used (Figure 3.1), and each provides different ecological functions:

- Sawgrass marsh (grading from dense to sparse)
- Wet prairies (two different kinds)
- Slough (pronounced "slew")
- Pond (alligator hole) and creek

Many species of marsh plants can be found in more than one community. Spikerush, for example, is common in wet prairies and sloughs, and it quickly invades areas where sawgrass has been killed by a soil fire or by salt water from a storm tide near the coast.[17] Spikerush leaves are nearly round in cross section. Initially green, they become brownish-orange upon aging, and early morning or late afternoon sunlight can produce a stunning visual effect in spikerush areas.

Recognizing certain kinds of plants, such as cattails and alligatorflag, can make hiking through Everglades marshes more pleasant. Most people recognize the tall, smooth cattail leaves, but fewer know alligatorflag, a plant with large, light-green oval leaves that may be over two feet long. The presence of either of these two species often indicates deeper water and soft muck soils. Venturing into these areas can be a rude surprise, resulting in stuck feet and a wet wallet (sometimes even if

---

[*] The main sources for this chapter are references 128, 145, 251, 252, 369, 402, and 452.

TABLE 3.1
Representative freshwater marsh plants of the Everglades[a]

Common name[b]	Scientific name[b]	Habitat[c]
widespread maiden fern, southern shield fern	*Thelypteris kunthii*	Sm
southern cattail	*Typha domingensis*	Da, Sm
bulltongue arrowhead (lanceleaf arrowhead)	*Sagittaria lancifolia*	Sl, Sm, Wp, Mp
hairawn muhly (muhly grass or hairgrass)	*Muhlenbergia capillaris*	Mp
maidencane	*Panicum hemitomon*	Wp
bluejoint panicum	*Panicum tenerum*	Mp
Jamaica swamp sawgrass (sawgrass)	*Cladium jamaicense*	Mp, Sm
gulf coast spikerush	*Eleocharis cellulosa*	Mp, Sm, Wp
slim spikerush	*Eleocharis elongata*	Wp
knotted spikerush	*Eleocharis interstincta*	Wp
starrush whitetop	*Rhynchospora (Dichromena) colorata*	Mp
spreading beaksedge	*Rhynchospora divergens*	Mp
narrowfruit horned beaksedge	*Rhynchospora inundata*	Wp
southern beaksedge	*Rhynchospora microcarpa*	Mp
Tracy's beaksedge	*Rhynchospora tracyi*	Wp
black bogrush (black sedge)	*Schoenus nigricans*	Mp
green arrow arum	*Peltandra virginica*	Sm
pickerelweed	*Pontederia cordata*	Sl
string-lily, seven-sisters (swamplily)	*Crinum americanum*	Sm, Wp
alligatorlily	*Hymenocallis palmeri*	Sm
alligatorflag, fireflag	*Thalia geniculata*	Da (deep peat)
spatterdock, yellow pondlily	*Nuphar lutea*	Sl
American white waterlily	*Nymphaea odorata*	Sl
marsh mermaidweed	*Proserpinaca palustris*	Sl, Sm
water cowbane (water dropwort)	*Oxypolis filiformis*	Wp
big floatingheart	*Nymphoides aquatica*	Sl
saltmarsh morningglory (Everglades morningglory)	*Ipomoea sagittata*	Sm
lemon bacopa, blue waterhyssop	*Bacopa caroliniana*	Sl
leafy bladderwort (flatstem bladderwort)	*Utricularia foliosa*	Sl
eastern purple bladderwort	*Utricularia purpurea*	Sl
rosy camphorweed	*Pluchea rosea*	Mp, Sm
semaphore thoroughwort	*Eupatorium mikanioides*	Mp

[a] Listing is in phylogenetic order, following reference 661.
[b] Common and scientific names follow references 222 and 661. Alternate South Florida vernaculars are in parentheses
[c] Habitat key: Da= disturbed areas; Mp=marl wet prairie; Sl=slough; Sm=sawgrass marsh; Wp=wet prairie on peat

it is in a shirt pocket). There may also be another, possibly dangerous surprise. These plants frequently fringe alligator holes. This thought can be disconcerting while you are struggling to catch your balance and your feet are stuck in the muck.

SAWGRASS MARSH

Sawgrass is not a true grass but rather a member of the closely related sedge family (Figure 3.2). It is named for the sharp, upward pointing teeth that line each edge of the leaf's stiff "V" cross section. Once its highly perishable seedlings become established, sawgrass is a very tough plant, well adapted to the rigors of the Everglades, and most of its subsequent reproduction is by lateral spread of its prolific root-like rhizomes. Its principal enemies are soil fire during drought, multi-year flooding deeper than about a foot, and fire followed by rising water. The

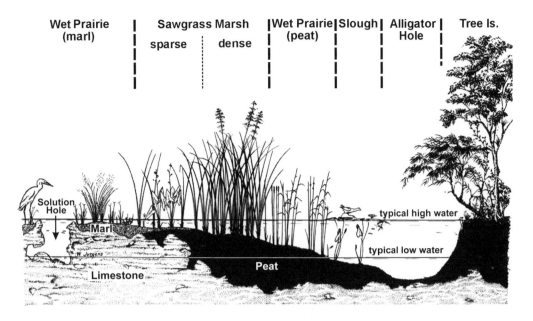

FIGURE 3.1 An idealized, greatly compressed section of typical Everglades plant communities. (Drawing by Wes Jurgens).

last of these conditions eliminates the oxygen supply to the roots, causing suffocation. As each sawgrass culm* matures, the older outer leaves die, and their conducting tissues become air tubes — tiny straws — that carry oxygen to the roots, which live in an oxygen-deprived environment when the soil is flooded. While sawgrass is recovering in the month or so after a burn, the roots are dependent on this oxygen supply, which would be cut off by flooding over the dead leaves. The implication to water-level management is obvious for dealing with too much or too little sawgrass. On its tougher side, essentially nothing eats sawgrass, although the abundant, late summer seeds are an important food for ducks.[19, 26, 201, 252, 272, 663]

Sawgrass marshes are areas overwhelmingly dominated by sawgrass, with little other conspicuous vegetation. Spacing of the plants grades from sparse to dense (Figures 3.3 and 3.4). Dense tall sawgrass originally covered most of the northern Everglades in an unbroken expanse called the "sawgrass plain." There it grew to over ten feet tall in deep, peat soils. Most of the sawgrass plain area is now in agriculture, mainly sugarcane, but some tall sawgrass remains in northern Everglades conservation areas. In the central and southern Everglades, sawgrass marsh typically occurs in patches; its stature is not so tall (usually three to five feet), and several other marsh species may occur together with the sawgrass.

The average hydroperiod** for sawgrass marsh is about ten months, but it ranges from less than six months to nearly continuous flooding, and typical wet-season depths range from one to one-and-a-half feet. The deeper water and longer hydroperiod support taller, more dense sawgrass, with the drier end of the range supporting more open, sparse growths. Dense sawgrass, especially, harbors little animal life, but is one of the habitats where alligators build their nests.[93, 360, 387]

---

* A culm is the vertical stalk and leaves (of grasses, sedges, and related plant families) that a lay person might call an individual "plant." Culms are interconnected by horizontal stems called *rhizomes* (not roots but roots may grow from nodes along the rhizome), making it a problem to define what is an individual "plant." As gardeners know, every time a rhizome is broken, two "plants" are made from one.

** Hydroperiod/hydropattern estimates are from References 250, 251, 322, and 452.

FIGURE 3.2 Sawgrass in bloom. (Photo by T. Lodge.)

Wet Prairies

Wet prairies are areas of marsh dominated by emergent plants other than sawgrass (and only superficially resembling prairie habitat of the Great Plains). The term *wet prairie* is not standardized in the Everglades literature and has been applied to two distinct communities. In common, these two communities have lower and less dense vegetation than sawgrass marsh, providing more open water for aquatic animals.[360]

FIGURE 3.3 Sparse sawgrass with periphyton over most of the surface, primarily growing on bladderwort. Water Conservation Area 3B. (Photo by T. Lodge.)

One type of wet prairie, termed *marl prairie*, occurs on thin calcitic soil (marl) over limestone bedrock, which may be exposed as jagged projections rising as high as one foot and called *pinnacle rock* or may be dissolved below the surface into pockets called *solution holes*. Such areas with pinnacle rock and solution holes are called *rocky glades*. Marl prairie hydroperiod is the shortest of all the marsh types in the Everglades, ranging between three and seven months, with a mean depth of only four inches. This community flanks both sides of the southern Everglades, where it was once much more extensive, especially in Miami-Dade County. It virtually always has scattered low sawgrass (less than three feet tall), but plant diversity is high. About 100 species occur there, thus the alternate term *mixed prairie*. Common species are beaksedges, spikerush, starrush whitetop, and muhly grass. The last of these can be recognized in summer by its pink inflorescence, sometimes giving whole vistas a pink glow. An interesting variation of marl prairie contains scattered, dwarf cypress trees, and is called *dwarf cypress prairie*, the most bizarre interior landscape in South Florida (Figure 3.5).

Certain marl prairies can be important short-hydroperiod feeding areas for wading birds, supplying prey early in the dry season (November into January) when water levels are initially receding. Rocky glades are also important in the early wet season as prey emerges from solution holes upon initial flooding. Snails, crayfish, small fish, and amphibians survive through the dry season in flooded underground cavities. Without solution holes that reach into the dry-season water table, marl prairies have less ecological value. As a cautionary note to hikers, solution holes may be hidden by marsh vegetation, especially in the wet season when covered by floating vegetation including mats of algae. For this reason, the rocky glades along the main Everglades National Park road between Taylor Slough and Pa-hay-okee Overlook are very hazardous (Figure 3.6).[314, 358, 359]

The other kind of wet prairie is a deeper-water marsh community that occurs on peat soil and has a long hydroperiod and lower plant diversity.[242, 251] Maidencane, Tracy's beaksedge, or spikerush usually dominate. These areas more commonly occur in the northern and central Everglades, where they characteristically lie between sawgrass marshes and sloughs. They are important for fish and

FIGURE 3.4 A wall of dense sawgrass borders a slough. The difference in soil elevation is typically a foot and a half. (Photo by T. Lodge.)

aquatic invertebrates, such as prawns, which require more permanent water. This habitat provides abundant prey for wading birds toward the end of the dry season, typically March and April. Soil evidence indicates that this peatland wet prairie was never common through Everglades history (see "Marsh Soils" section below).

FIGURE 3.5 Dwarf cypress prairie near Pay-hay-okee, Everglades National Park. Note the cypress dome in the background. The photograph was taken in mid-February, 2002, while the cypress were still in winter, defoliated condition. (Photo by T. Lodge.)

SLOUGH

These deepest marsh communities are the main avenues of water flow through the Everglades. The average annual hydroperiod is about 11 months, and flooding may be continuous for several years. Water depth in the rainy season may be over three feet, and the annual average is now about a foot, which is significantly deeper than the other marsh habitats. Depths prior to hydrologic modifications of the Everglades were greater.[402] Like the longer hydroperiod type of wet prairie, sloughs occur over peat soil and support an abundance of fishes and aquatic invertebrates.[357, 393, 477, 604] The dominant vegetation includes submerged and floating aquatic plants such as bladderworts, white waterlily, floatingheart, and spatterdock (see Figures 3.4 and 3.7). Normally, there is little or no sawgrass, but maidencane may be abundant.

Areas where slough communities are abundant are dotted with tree islands and contain large, elongated patches of sawgrass marsh. This configuration is called *ridge-and-slough* landscape because the sawgrass habitat is on elevated "ridges" (relative to the foot-and-a-half deeper sloughs). Tree islands (see Chapter 4) are on still higher ground. The Shark River Slough is a major Everglades flow-way where slough habitat is abundant. Historically, it was over 20 miles wide where the Tamiami Trail crossed it, narrowing to about six miles southward in Everglades National Park. Its waters flow into several tidal rivers in the coastal area at the park (one of which is the Shark River, thus the slough's name). Other flow-ways that carried Everglades water south and southwest were Lostmans Slough along the border of the Big Cypress Swamp (separated from the Shark River Slough by a higher area of marl soils) and the much smaller Taylor Slough, which is on the east side of the southern Everglades (see Color Figures 2 and 3).

FIGURE 3.6 Solution hole in the rocky glades west of Taylor Slough, Everglades National Park. Photographed late in the dry season, the pool of water still contained several small fish. (Photo by T. Lodge.)

RIDGE-AND-SLOUGH LANDSCAPE AND WATER FLOW [40]

A glance at an aerial photograph of well preserved Everglades landscape reveals a pattern that appears obviously related to water flow (Figure 3.8, also see cover). Once more widespread, ridge-and-slough landscape now characterizes a protected part of the northeastern Everglades (the Arthur R. Marshall Loxahatchee National Wildlife Refuge, historically called the Hillsborough Lake area

FIGURE 3.7 Aerial view of sawgrass patches amid more open water slough with abundant lily pads, Water Conservation Area 3A. (Photo by T. Lodge.)

FIGURE 3.8 Aerial view of ridge-and-slough landscape. Looking south, Water Conservation Area 3A. (Photo by T. Lodge.)

because of the abundance of open-water slough habitat) and the Shark River Slough of the central and southern Everglades. It has recently been determined that the central Everglades between these two areas was also ridge-and-slough habitat (see Color Figure 2), but had been degraded by drainage canals (constructed between 1910 and 1920) so that vegetation mapping in the early 1940s[145] did not recognize this earlier landscape feature.[402] The orientation of the elongated sloughs and sawgrass ridges, and of tree island tails (see Chapter 4), are clearly in the direction of historic water flow. Bedrock highs and lows have no relationship to ridges and sloughs, but high points are often the basis of tree islands (Figure 3.9).

The origin and maintenance of the flow pattern are not understood. The obvious hypothesis is that water flow is involved because other explanations, such as those involving wind and fire, would not create such directional features. The intriguing question is whether average-year flows are sufficient or are rare excessive flows required, such as those that occurred in 1947 by two hurricanes on top of an abnormally rainy wet season. The first puzzle is the initial formation of slough habitat. Did the pattern form, for example, from sawgrass marsh by deep soil fire scars that became sloughs, subsequently shaped by flow, or is there another explanation? Second, what water-flow velocities are required for initial formation and for subsequent maintenance? Another question is: why was there no slough habitat in the historic sawgrass plain of the northern Everglades? At the heart of scientists' consideration of these questions is the assumption that flow carries light, flocculent sediments, eroding them from, and preventing their settling in, slough habitat and selectively causing them to settle in sawgrass habitat where flow is slowed by abundant plant stems. Various questions are easy to ask, but demonstrations and proof are difficult, presenting obvious areas for research. The answers will likely have important applications in restoration of the Everglades (see Chapter 20).

FIGURE 3.9 A hypothetical oblique section showing Everglades ridge-and-slough landscape with tree islands. It shows that the bedrock is the same below sawgrass ridges and open sloughs, but may have a prominence below tree islands (Redrawn by W. Jurgens from Ref. 40).

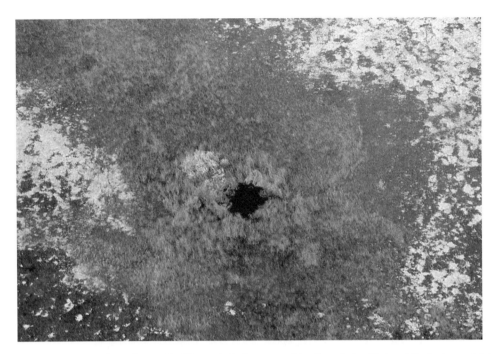

FIGURE 3.10 Aerial view of a small alligator hole in wet prairie, Everglades National Park. A coastalplain willow is at upper left of the pond, which is only about 10 feet across and mostly bordered by sawgrass. The light color areas in the marsh at left and upper right, extending down the right side, are periphyton. (Photo by T. Lodge.)

POND (ALLIGATOR HOLE) AND CREEK [93, 200]

Small open-water areas are scattered through most of the Everglades (Figure 3.10; also see Figure 17.6 and Color Figures 8 and 9). They are flooded continuously except in the most severe droughts. They occur mostly in marshes having longer hydroperiod and not in marl prairies. Most of them probably originated where alligators excavated an existing depression, such as where fire burned a pocket in peat soil during an extended drought. Whatever the origin, alligators perpetuate these ponds by their activity, most notably where alligators forage in these ponds during low water and dislodge rooted plants in the process (see Figures 17.7 and 17.8). Thus, active alligator holes are relatively free of vegetation. There may be scattered spatterdock and other species of aquatic and floating plants like those of the sloughs, and wetland trees take root on the banks. Alligator holes (see the alligator section in Chapter 17) are very important as *refugia*[*] for aquatic life during the dry season. In the extreme southern Everglades, where freshwater and tidal water meet, there are numerous creeks (see Chapter 8) that function like alligator holes and are maintained by alligator activity. The creeks perform the same important function for wildlife as alligator holes. [360]

PERIPHYTON [82, 235, 360, 393, 584]

Although rooted wetland plants dominate the appearance of the Everglades, algae may be the more important producers at the base of the food chain. Algae in the Everglades, as in aquatic habitats everywhere, grow on any submerged surfaces wherever adequate sunlight is available. Plant stems

---

[*] The term *refugia* (plural of the Latin *refugium*) is used to avoid confusion with various uses of the word "refuge," such as a national wildlife refuge. In Everglades jargon, refugia are habitats where aquatic organisms survive temporary drydowns of Everglades marshes and include solution holes in the rocky glades and alligator holes, but also include sloughs in years when they remain continuously flooded.

FIGURE 3.11 Abundant periphyton on plant stems, Shark Valley, Everglades National Park. (Photo by T. Lodge.)

and the soil surface of the marsh are covered with a complex association of numerous kinds of algae in a community called *periphyton* (Figure 3.11). Where it grows on the soil surface, it is commonly called *algal mat*. Snails, many small crustaceans and aquatic insects, and herbivorous fishes graze on this very important food source. Periphyton is least abundant where there is heavy shading, as in dense sawgrass, and most abundant where sunlight is strong as in sparse sawgrass, wet prairies, and sloughs that have abundant floating vegetation, such as bladderwort. In addition to supporting more aquatic animal life, abundant periphyton produces higher dissolved oxygen because of photosynthesis. Limited periphyton is the main reason that dense sawgrass communities do not support an abundant aquatic life.

Two general kinds of periphyton occur in the Everglades. In more acid or soft water, with little dissolved calcium, which normally occurs in areas of peat soils such as sloughs and deeper wet prairies, the periphyton is a thin association of mostly green algae. This type has been shown to be a nutritious food for grazing invertebrates and fishes, such as grass shrimp, applesnails, flagfish, and sailfin mollies.

In areas of harder water, which contain abundant calcium from contact with limestone bedrock, the periphyton looks and feels very different. It contains relatively more blue-green algae (technically *cyanobacteria*), is usually a light yellowish-brown, and forms a thick mass with a spongy texture. Around thin plant stems, it may be a half inch in diameter, and over the bottom an inch or more thick. The texture is due to minute calcium carbonate crystals that are formed as the algae remove carbon dioxide from the water during photosynthesis. This type of periphyton is also consumed by grazers, but is not as nutritious as the other type, possibly because of the calcium carbonate particles (which make it "crunchy"). It typically occurs in areas of short hydroperiod, notably in the wet prairie community on marl soil.[233, 358, 540]

Periphyton has been shown to be sensitive to changes in water chemistry, including nutrients, as evidenced by shifts in the types of algae that compose it. Typical periphyton in the Everglades is adapted to water containing low levels of nutrients, and the addition of nutrients changes the assemblage. Certain types, notably species of cyanobacteria, can "fix" nitrogen, and they become more common when the water is enriched in phosphorus.[108]

MARSH SOILS [111, 128, 233, 237, 563]

Two kinds of soil occur in the Everglades: marl and peat. Both are mentioned above in the context of the plant communities. Marl is a product of periphyton. During the dry season, the organic material (dead algae and cyanobacteria) in the periphyton mass oxidizes, leaving the calcium carbonate particles as this light-colored soil. Marl is descriptively called *calcitic mud* and is the main soil of the short-hydroperiod wet prairies near the edges of the southern Everglades, where the bedrock lies close to the surface. It is the primary soil type along much of the main park road to Flamingo.[514]

Most marsh peat (and related muck) is a product of long-hydroperiod wetlands and usually occurs in areas where the bedrock lies deeper. Peat is composed of the organic remains of dead plants. Examination, requiring magnification, can reveal the kinds of plants involved if the soil has remained undisturbed. Two common types of peat occur in Everglades marshes: *Everglades peat* and *Loxahatchee peat*. Everglades peat is composed almost entirely of the remains of sawgrass and is often descriptively called *sawgrass peat*. It is dark brown to black in color, the darkness attributed to charcoalized matter from frequent fires. Everglades peat is the more widespread type, especially in the northern and central Everglades, and it originally covered about 1700 square miles of the Everglades. Loxahatchee peat occurs in deeper marsh areas, typically the northeastern Everglades and the Shark River Slough. It is lighter colored (fire is rare in sloughs) and contains the remains of slough vegetation, especially the roots, rhizomes, leaves, and seeds of waterlilies. It covered about 1140 square miles of the historic Everglades. Oddly, no peat soil type has been found that represents the peatland wet prairie community. Because of the lack of wet-prairie peat, it is assumed that this community was relatively rare through Everglades history, perhaps merely a short-term transitional community between slough and sawgrass.[581]

If the peat soil has become too finely divided (often due to animal activity by everything from earthworms to alligators), the soil is commonly called muck. This term also applies to black peat soil combined with inorganic material.[87] The former Everglades soils near Lake Okeechobee ("consumed" by agricultural development by the mid-1900s) contained substantial sediments originally from the lake and is thus called muck or peaty muck. Closest to the lake was *Okeechobee muck*, followed southward by a band called *Okeelanta peaty muck* that lay just north of the widespread Everglades peat. Together they covered about 90 square miles.

Coal is fossilized peat and muck soils, and coal even hundreds of millions of years old often contains fossil imprints of the plants that flourished in those ancient wetlands. Most of the peat in the Everglades is the remains of marsh vegetation, primarily roots, and it occurs where the historic hydroperiod was long enough to keep the soil saturated throughout most years. This condition occurred historically in sloughs, long hydroperiod wet prairies, and deeper sawgrass marshes.

Water is required for the formation of peat soil because it restricts atmospheric oxygen from the soil. Without oxygen, microorganisms cannot decompose dead marsh plants as fast as they accumulate. Through the years, ever thicker layers of peat are deposited until an elevation is reached where the surface dries enough that oxygen-dependent decay (the opposing aerobic process), or fire, prevents accumulation.

Near the turn of the century in the northern Everglades, the peat and muck soils were commonly 14 or more feet deep over the rock base when drainage activities began. In the southern Everglades, peat soils were much thinner, but in the deepest parts of the Shark River Slough, there were (and still are) several feet of peat. Look back to the northern Everglades, where most of the area has been developed for agriculture. Those soils were coveted because they were enriched by sediments from Lake Okeechobee, but agricultural drainage and soil tilling practices stimulated their loss. Through the decades, these soils, supporting some of the most productive agriculture in the world, literally disappeared. The process, called *subsidence*, has resulted from compaction (by drying), from fire, and from aerobic decay. Like coal, dried peat will burn, and fires have been responsible for substantial soil losses. However, the most persistent factor is aerobic decay by microorganisms that degrade the peat into carbon dioxide and water. In large areas, more than eight feet of soil depth has been lost. Visual evidence has been provided by houses built on pilings, but with their frames originally located on the existing ground surface. Years later these houses stood high in the air, requiring long stairways to reach the doors.[87, 563, 575]

Marl and peat soils are essentially opposites, requiring aerobic and anaerobic soil conditions, respectively. Peat does not accumulate in shorter hydroperiod marshes where marl persists, and acid conditions within a peat soil dissolve marl, preventing its accumulation. However, there are large areas in the northern Everglades where a thick layer of marl *underlies* peat. This occurrence stands as evidence that the hydroperiod increased substantially in those areas as the Everglades ecosystem evolved.

WATER QUALITY [108, 149, 393]

The appearance of the water in the Everglades marshes surprises many first-time visitors. They frequently expect the water to be unpleasant in some way (perhaps turbid and foul smelling), but it is clear. The water ranges from very soft and slightly acidic where it is derived from rainwater that has only contacted vegetation and peat soils, to hard and alkaline (containing abundant dissolved calcium and other ions) where it has contacted the limestone bedrock. Changes in Everglades water quality have become highly controversial. Relevant issues including urban and agricultural activities, water routing by canals, and even man-altered rainwater chemistry are addressed in Chapter 21. The "original" conditions are covered here.

Historically the water of the central and southern Everglades has been very low in the nutrients that promote plant growth, a condition called *oligotrophic* by ecologists. Concentrations of nitrogen and especially phosphorus, which are principal components in fertilizers, were notably low. Most ecosystems are nutrient-limited. Of the many nutrients required by plants, the one that is in least supply relative to plant growth responses is defined as the "limiting nutrient." Forms of nitrogen or phosphorus that are usable by plants are common limitations. Phosphorus, at levels of or below about 10 parts per billion (ppb) or less, has been demonstrated to be the limiting nutrient in most of the Everglades. The historic level of phosphorus in Lake Okeechobee, although limiting there also, was substantially higher (20 ppb or more) than in the Everglades, apparently because of natural sources of phosphorus within its watershed. The nutrient condition of the historic lake is described as *eutrophic* (see Chapter 11). The original swamp forest around the south shore of Lake Okeechobee and the dense, tall sawgrass of the northern Everglades sawgrass plain were probably stimulated by phosphorus from the lake's overflow waters. In turn, these communities extracted the phosphorus, enriching the northern Everglades soils and leaving the downstream Everglades in its oligotrophic condition. Important sources of phosphorus for the central and southern Everglades were small inputs from rainfall and airborne dust because little arrived in the flowing water.

The characteristic Everglades marsh plants thrive in this low-nutrient environment, but this condition accounts for the rather sparse, open character of much of the marsh and for the abundant periphyton, which thrives in the oligotrophic environment. In turn, the open character and periphyton are conducive to the kinds of aquatic animals (fish and invertebrates) that are present. The sparsely vegetated water surface promotes gas exchange with the air and periphyton provides oxygen within the water. These features maintain a well-oxygenated aquatic environment. In marshes with nutrient-rich waters, other kinds of vegetation predominate, sometimes in such thick profusion that there is little open water and little periphyton. Such situations promote poorly oxygenated conditions and inferior habitat for most amphibians and fishes, although certain aquatic insects and other invertebrates may be abundant.

Cattails are among the marsh species that compete successfully where there are higher levels of nutrients, or where there has been soil disturbance. They occur naturally where soil fires removed competing marsh plants (and released nutrients) or in and around disturbed and enriched depressions such as alligator holes. Sediments in alligator holes contain elevated concentrations of phosphorus from excrement of wading bird- and alligator-use during low water, and from aquatic life that perishes there during severe droughts.[350] Two species of catttails occur in the Everglades. The broadleaf cattail (*Typha latifolia*) is relatively rare, but the southern cattail (*T. domingensis*) has invaded sawgrass and other marshes since the 1970's, apparently due to nutrient enrichment.[149, 393, 488, 540] Cattails do not normally occur even in the deeper marsh communities, such as peat-based wet prairies* and sloughs, where the depth should not restrict their invasion — they tolerate deeper water than does sawgrass.[19] They are mentioned only as a minor element in descriptive literature of natural Everglades plant communities.[145, 369] Their presence during the initial formation of the Everglades has been confirmed by paleobotanists who have found cattail pollen in certain older peat deposits, but peat composed of cattail roots and leaves has not been found in the Everglades, confirming that cattails were not common.[237]

One final thought on the importance of Everglades water quality (as well as quantity): water from the Everglades directly enters the groundwater supply of Florida's metropolitan southeast coast. The same highly porous limestone bedrock, the Biscayne Aquifer, is continuous from the central and southern Everglades to these areas. Wells in and near urban centers are used for the region's water supply.[316, 416]

WEATHER AND FIRE [106, 128, 253, 271, 281, 623, 657]

Any study of the Everglades requires an understanding of the role of fire, without which the Everglades would be very different. Fire is an important controlling agent for most habitats of southern Florida, as it is in all seasonally dry areas of the earth where lightning occurs. Without fire, wetland shrubs and trees would invade Everglades marshes, especially shorter hydroperiod communities including marl prairie and sawgrass. The process is called *succession* (see Chapter 20). Fire sets succession back and has the positive effect of releasing and redistributing nutrients that become "locked-up" in plant tissues during their growth. Death and decay would eventually release the nutrients, but fire is faster. Thus, in weeks following "normal" fires, new growth rejuvenates the marshes. However, defining "normal fire" requires an understanding of the region's seasonal weather patterns.

It is during the summer rainy season that nature's fire starter — lightning — plays an important role. The summertime frequency of lightning in peninsular Florida is very high. In South Florida, lightning strikes the ground about 25 times in each square mile annually, and most of these strikes occur in June through September. [251, 271, 440, 657] The fact that lightning starts wildfires is well

---

* Certain established species such as knotted spikerush appear to inhibit cattail invasion,[348] but not sawgrass. Cattails are well known for their rapid invasion of a wide variety of disturbed habitats and are often cited as a problem in wetland mitigation.[140, 187, 221, 441]

documented,[504] and in an eye-opening demonstration, a lightning strike about 100 feet from the author made a cabbage palm seemingly explode into flame.[348] Summertime fires are usually small in area and do not damage soil or roots because abundant rainfall and higher water levels increase soil moisture. Marsh vegetation is reduced in stature, invading shrubs are set back, and dead portions of plants are consumed. Most of these effects are beneficial, short-term changes. Sawgrass marsh, for example, is perpetuated by most wet-season fires (Figure 3.12).

Dry season fires can change Everglades landscapes extensively, especially during severe droughts when peat soils are dry enough to ignite. At those times, roots and soils both go up in smoke, along with the rest of the plants. Thunderstorms that immediately precede the rainy season in April or May, can start fires that sweep over huge areas of the still-dry Everglades. Such catastrophic fires have always been part of the natural regime, and their charred evidence is found in buried layers of peat soils.[110, 237] These fires reduced accumulated peat levels, setting succession back and thus perpetuating long-hydroperiod marsh habitat. Documented examples include the conversion of sawgrass marsh to spikerush wet prairie.[253] Peat fires occur infrequently, however, perhaps once in a decade; thus, in the long intervening periods, Everglades wildlife has, in the past, prospered. Since the early 1900's, however, man's unmanaged dry season fires — due to arson or

FIGURE 3.12 Sawgrass recovering after a wet-season fire, Water Conservation Area 3A. (Photo by T. Lodge.)

accident — have been a problem of great magnitude. Many are set in the mid-dry season (January through March) when cooler temperatures are inviting for people to explore the wilderness. Such timing is highly unnatural for South Florida landscapes.[*]

Despite grave concerns about unmanaged wildfires, since the 1950's there has probably been more alteration of Everglades marshes by too little fire than by too much fire.[7] Roads and canals act as firebreaks, reducing the spread of fires, and fire fighting has traditionally endeavored to suppress or extinguish wildfires without regard to ecosystem benefits, especially (and understandably) near highways and residences. To improve this situation, wildlife and ecosystem scientists have increasingly used fire for regional natural-area management. Prescribed burning in the Everglades and Big Cypress Swamp is now an important tool in an evolving science that strives to mimic the natural effects of fire. For example, the ability to perpetuate or terminate sawgrass marsh by coordinating fire and water levels is well known (see Sawgrass Marsh section, above). The uses and effects of fire on other plant communities are discussed in the succeeding chapters.

---

[*] Native Americans in the Everglades region used fire, primarily as an aid in hunting, and the region was altered as a result. However, what impact these native humans had still resulted in an ecosystem that provided varied habitats, including fire-sensitive plant communities and abundant wildlife.[248, 504] See also Chapter 21.

4 Tree Islands*

In South Florida, the term *tree island* is most often used to depict the island-like appearance of a patch of forest in an Everglades marsh, the topic of this chapter (Figure 4.1). The term is sometimes extended to mangrove islands in Florida Bay and to broad-leaf forest patches (hammocks) in pinelands, but such alternate uses are rare. The "island" part of the term is its superficial resemblance. In the Everglades, if you are looking for relief during a summer adventure, you are in for an unpleasant surprise: most of the tree islands have no dry land during ordinary wet-season water levels.

Several forested plant communities occur as tree islands in the Everglades. Wetland shrubs and trees found throughout the region's swamps occur on all Everglades tree islands. In addition, those tree islands having ground rising above ordinary flooding elevations support upland tree species that only survive in aerobic (unsaturated) soils. The upland forest community is called a *hammock* and is covered in Chapter 5. This chapter deals with the wetland forest communities of tree islands. In contrast to tropical domination in the hammocks (at least those of the southern Everglades), most of the wetland species are of temperate, North American origin, and are common in swamps throughout the southeast. Common wetland tree island species in the Everglades are listed in Table 4.1.

Tree islands occur in many wetlands around the world. In this hemisphere, wetlands with tree islands range from the extreme cold regions of northern Canada and Alaska through temperate climates from Minnesota to the Okefenokee Swamp of Georgia, to the tropics of South America. Although the tree islands in each region may have unique features, there are more similarities than differences. Common processes account for their formation, and they are universally important to the surrounding marsh by providing habitat diversity that supports many added wetland functions. In the Everglades, tree islands provide nesting sites for turtles and alligators, protective cover for wildlife such as deer and otters, and roosting and nesting locations for many birds, notably snail kites, anhingas, and wading birds. Abundant evidence also shows that Native Americans used the high ground on tree islands through a long history (see Chapter 21). Without tree islands, marsh habitat would have far less value to an ecosystem.[77, 97, 210, 410, 645]

SIZE AND LOCATION OF EVERGLADES TREE ISLANDS

Everglades tree islands range from as small as one hundredth of an acre to hundreds of acres. The largest examples are found in the Shark River Slough north of the Tamiami Trail, where many have interesting names like Nut House Island, Tommy Tiger's Camp, Cumpleaños Perdido Island, and Skinner's Island. Skinner's Island is by far the largest in the Everglades with a swamp-forest area of about 650 acres; it can be easily identified on satellite images (see Color Figure 3). The names are often for a camp on a small area of high ground typically on the island's north end. Huge numbers of tree islands without high, dry ground have not been used for camps and, thus, usually remain nameless.

Tree islands occur both in peatlands of the Everglades "mainstream" and in peripheral marl prairies of the southern and southeastern Everglades. Those in the peatlands are the more numerous and have received the most attention. Today, most peatland tree islands are in the northeastern

---

* The main sources for this chapter are references 128, 145, 452, and 538.

FIGURE 4.1 A bayhead in the southern Everglades near Pah-hay-okee, Everglades National Park. A cabbage palm stands prominently at the edge. (Photo by T. Lodge.)

Everglades (the "Loxahatchee," short for the Arthur R. Marshall Loxahatchee National Wildlife Refuge created to preserve it) and in the Shark River Slough of the central and southern Everglades. In both areas, the tree islands are part of the ridge-and-slough landscape. However, ridge-and-slough landscape was historically contiguous between Loxahatchee and Shark River Slough (see Chapter 3 and Color Figure 2). The gap that exists there today lost its historic ridge-and-slough characteristics and most tree islands as a result of early drainage and attendant soil fires.[402]

KINDS OF TREE ISLANDS [34, 150, 233, 237, 369, 623]

For most people familiar with the Everglades, kinds of tree islands are recognized for their appearance, related to the kinds of trees that dominate them regardless of how they formed (see next section). The word "head" is the standard jargon for a tree island. Ask a hunter or fisherman about tree islands, and your answer will be names like "bayhead," "willow head," and "cypress head." These common terms are described as follows:

BAYHEADS

Bayheads (see Figures 4.1 and 4.2) are by far the most common type of tree island. They are named for swamp bay* but may often contain other species that superficially resemble swamp bay, such as sweetbay. In fact, swamp bay may not even dominate a bayhead. A good example is the strand-type tree island of Loxahatchee, which is nearly always dominated by dahoon holly yet still commonly called a bayhead. Bayhead hydroperiods vary greatly. Dryer types dominated by swamp bay range from one to four months, but those dominated by dahoon holly may be up to 10 months. Even tree islands with an upland area are regularly called bayheads although they support a diverse assemblage of tropical hardwood trees called a hammock (see Chapter 5). Hammock species grow to a much greater height than bays and would not flood in normal years.

---

* Widely called red bay. See Table 4.1 for comment on swamp bay.

TABLE 4.1
Representative wetland trees, shrubs, and understory plants of Everglades tree islands[a]

Common Name[b]	Scientific Name[b]
Trees	
pond-cypress	*Taxodium ascendens*
bald-cypress	*Taxodium distichum*
strangler fig	*Ficus aurea*
sweetbay	*Magnolia virginiana*
pond apple (custard apple [c])	*Annona glabra*
swamp bay[d]	*Persea palustris*[d]
red maple[e]	*Acer rubrum*
Carolina ash, water ash, pop ash[e]	*Fraxinus caroliniana*
Small Trees/Shrubs	
cabbage palm	*Sabal palmetto*
Carolina willow, coastalplain willow	*Salix caroliniana*
wax myrtle, southern bayberry	*Myrica cerifera*
coco plum	*Chrysobalanus icaco*
dahoon (dahoon holly)	*Ilex cassine*
common buttonbush	*Cephalanthus occidentalis*
American elder, elderberry	*Sambucus canadensis*
Understory and Ground Cover Plants	
royal fern	*Osmunda regalis*
giant leather fern	*Acrostichum danaeifolium*
swamp fern, toothed midsorus fern	*Blechnum serrulatum*
sword fern, wild Boston fern	*Nephrolepis exaltata*
hottentot fern (shiny maiden fern)	*Thelypteris interrupta*

[a] Selection of species is based on information provided in references 77, 128, 269, 369, 452, and 652.

[b] Common and scientific names follow references 222 and 661. Alternate South Florida vernaculars are in parentheses.

[c] The name "custard apple" has been used for pond apple, especially around Lake Okeechobee, causing confusion with a related tropical American fruit tree, *Annona reticulata*, which is also known as custard apple and has been introduced and grown in South Florida for its sweet, palatable fruit (unlike pond apple).

[d] Everglades literature usually lists red bay (*Persea borbonia*), not swamp bay. The reason stems partly from their similarity, but mostly from taxonomic literature (e.g. Tomlinson[599]) that treats the two as varieties of red bay (*P. borbonia*). By treating red bay and swamp bay as a separate species, recent work, including examination of herbarium specimens, has shown that swamp bay is the species in Everglades wetlands, and that red bay occurs coastally on higher ground, including dunes. Treated as separate species or as varieties, the important point is that they are ecologically distinct. Swamp bay occurs in wetter habitats.

[e] Red maple and pop ash are good examples of the reduced temperate flora in extreme southern Florida. They are common in tree islands of the northern Everglades and Big Cypress Swamp but decline southward and are rare in the southern Everglades.

WILLOWS AND WILLOW HEADS

Coastalplain willow, a low tree that colonizes rapidly with wind blown seeds, may form discrete willow heads but is associated with a variety of disturbed habitats. It prefers lower, wetter elevations than most species of the bayhead community and is a good marker for an active alligator hole (see Chapter 17 and Figure 3.10). Willow heads often mark the location of a previous bayhead that was completely destroyed by fire, especially where fire has burned the peat soil, reducing the elevation. Willows have also overtaken extensive areas of the Everglades where the combination of reduced

FIGURE 4.2 Abundant bayheads in Loxahatchee. The larger bayhead at upper left is a strand-type tree island developed on a sawgrass ridge. (From Ref. 581, with permission from Kluwer Academic Publishers.)

water level and fire have eliminated sawgrass. However, in the long term without fire, willows are typically displaced by other tree species or by sawgrass, depending on the flooding conditions and fire regime.

CYPRESS, CYPRESS DOMES, AND CYPRESS HEADS[188]

Cypress forests of various kinds occur widely in the Big Cypress Swamp (see Chapter 7) but are very limited in the Everglades. Historically, a narrow cypress swamp bordered the eastern edge of the Everglades in Palm Beach and northern Broward counties, south to the New River at Fort Lauderdale, and small vestiges of this former community are scattered throughout developed areas of those counties. Examples are found along Sample Road between State Road 7 and Rock Island Road in northern Broward County, and at the Loxahatchee National Wildlife Refuge visitors' center, located in the former headwaters of the Loxahatchee Slough. The latter has a boardwalk that affords an excellent view of this beautiful remnant.

Areas of cypress in the southern Everglades include the Long Pine Key area of Everglades National Park, where most of the cypress exist as dwarf cypress in marl prairies. Occasional depressions in marl prairies contain dense cypress communities that are called cypress heads or domes. The depressions provide better growth conditions for cypress (Figure 4.3).

The roots of cypress trees produce growths that protrude far above the soil and look like sawed-off trees that healed over. These structures, called knees, may range from a few inches to more than six feet tall, apparently in response to local water level fluctuations. Cypress knees are thought to be "breathing" organs for the roots, where gases are exchanged with the atmosphere. Cypress trees typically grow in hydric, oxygen-deficient soils. Roots require oxygen, just as other plant tissues do. Only the leaves produce oxygen, and then only during daylight. Thus, the knees are an adaptation to anaerobic soils. Such adaptation is described for sawgrass (Chapter 3) and is also important to mangroves (Chapter 8).[599]

FIGURE 4.3 Cypress dome amid marsh habitat in the Big Cypress National Preserve on the Tamiami Trail. Note the precise, graph-like arrangement of tree height increasing toward the center. Similar small domes occur in the Everglades National Park along the main park road between Pay-hay-okee and Mahogany Hammock (Photo by T. Lodge.)

Another characteristic of cypress is fire resistance. Like pine trees, cypress can survive fire with little apparent damage. However, fire can be decimating during an extreme drought if the soil has dried. Dry peat will burn; causing root damage that kills the trees. Cypress can grow in uplands, but fire restricts their normal distribution to perennially saturated soils. Cypress heads that have been damaged or destroyed by fire are often invaded by willow.

TREE ISLAND FORMATION [97, 539, 581]

In its early history the Everglades was nearly devoid of tree islands. An older documented example began to form about 3200 years ago, and most tree islands are younger than 2000 years. Tree islands of the Everglades peatlands have mainly formed by three processes. They may form on a "fixed" high point of rock, on a peat "pop-up," or on a sawgrass "strand," all discussed below. Once established, generations of trees and shrubs produce accumulations of roots, twigs, branches, and leaves that become woody peat, called *Gandy* peat, which is more resistant to aerobic decay than marsh peat because of tannins and other decay inhibiting chemicals produced by woody plants. Thus, tree island elevations stay above the surrounding marsh as Gandy peat accumulates.

FIXED

This tree-island origin involves a high spot of the rock base: a true island in the marsh. In the Shark River Slough, where only about three to five feet of peat overlies most of the bedrock, hundreds of such rock humps are high enough to protrude to or above the normal wet-season water level. There, they provided a surface that became colonized by shrubs and trees, including upland trees if the elevation is high enough. These upland "heads" of fixed tree islands contain by far the largest trees found in Everglades tree islands. From an initial nucleus provided by the topographic high spot or "head," fixed tree islands could grow to enormous sizes. The expansion always proceeds in a downstream direction from the head, producing a tear-drop-shaped tail (some are better described as "comet-shaped"). Dating of peat on these islands shows that tail development takes 1000 to 2000 years after the head becomes established. There is evidence that bird use of fixed tree islands, which support larger trees on the stable rock head, provides a source of nutrients that stimulates island elongation. Phosphorus levels can be about six times higher in fixed tree island heads than in the surrounding marsh. The near tail also has elevated phosphorus and is composed of shrubs and trees that tolerate seasonal flooding, grading downstream into shrubs mixed with sawgrass, then followed by sawgrass. Sawgrass also forms a border around most of the tree island but is very narrow or missing at the head (Figure 4.4, also see Figure 3.9). The tree islands, having human habitation and associated names such as those mentioned above, mostly formed from fixed heads. In addition, prehistoric human use is known to have elevated the heads of certain fixed tree islands by midden deposits.[97] Human habitation undoubtedly enhanced downstream development just as bird use has (see Chapter 21 and *Other Processes* below).

POP-UP OR BATTERY

The most common process, accounting for the vast majority of Everglades tree islands, is peat "pop-up." This process occurs in sloughs where waterlilies are abundant. Waterlily roots and rhizomes commonly fill with gas, and additional gas bubbles, mostly methane, are produced in the peat by decomposition. As a result, slough-bottom peat can become very buoyant. If an area of peat is large enough that its buoyancy exceeds the peat's strength to hold together, then it will tear free and rise to the surface: a *pop-up* or *battery* (also called *sudd* elsewhere). Pop-up sizes are highly variable, but 30 to 90 feet across and about a foot thick is common. They float freely until lodged against another surface, typically over the slough bottom or on a sawgrass ridge, depending on water levels and pop-up thickness. Although pop-ups do not look substantial, their strength of interwoven roots can usually support the weight of a person (Figure 4.5) — but walk carefully and do not jump!

FIGURE 4.4 Aerial view of a large fixed tree island in the Shark River Slough, Water Conservation Area 3A. looking south. A tropical hardwood hammock is on the upstream end at left, with bayhead vegetation behind, grading southward to sawgrass in the distance. (Photo by T. Lodge.)

FIGURE 4.5 A peat pop-up amid slough habitat in Water Conservation Area 3A, easily supporting wading-bird researcher, Lisa Borgia on dry land. (Photo by T. Lodge.)

The higher surface of the pop-up provides a different habitat: a soil surface slightly above the water level. The new surface is soon colonized by shrubs that could not endure the adjacent marsh's depth and hydroperiod. Peat core analyses show that wax myrtle is most common at first, but new pop-ups have been observed to have — in addition to wax myrtle — swamp bay, dahoon holly, coastalplain willow, buttonbush and even strangler fig, which is normally an upland species. Succession leads to swamp bay (thus the name, "bayhead," described below) and dahoon holly in the interior, which never become very large, with wax myrtle and cocoplum shrubs on the edges, often skirted on the outside by sawgrass and ferns. After shrubs and small trees become established, their weight and roots may anchor the pop-up, making it a stationary feature of the marsh, but some continue to slide about in response to strong winds even with small trees. Tree islands that have formed from pop-ups remain small and roughly circular (Figure 4.6) but may be accidentally joined by other pop-ups to make larger, irregular shapes. Mature pop-up tree islands may have interior elevations of about three feet above the surrounding marsh. In Loxahatchee where they are most common, there are about 4000 such bayheads (see Figure 4.2). The woody peat of several has been dated and found to be relatively young — less than 1500 years old — whereas the Everglades at that location is almost 5000 years old. Thus, the early Everglades may have been devoid of these tree islands for its first 3500 years.

STRAND

This third type of tree island formation process has only recently been discovered. Certain tree islands in Loxahatchee are huge compared to the rounded pop-up tree islands (see Figure 4.2). Strand islands are relatively narrow but greatly elongated, the largest being up to a mile long and a thousand feet wide. Their configuration is similar to large sawgrass ridges (sometimes called "strands") in ridge-and-slough areas. While they are fixed in location, they do not have the characteristic teardrop shape of fixed tree islands. Investigation of the peat strata in strand tree islands reveals Gandy peat underlain by sawgrass peat (Everglades peat), with no relationship to the topography of the underlying

FIGURE 4.6 A still fully floating peat island covered with swamp bay and other tree island vegetation, in Loxahatchee National Wildlife Refuge. The gap of open water to its right is the origin of the pop-up. (From Ref. 581 with permission from Kluwer Academic Publishers.)

rock surface, which is covered by 7 to 11 feet of peat in Loxahatchee. Thus, it is apparent that they evolved from sawgrass strands by shrub invasion. The abundance of slough habitat in Loxahatchee helps protect them from fire, which would inhibit or eliminate shrub invasion. The dominant species of strand islands is dahoon holly, a shrub or small tree that tolerates a relatively long hydroperiod. Over a hundred of these low, wet, strand tree islands exist in Loxahatchee, the only part of the Everglades where they occur. Like pop-up tree islands, strand islands are thought to be much younger than the Everglades, but age determinations have proved difficult.

OTHER PROCESSES

Prehistoric man is known to have enhanced numerous tree islands, such as the fixed tree islands discussed above, but in addition, some tree islands that resemble fixed types have heads composed only of midden material, not rock, indicating that the continuity of prehistoric human habitation perpetuated their existence as the Everglades evolved.[97] Still other tree islands are associated with depressions in the underlying bedrock. Their origin is still in question, but evidence points to an origin from wetland forest pockets that existed when the Everglades began forming, making them the oldest tree islands in the ecosystem. In the modern Everglades, shrubs may root on any elevated surface. Alligator activity — nesting or working around their ponds (see Chapter 17) — may produce a suitable surface. Round-tailed muskrats make homes of piled plant debris that could act as a tiny tree-island nucleus. Coastalplain willow, with its wind-blown seeds, is quick to colonize such surfaces, often producing a small but usually temporary tree islands. However, if by chance a tree or shrub grows large enough to support nesting birds, the resulting enrichment from their droppings could start a larger tree island.

In marl prairies, tree islands have a very different relationship to their environment than in peatlands. Peat deposition is not possible on top of the marl prairie elevation. Low nutrients and the lack of natural firebreaks in marl prairie also restrict the development of tree islands. The few invading shrubs are regularly burned back. However, the marl prairies of the southeastern Everglades such as near U.S. 1 south of Homestead, contain drainage depressions that are deep enough, therefore wet enough, for peat deposition; and small tree islands occupy these depressions. These tree islands have shallow moats around them that appear to be the result of acids from peat soil (and leaf-litter decomposition) dissolving the adjacent marl, but probably aided by deer and other wildlife walking around the bayheads (see section on moats below). The flow pathway downstream of these tree islands also follows a depression, which probably results from the tree island acids dissolving the marl (Figure 4.7). In many locations, there are lines of tree islands in these drainage-ways. Thus, the bayheads in marl prairie occupy depressions filled with peat whereas the bayheads in the Everglades peatlands occupy areas that are substantially elevated above the surrounding marsh.

Deeper depressions in marl prairies may also provide a wetter environment conducive to cypress trees. While a resulting localized cypress forest could be called a tree "island," it is actually a deeper area than the surrounding marsh, and could aptly be called a cypress pond (see description below).

TREE ISLAND MOATS[127, 128]

A mostly former feature of many peatland bayheads, especially those of fixed tree island origin, was a zone ranging from 10 to 20 feet wide, that was both deeper and less densely vegetated than the surrounding marsh. Called "moats" or "channels," these areas were heavily used by alligators, fish, and other aquatic life. One or a combination of factors may explain the existence of bayhead moats: water flow and wildlife activity are suspected. The shape of bayheads is related to water flow, and it is hypothesized that flow, accelerated by passing around islands, scoured the edges. Accordingly, moats were most prominent on the upstream end and sides of the fixed tree islands, often disappearing along the sawgrass tails. Whatever the cause, moats are not as apparent today as they were to observers several decades ago.

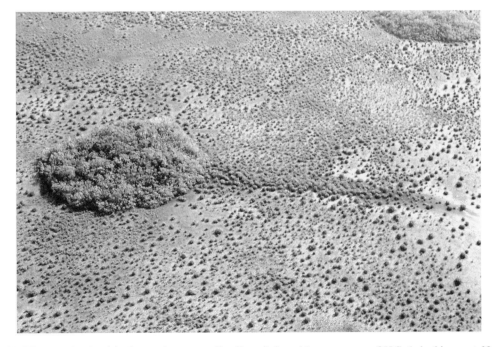

FIGURE 4.7 A bayhead in the southeastern saline Everglades white zone west of U.S. 1, looking east. Note the downstream "tail" developed by humic acid dissolution of the surface marl. (Photo by T. Lodge.)

The wildlife hypothesis for bayhead moat origin considers deer and alligator activities. Deer regularly walk around tree islands, browsing on vegetation at the edge. Alligators are commonly associated with bayheads (see Chapter 17) and patrol the edges where the chance of ambushing prey is higher than in the open marsh. Such wildlife activity would dislodge soil and vegetation around the edges of tree islands. Combining wildlife activity with flow, which would carry the dislodged material away, may explain the former existence of moats. Whatever the cause, the moats helped prevent fires from entering tree islands, so their maintenance is important to their integrity of respective bayheads. Moats and their origin have received very little attention, and they appear to have degraded in the modern Everglades. Sediment accumulation and encroaching sawgrass and other marsh vegetation may be erasing them.[7]

POND APPLE/"CUSTARD APPLE"

The most peculiar tree associated with tree islands is the pond apple (Figure 4.8), a species that thrives in a longer hydroperiod and deeper water than most. The edge of a bayhead moat and the interior of a cypress dome are ideal for the pond apple. Though seldom taller than 35 feet, it develops a massive trunk and buttressed roots. Together with the robust load of epiphytes that typically thrive on its rough, dark-brown bark, it presents a bizarre spectacle. Due to its common location in or next to open water, and to its convenient configuration of branches, nesting waterbirds such as the anhinga, regularly use the pond apple. The greenish-yellow to yellow fruit resembles a large apple and is eaten by animals, which then serve to distribute the seeds. It has an unpleasant musky odor and most people describe its taste as sour (thus the alternate name "custard apple" see footnote, Table 4.1). Politely stated, it is undesirable.

Pond apple forests occur in some locations where deeper water is sustained most of the year. Historically, there was a forest of pond apple at the head of the Everglades along the south shore of Lake Okeechobee (see Chapters 2 and 11). Other pond apple forests have been called "pond-apple sloughs" including a remnant in the headwaters of the South Fork of the New River near Fort Lauderdale, where historic Everglades overflows and groundwater seepage kept the area flooded with freshwater most of the time.[649] The Fakahatchee Strand in the Big Cypress Swamp also has deep-water areas dominated by pond apple (see Chapter 7).

ROLE AND INTEGRITY OF TREE ISLANDS [26, 76, 127, 210, 369, 540, 623]

Tree islands greatly enhance the ecological value of the Everglades, especially where they support wading bird rookeries and other wildlife. Their survival depends on a balance between fire and water levels, and alligator activity may also be important. Hydrologic alteration of the Everglades (see Chapter 21) and probably the reduced numbers of alligators, have caused many complex changes in tree islands, most of which have been deleterious, even their complete destruction.

FIGURE 4.8 A large pond apple at the Anhinga Trail, Everglades National Park, supporting two active anhinga nests, adults, and visiting double crested cormorants. (Photo by T. Lodge.)

5 Tropical Hardwood Hammocks*

A hammock is a localized, mature hardwood forest. By *hardwood*, the distinction is made that broad-leaf trees are prevalent, as opposed to pines, which normally have softer wood. The term *hammock* is of uncertain origin, but it may have been derived from a Seminole Indian word that means house or home, and many hammocks were used by Seminoles as encampments. In southern Florida, hammocks occur in marshes, pinelands, mangrove swamps, and the upstream heads of the larger tree islands of the Shark River Slough. In order for hammocks to exist, the ground must be high enough so that seasonal flooding does not occur.

The flora of hammocks is markedly distinct from the wetland tree islands, which are dominated by a temperate swamp flora. In the northern Everglades region, hammocks are dominated by trees of temperate climate origin such as the Virginia live oak or hackberry, and the occurrence of tropical trees is highly restricted by cold weather. From the latitude of Miami southward, however, nearly all of the trees are of tropical origin, with Virginia live oak being the only common temperate representative. For this reason, the term *tropical hardwood hammock* is commonly used for any hammock in Everglades National Park, the Florida Keys, and in proximity to the Atlantic coast northward, historically, to Pompano Beach.[13]

Early botanists and plant explorers in southern Florida were greatly impressed by tropical hardwood hammocks.[190, 515, 543, 572] The types of trees in these hammocks have unusual names for anyone unfamiliar with the West Indian region of the American tropics. Some of the prevalent trees and shrubs of the southern Everglades hammocks are listed in Table 5.1 together with their origins. Of the four temperate North American species listed, hackberry and red mulberry are infrequent. Virginia live oak and cabbage palm (see Figure 4.1) are common, but they are overwhelmed by the diversity and abundance of tropical species. Among the tropical trees and shrubs, nearly all have seeds that are dispersed by birds, which probably explains how they arrived in South Florida from the West Indies where all of them occur naturally (see Chapter 12). The West Indian mahogany is among the few tree and shrub species having wind-blown seeds.

With the exception the Fakahatchee Strand (see Chapter 7), the hammocks of the Florida Keys harbor the most sensitive and rare tropical species, to the intrigue of botanists. Tropical hammocks in the keys are now generally protected by Monroe County regulations and by state preserves including the Key Largo Hammock Botanical State Park,** Windley Key Fossil Reef Geological State Park, and Lignumvitae Key Botanical State Park. The last of these is named for an important and now rare small tropical tree of the keys hammocks: lignumvitae (*Guajacum sanctum*), which is renowned for its extremely hard wood. So dense is its wood that even cured boards do not float, and the early shipbuilding industry revered it for shaft bearings. Other desirable woods were West Indian mahogany and black ironwood (*Krugiodendron ferreum*). Lignumvitae and black ironwood are not included in Table 5.1 because of their absence and rareness, respectively, in the Everglades hammocks. Desirable specimens of these three species were logged out of the keys and other accessible South Florida locations by the late 1700's.[582] West Indian mahogany, relatively much faster growing, continued to be sought and harvested for its fine furniture value to about 1940, so

---

* The main sources for this chapter are references 12, 14, 16, 131, and 452.
** Now named in honor of the late Florida Keys environmentalist, Dagny Johnson.

TABLE 5.1
Representative trees and shrubs of the tropical hardwood hammocks of the southern Everglades

Common name[a]	Scientific name[a]	Tropical origin
Canopy trees		
Florida royal palm	*Roystonea regia*[b]	X
Virginia live oak	*Quercus virginiana*	-
sugarberry, hackberry	*Celtis laevigata*	-
strangler fig	*Ficus aurea*	X
red mulberry[c]	*Morus rubra*	-
pigeon-plum	*Coccoloba diversifolia*	X
lancewood	*Ocotea coriacea*	X
West Indian cherry	*Prunus myrtifolia*	X
false tamarind (wild tamarind)	*Lysiloma latisiliquum*	X
Jamaican dogwood, Florida fishpoison tree	*Piscidia piscipula*	X
paradisetree	*Simarouba glauca*	X
gumbo limbo	*Bursera simaruba*	X
West Indian mahogany	*Swietenia mahagoni*	X
inkwood, butterbough	*Exothea paniculata*	X
false mastic (mastic)	*Sideroxylon foetidissimum*	X
willow bustic	*Sideroxylon salicifolium*	X
Shrubs and understory trees		
cabbage palm	*Sabal palmetto*	-
tallow wood, hog plum[d]	*Ximenia americana*[d]	X[d]
wild lime, lime pricklyash	*Zanthoxylum fagara*	X
poisonwood, Florida poisontree	*Metopium toxiferum*	X
white stopper	*Eugenia axillaris*	X
Spanish stopper	*Eugenia foetida*	X
marlberry	*Ardisia escallonioides*	X
myrsine, colicwood	*Rapanea punctata*	X
wild coffee	*Psychotria nervosa*	X

[a] Common and scientific names follow references 222 and 661. Alternate South Florida vernaculars are in parentheses
[b] The royal palm is treated here as the same species as found in Cuba, following Tomlinson[599] and Wunderlin.[661] Other literature may recognize it as an endemic species, the Florida royal palm (*Roystonea elata*), distinct from the Cuban royal palm (*R. regia*).
[c] Red mulberry is a temperate tree of the central and eastern U.S. Its occurrence in southern Florida — mostly in hammocks — often correlates with sites of aboriginal occupation. Thus, it was probably transported by Native Americans for its sweet, blackberry-like fruit.[553]
[d] Found in hammock margins as well as pinelands, tallow wood or hog plum is among the most cold-tolerant of tropically derived species, being found throughout the Florida peninsula.

that the presence of huge West Indian mahoganies in Everglades National Park's Mahogany Hammock is credit to the former remoteness of that location. [9]

A tropical species in Table 5.1 that is very limited in South Florida distribution is the Florida royal palm, included because of its magnificent stature and because there is a prominent group of them at Royal Palm (Figure 5.1), a popular visitor location in Everglades National Park. They have a massive light-gray trunk that looks like concrete. The species is also abundant in some locations in the Big Cypress Swamp. It is now considered to be the same species as the Cuban royal palm,[599] widely planted as an ornamental, although the native South Florida population appears to be more tolerant of hydric soil conditions. Wherever royal palms grow, they usually stand conspicuously above the other hammock trees, sometimes reaching well over 100 feet.[543]

FIGURE 5.1 A royal palm at Royal Palm, Everglades National Park, towers over the hammock. (Photo by T. Lodge.)

THE HAMMOCK ENVIRONMENT

The feeling inside a tropical hardwood hammock is special. For those familiar only with temperate forests, the hammocks are full of strange sights and sensations. A first impression is the smell, which is unlike any northern forest. It reminds most people of a faint skunk odor, yet is not unpleasant. This aroma is emitted by one of the most ubiquitous of the hammock species, a small tree called the white stopper, named "stopper" for its use as in treating diarrhea. The protection afforded by the cover of trees also creates acoustic effects, with an eerie silence prevailing.

Because most of the foliage is overhead, the inside of a mature hammock is easy to explore, with the dense shade preventing much growth at ground level. The forest floor, on the other hand, may present obstacles. There are numerous rough limestone rock outcroppings and frequent solution holes or "sinks," some of which are deep enough to contain a pool of water. In general, hammocks developed where the surface rocks were slightly harder than in surrounding areas. This condition resulted in slower weathering of the rock, leaving the elevation higher. However, chemical weathering of the limestone underneath these areas continued until, here and there, the surface collapsed, producing a sink. Exposed groundwater within the sinks contributes to higher humidity, promoting luxuriant growths of mosses and ferns on the walls. Most of these species are also of tropical origin. More ferns, such as the resurrection fern, are found on the trunks and limbs of the hammock trees, in addition to air plants and orchids.

The dense shade provided by the trees of a mature hammock, and the exposed ground water help keep the inside temperature several degrees cooler during hot summer weather. Conversely, during severe winter cold fronts, tropical hammock plants are usually spared by the higher humidity and protection from wind, while the same kinds elsewhere perish.

TREE HEIGHT

Although the trees protect plants inside the hammock from cold weather, the tallest trees are frequently damaged at their tops and effectively pruned by the cold. The cold, together with hurricanes and lightning, explains why hammock trees do not grow very tall, only a few reaching as much as 55 feet.[452] Any tall tropical hammock tree in the path of a hurricane is sure to be uprooted, because the roots of these species are surprisingly shallow (Figure 5.2). The summertime incidence of lightning strikes in peninsular Florida is very high (Chapter 3), and the tallest trees are prime targets. Without these limitations, most of the tropical hardwoods would be much taller.

THE STRANGLER FIG

One of the common hammock trees is so strange in its growth form that it deserves special recognition: the strangler fig (Figure 5.3). There are two native figs, both from the West Indian tropics, but the strangler is more common. This species produces a berry-like fruit commonly eaten by birds. The seeds within the fruit are not digested and remain viable for dispersal in the birds' droppings. Although the seeds can start growing wherever there is sufficient moisture, one of the most common places is in the "head" of a cabbage palm. Thus, a fig seed sprouting there begins life as an epiphyte. Like orchids, ferns, and air plants, figs are not parasites because they do not derive nutrients from the host tree; but the end result is the same. As the fig grows, it proliferates roots downward, toward the forest floor. On reaching soil, they expand greatly in size and form a network that completely entwines the host, which is eventually killed by the fig tree's dense shade. The strangler fig literally takes the host's place in the forest. Strangler figs are not choosy, they engulf stone walls, abandoned cars, and even household chimneys if not removed.

HAMMOCKS, FIRE, AND SUCCESSION

A common characteristic of all tropical hardwood hammocks is intolerance of fire. Fire actually helps maintain many other kinds of plant communities in the Everglades, especially marshes and pinelands, but it can destroy hammocks. Fortunately, the very characteristics that make hammocks such interesting places also help prevent fire from entering. These characteristics are the open understory, with little accumulated fuel, and the high degree of shade, which keeps the inside

FIGURE 5.2 Chaos at Royal Palm five months after Hurricane Andrew. A gumbo-limbo (on its namesake trail in Everglades National Park) rests against another. In response to the newly opened habitat, vines have advanced up most trees, but not on gumbo-limbos, which have peeling bark that may be an adaptation to vine encroachment. (Photo by T. Lodge.)

climate relatively cool and moist. Thus, the tropical hardwood hammock is the climax community — the ultimate final stage of succession — where fire is rare or absent (see Chapter 20). In very dry years, however, hammocks are highly vulnerable to fire, which may then destroy the humus soil. Soil fires burn slowly but persistently, killing the trees by root damage. Hammocks may survive

FIGURE 5.3 Strangler figs grip over the trunk of a long dead and toppled tree on the Gumbo Limbo Trail, Royal Palm, Everglades National Park. (Photo by T. Lodge.)

a crown fire and look almost untouched after a few years, as did Mahogany Hammock following a fire in the early 1970's, but soil fires have completely eliminated many hammocks.

UNPLEASANT ASPECTS OF HAMMOCKS

So far, a wondrous picture of the tropical hardwood hammock has been painted here, but there are also drawbacks. First is the problem of just getting inside. While it is easy to walk around inside a mature hammock, it is not easy to enter it from the outside. The edge of a hammock does not have the tall trees that produce dense shade under which little grows at ground level. On the contrary, the edge is nearly impenetrable. The outermost edge is sometimes dominated by the saw palmetto (*Serenoa repens*), a low, sprawling species of palm, which is common in the pinelands. This species seldom grows over about six feet tall, and its trunks lie prostrate along the ground. The leaf stem (properly called the petiole) is armed with upward-pointing teeth that make those of sawgrass look and feel like child's play. The saw palmetto's stiff, toothed petioles can cut your legs and ankles and the trunks can make walking difficult, while the large broad leaves may hide impending dangers. This exciting adventure is intensified by the possibility of encountering the sometimes huge (up to about seven feet long) eastern diamondback rattlesnake that may lie concealed among saw palmettos. The species is not common, but the thought is still frightening.

The area of saw palmettos is generally the easy part, and they are not even present at the edges of all hammocks. The next rim of fringing vegetation, which is always present, is a tangle of vines and shrubs that defy intrusion. Saw greenbrier (*Smilax bona-nox*) is a common vine with very tough stems and strong curved spines. Added to this difficulty is a shrub or small tree common in the hammock fringe: poisonwood. It is a close relative of poison ivy and similarly produces an itchy, unpleasant skin rash. The leaves of poisonwood are deceptively similar to those of gumbo limbo, a harmless hammock species, and many people are fooled by the resemblance. Once through the hammock's edge, you have paid the price and are free to walk through the hammock forest. Fortunately, several hammocks have been "conquered" for public access with pathways and boardwalks (Figure 5.4). Examples are Miami-Dade County's Castellow Hammock and Everglades National Park's Gumbo Limbo Trail at Royal Palm and the Mahogany Hammock board walk.

The second unfortunate feature for visitors to a hammock takes only one word to describe: mosquitos. Mosquitos can be unbearable under the shade of hammock trees, particularly near salt water, during the warmer, wetter months of the year, generally May through October. If a hard-blowing cold front has not occurred for a month or so, mosquitoes may be a problem even in winter months. The answer is to use plenty of insect repellent.

FIGURE 5.4 Boardwalk in Mahogany Hammock, Everglades National Park. (Photo by T. Lodge.)

HAMMOCKS AND WILDLIFE

Hammocks that occur inside wetland tree islands, or alone in an area of marsh, may support a wading bird rookery. However, most hammocks accessible to people lie in upland communities and would not normally be used by colonial nesting birds, which seek the protection afforded by surrounding water. Thus, there is an element of disappointment for many visitors who expect to see abundant animal life. Hammocks are not a showcase for wildlife. Some of the interesting Florida tree snails, a few lizards (most commonly the southeastern five-lined skink, a sleek lizard with an electric-blue tail), occasional owls (barred owls are the most common), and a raccoon can sometimes be seen; but more often, only patient, skilled observers are rewarded. Hammock wildlife is not rare but is difficult to see. The spectacular vegetation is the main attraction.

6 Pinelands[*]

Many visitors to Everglades National Park are surprised to find that one of the most extensive habitats appears to be the open pinelands. The dominant species, and the only true pine present in and around the Everglades, is slash pine, a handsome, long-needled species. The name *slash* is derived from the once widespread practice of extracting sap from this species by cutting diagonal slash marks in the trunk. Sap draining from the cuts was collected in cups and used to make turpentine and other products.[66] Slash pine is abundant throughout Florida and neighboring states of the southeast coastal plain where it is the most important species in the paper (pulp) industry. In northern Florida, enormous acreages of row-planted slash pines now replace the natural pinelands. Patches or blocks of trees are clear-cut in about 20 to 25 years and replanted. Where natural pinelands occur on flat sandy soils, from the southeast into southern Florida, they are commonly called *flatwoods*, formerly the most widespread plant community in Florida. Longleaf pine (*Pinus palustris*) and pond pine (*Pinus serotina*) were also important in the flatwoods of northern and central Florida, but slash pine was most abundant.

In southern Florida, from the approximate latitude of Orlando and southward, there is a special variety of slash pine: *Pinus elliottii* var. *densa*. It is locally known as Dade County pine.[599] Older *densa* variety trees have stored unusually large amounts of resins, making the wood very hard (dense), unlike most pines. The early lumbering industry of the Miami area (formerly Dade County) used this once abundant resource, and houses made with this wood have a time-tested value: termite resistance. The only drawback to the wood is that it is difficult to drive a nail into it.[128] The concentration of pine resins in older Dade County pines was so great that an industry was based on digging up long-dead stumps because they could be processed into the same products as derived from the sap of the more northern trees.

Slash pine prospers in a range of elevations from "high and dry" to levels that are only slightly (inches) above marsh habitats. Soil saturation and patchy, shallow flooding may occur for two or three months during the height of the summer rainy season. Pinelands originally dominated the Atlantic Coastal Ridge upon which most of the metropolitan areas of Miami and Fort Lauderdale have been built. On the southern part of this ridge, from Miami southward, the pinelands grow on very rough limestone, full of holes and cavities. It provides a very different ground surface for the local "rocky" pinelands, often termed *pine rockland*, compared to the sandy soils common to the flatwoods that border the northern Everglades and occur widely in the remainder of the state. The southern end of this ridge curves westward into Everglades National Park as a large area or pine rockland called Long Pine Key, which recedes in elevation westward, breaking into pineland islands that continue to Mahogany Hammock (one of the best known of the park's tropical hardwood hammocks). The main road through Everglades National Park follows this pine rockland habitat for many miles, giving the impression that much of the park is pineland. Actually, the pinelands occupy only a small percentage of the total mainland area of the park. Pine rocklands also occur in the lower Florida Keys and in the Bahamas where a related pine occurs instead of slash pine, the Caribbean pine (*Pinus caribaea*).[117] A list of plants found in South Florida pine rocklands is presented in Table 6.1. Many of them also occur in the Bahamas.

---

[*] The main sources for this chapter are references 1, 367, and 565.

TABLE 6.1
Representative native pine rockland plants[a]

Common name[b]	Scientific Name[b]	Status[c]
Canopy[d]		
(South Florida) slash pine, Dade County pine[e]	*Pinus elliottii* var. *densa*	
Small trees and shrubs[d]		
Florida silver palm	*Coccothrinax argentata*	Ft
cabbage palm (sabal palm)	*Sabal palmetto*	
saw palmetto	*Serenoa repens*	
lusterspike indigobush (crenulate leadplant)	*Amorpha herbacea* var. *crenulata*	Ens
Long Key locustberry	*Byrsonima lucida*	
winged sumac (southern sumac)	*Rhus copallinum*	
varnishleaf, Florida hopbush	*Dodonaea viscosa*	
Florida clover ash (West Indian lilac)	*Tetrazygia bicolor*	
snowberry, milkberry	*Chiococca alba*	
rough velvetseed	*Guettarda scabra*	
white indigoberry	*Randia aculeata*	
Herbaceous species/ground cover		
lacy bracken (lacy bracken fern)	*Pteridium aquilinum* var. *caudatum*	
Florida arrowroot, coontie	*Zamia pumila*	
running oak, bluejack oak	*Quercus pumila*	
eastern milk pea	*Galactia regularis*	
Small's milkwort, tiny polygala	*Polygala smallii*	Ens, Ue, Fe
Redland sandmat (deltoid spurge)	*Chamaesyce deltoidea*	Ens, Ue, Fe
Garber's spurge	*Chamaesyce garberi*	Ens, Ue, Fe
pineland croton	*Croton linearis*	
fivepetal leafflower	*Phyllanthus pentaphyllus*	Ens
pineland spurge (pineland poinsettia)	*Poinsettia pinetorum*	Ens, Fe
Christmasberry (quail berry)	*Crossopetalum ilicifolium*	Enf, Ft
corkystem passionflower	*Passiflora suberosa*	
pineland golden trumpet (pineland allamanda)	*Angadenia berteroi*	Enf, Ut
pineland clustervine, pineland jacquemontia	*Jacquemontia curtisii*	Ens, Ft
rockland shrubverbena, pineland lantana	*Lantana depressa*	Ens, Fe
redgal (yellowroot, cheeseshrub)	*Morinda royoc*	
Everglades key false buttonweed	*Spermacoce terminalis*	Ens, Ft
snow squarestem (pineland blackanthers)	*Melanthera nivea (parvifolia)*	Ft

[a] Selection of pine rockland species is mostly based references 49 and 565.
[b] Common and scientific names follow references 222 and 661. Alternate local vernaculars are in parentheses.
[c] Key: Enf=endemic to Florida; Ens=endemic to South Florida; Ue=endangered, and Ut=threatened on USFWS listing; Fe=endangered, and Ft= threatened on Florida Department of Agriculture listing.
[d] Trees and shrubs that occur in tropical hardwood hammocks nearly all occur in pine rocklands but are regularly killed-back by fire. Live oak, wax myrtle, poisonwood, willow-bustic, and tallow wood (or hog plum) are common examples but not listed to avoid redundancy with Table 5.1.
[e] Wunderlin[661] does not recognize varieties of *P. elliottii* and thus uses "slash pine" as the only common name. The name "South Florida slash pine" is used by Austin[46] for var. *densa*, and the corresponding historic name, "Dade County pine," are still widely used.[599]

PINELANDS AND FIRE

Pinelands are called a "fire climax" by ecologists, meaning that the pineland community is the final vegetation that develops (on acceptable elevations) if fire is common (see Chapter 20). The pineland community is highly resistant to fire and is actually maintained or perpetuated by fire (Figure 6.1). In the pine rocklands, slash pine is the only tree that reaches canopy stature in regularly burned areas, and its long needles provide an envelope of air that helps protect the growing tips of branches from heat damage (although the lower limbs are often killed by fire). At the shrub level, two obvious fire-adapted species are palms. The saw palmetto, a low-growing, sprawling species, is the more abundant both in pine rocklands and in the flatwoods farther north. The taller cabbage palm (see Figure 4.1) is commonly dwarfed (less than 20 feet) in pine rocklands but often grows 50 or more feet tall in better soils. There are several normal shrubs that characterize the pine rocklands, notably West Indian lilac,* rough velvetseed, and varnishleaf. Herbaceous plants, however, are the most diverse and botanically interesting component of the pine rockland community. Most herbaceous species are southernmost variations of temperate plants with relatively few tropical representatives. Examples of particularly colorful plants are the pineland allamanda and the pineland spurge, a small species of poinsettia.

The rocky pinelands also harbor numerous hardwood shrubs and small trees. Wax myrtle and live oak are abundant, but also most of the tropical hardwood species that dominate the hammocks encroach rapidly. They are kept in check by fire. Pinelands that have burned within the past two or three years are very open with almost nothing blocking the view between the tops of the palmettos (usually about four to five feet) and the lower limbs of the pines, which may begin over 20 feet up the trunk. Pinelands that have not burned within four or five years begin to reveal the vision-blocking invasion of the hardwood species. If this invasion continues unchecked for several more years, a subsequent fire can be devastating, especially when very dry. Not only are the hardwoods

FIGURE 6.1 Pine rockland in Everglades National Park. Note the burn marks on the trunks from a fire within the last year, judging from the open understory, and the abundant saw palmettos. (Photo by T. Lodge.)

---

* Often overlooked by landscape architects, this species is an excellent and beautiful landscape shrub for South Florida locations.

killed-back, but with so much fuel, the pines themselves may be killed, resulting in a nearly barren landscape with dead pines standing like tombstones. However, most cabbage palms and palmettos survive. Managers striving to maintain the natural character of pine rocklands recommend fire at three- to seven-year intervals.

In the absence of fire for two to three decades, a pineland will take on the characteristics of a hammock in the process of succession. Castellow Hammock, a Miami-Dade County park located on the coastal ridge between Miami and Homestead, is an example where tropical hammock vegetation overtook pine rocklands. Between 1940 and 1963, this hammock expanded into adjacent pinelands, causing it to merge with a neighboring hammock.[16] The dense shade produced by the encroaching hardwoods prevents seedling survival by all but certain shade-tolerant hammock species, such as pigeon plum. A "young" hammock can be recognized by the remaining presence of tall pines, such as in the respective part of Castellow Hammock.* Once established, the hammock itself becomes fire resistant (see Chapter 5). Thus, while pinelands are a fire-climax, ecologists consider the tropical hardwood hammocks to be the climax community of upland succession in southern Florida (see Chapter 20).

ENDEMIC PLANTS AND THE ROCKY PINELANDS

Botanists are especially interested in the rocky pinelands of southern Florida. Other plant communities of the region are populated almost entirely by plants that occur in other areas of the world and have invaded the region only recently. The rocky pinelands, however, harbor many "endemic" species (found only in a certain region and nowhere else on earth). The rocky pinelands of the southern Everglades and nearby portions of the Atlantic Coastal Ridge harbor at least 20 endemic species** that are not found even in the pine flatwoods of the sandy, acid soils in the remainder of the state. Examples include eastern milk pea and pineland spurge. The occurrence of so many endemics in this community indicates isolation and long-term continuity of conditions favorable to their survival. As a result, they have evolved into new distinct species or varieties. Like slash pine, the endemic pineland species are adapted to fire, not by physically withstanding the heat, but by other mechanisms that involve underground tubers and rhizomes that quickly resprout after fire and flowering/seeding cycles that respond rapidly to fire.

If the changes in southern Florida through the glacial cycles could be viewed, the perpetuity of the pinelands would probably be observed, while the tropical vegetation and freshwater wetland communities would come and go. Support for this theory has recently come from the discovery of pineland forest remains, complete with charred evidence of fire, dated at 8500 years ago in 40 feet of water about 35 miles west of Key West.[164, 379] It is most likely that the pine rockland community followed marginal wetland conditions that would have been in lowlands near the coast. The higher elevations of the peninsula during low stands of sea level would have been dry because limestone bedrock is so leaky. In fact, fossil pollen and other records indicate that communities dominated by slash pine were restricted in distribution during glacial times and the peninsula was dominated by xeric (dry land) plant communities, like the sand pine scrub community found today on dry, sandy soils.[642] Nevertheless, one is impressed that the really characteristic plant community of southern Florida, through time, is the pine rockland.

---

* Hurricane Andrew (1992) killed most of these old, remnant pines.

** The number of endemic plants varies considerably among sources (i.e. reference 128 states 100 "plants" and reference 565 lists just over 30 species). Much of the variation comes from whether only "species" are counted or "taxa;" the latter term includes species, subspecies and even named varieties. Also, recent discovery reveals that many rocky pineland plants formerly considered endemic to southern Florida also occur in the Bahamas thus reducing the list.[51]

7 The Big Cypress Swamp*

Traveling west on either of the two highways across the Everglades, one enters very different terrain at Fortymile Bend on the Tamiami Trail (U.S. 41) or at the Miccosukee Service Plaza at Snake Road on Alligator Alley (I-75) (Figure 7.1 and Color Figure 3). Called the Big Cypress Swamp (often shortened to just "the Big Cypress"), this mostly forested region reveals a greater variety of plant communities than the Everglades.

Like all of southern Florida, the terrain of the Big Cypress Swamp is nearly flat so that small variations in elevation make large differences in vegetation. Slightly higher than the Everglades, it has a maximum elevation of about 22 feet in the north, sloping almost to sea level in the south, over a typical distance of 35 miles. More than in the Everglades, differences in elevation here are directly related to local variations in the underlying geology, which greatly affect hydrology. The result is a range of plant communities from deep sloughs dotted with open ponds to cypress swamps, to regions of countless cypress domes, to open marshes, pinelands, and hammocks. Variations in forest cover often make the region appear to have hills, in contrast to the flatness of the Everglades. It is obvious that the key to the variety of plant communities in the Big Cypress is its more complex and shallow bedrock compared to the relatively simple Everglades bedrock, which mostly lies buried under peat (Figure 7.2).

While the name "Big Cypress" implies that the cypress trees are large,[423] it has come to denote that the region is big[428] — 1200 square miles by the most limited definition and nearly 2500 square miles using its hydrologic boundary, which includes now-developed lands near Naples. However, there once were abundant large cypress trees in restricted habitats. The best-known area was the Fakahatchee Strand, located west of State Road 29. Locals historically referred to that area as "the big cypress country" because of its huge cypress up to 130 feet tall. They called more easterly areas, now in the Big Cypress National Preserve, "the glades" because of the many interspersed marshes and dwarf cypress prairies that resembled parts of the Everglades.** So, it is not clear how the name "Big Cypress Swamp" evolved or what lands it designated. The historic areas having huge cypress were nearly all logged in the 1940's and early '50's and hauled away by steam locomotives. Only remnants survive today. But even without those historic giant cypress, the Big Cypress is famed for its spectacular variety of plant life and intriguing wilderness.

PUBLIC CONSERVATION UNITS AND NATIVE AMERICAN LANDS

Because of its botanical, wildlife, and wilderness values, extensive areas of the Big Cypress Swamp are now protected in the following preserves listed generally from northwest to southeast (see Figure 7.1):

- Corkscrew Swamp Sanctuary
- Florida Panther National Wildlife Refuge
- Picayune Strand State Forest
- Collier Seminole State Park
- Ten Thousand Islands National Wildlife Refuge

---

* The main sources for this chapter are references 176 and 399.
** Oral history, Taylor Alexander talking with Fred Dayhoff: see "Recollections."

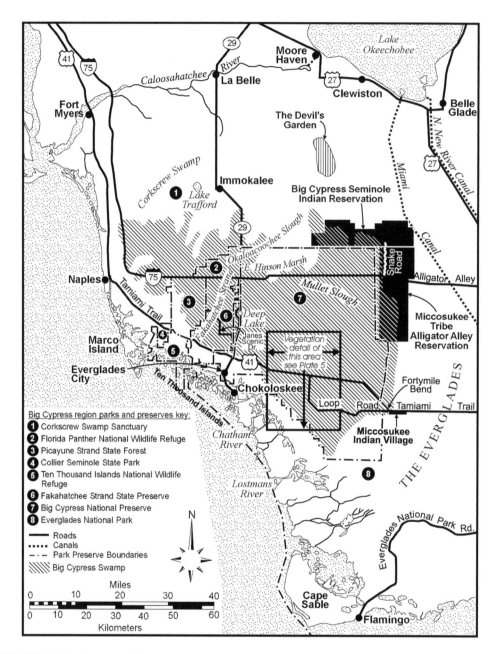

FIGURE 7.1 Location map of the Big Cypress Swamp region, covering coastal areas from Charlotte Harbor (upper left) to Florida Bay (lower right). (Selectively redrawn from USGS State of Florida map, scale 1:500,000, compiled 1966, 1967 edition, with details from Ref. 428. Big Cypress Swamp location from McPherson.[399] Alligator Alley/I-75 alignment interpolated from various sources.)

- Fakahatchee Strand State Preserve
- Big Cypress National Preserve
- Everglades National Park

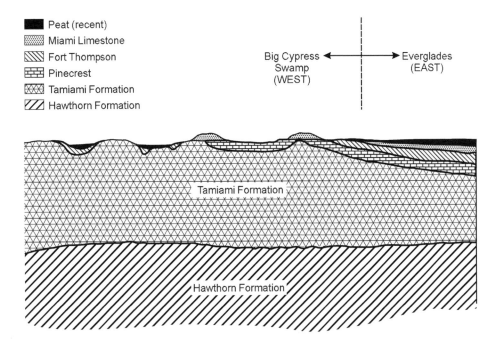

FIGURE 7.2 Eastern Big Cypress Swamp/western Everglades typical geologic cross section near the Tamiami Trail. The Tamiami Formation here is about 50-feet thick. Surface layers are shown proportionally thicker for presentation. Most of the surface layers are a maximum of about 3-feet thick, including the two outlier-mounds of Miami Limestone, which, in the eastern Big Cypress, typically provide the higher elevation surface where hammocks occur. Topography in the western Big Cypress is nearly all the result of undulations in the surface of the Tamiami Formation, with depressions, including solution holes, containing remnants of younger limestone and superficial peat of recent formation. (Redrawn and modified from Ref. 176.)

In addition to the conservation lands, two large Native American lands lie in the northeastern Big Cypress:

- Big Cypress Seminole Indian Reservation
- Miccosukee Tribe Alligator Alley Reservation

BIG CYPRESS BOUNDARIES AND SURFACE HYDROLOGY

Except for its easily observed eastern transition with the Everglades, the Big Cypress Swamp does not have clear boundaries. The least defined boundary is to the north, against an elevated area sometimes called the Immokalee Rise. The latter is a region of sandy soils west of the northern Everglades and south of the Caloosahatchee River, occupying parts of Hendry, Collier, and Lee counties. With its maximum elevation of about 40 feet, the Immokalee Rise is substantially higher than the Everglades and Big Cypress. Agriculture — row crops, cattle, and (since the 1980's) large tracts of citrus — is an important land use in the eastern part of the Immokalee Rise, and a hub of the agricultural activity is its namesake town of Immokalee. The original eastern uplands, prior to clearing, were dominated by pine flatwoods and "dry prairies." The latter are areas of grasses and saw palmetto where conditions were dry enough that wildfires were frequent and intense — too severe even for the slash pine. And while the region is relatively high, it is flat and thus drains slowly so that there are numerous pockets of wet prairie and cypress swamp. The eastern portion of the Immokalee Rise, nearer the Everglades, was dominated by wet prairies but also contained islands of pine and deeper areas of cypress in a region so complicated that it was given the name, "The Devils Garden." Much of that area has been drained for agriculture.

Now largely replaced by canals, the natural surface drainage from the Immokalee Rise flowed to all neighboring areas. The northeastern portion followed sloughs into streams such as the Orange River, northward to the Caloosahatchee River. However, the historic Caloosahatchee River prior to 1881 did not flow like today's canalized river. The upper portion, east of LaBelle, may have partially overflowed southward toward the Devil's Garden, where the waters joined flows coming from the Immokalee Rise. The Devil's Garden flowed partly eastward to the Everglades and partly southward through the present Big Cypress Seminole Indian Reservation, through Kissimmee Billy Strand of the northern Big Cypress to Mullet Slough. Further west, two drainages flowed southward to the Big Cypress. One, the Okaloacoochee Slough, flowed from the mid-southern edge of the Immokalee Rise into the most botanically rich area of the Big Cypress, the Fakahatchee Strand. Still farther west, another drainage from the Immokalee Rise flowed southwest to Corkscrew Swamp, near Lake Trafford, one of the few natural lakes south of Okeechobee. Corkscrew and Lake Trafford are at the northwestern edge of the Big Cypress Swamp (see Figure 7.1).

Looking at surface-water flows of the Big Cypress, it is impressive that the plant community patterns of the southern Big Cypress Swamp are generally oriented northeast to southwest. This trend begins with water coming from Mullet Slough, initially flowing eastward toward the Everglades. At the edge of the Everglades this flow pattern turns abruptly south and then southwest, into many sloughs and "strands" (a "strand," further defined below, is a common plant-community term in the Big Cypress and denotes an elongated corridor of swamp forest). This pattern in the southern Big Cypress signifies the southwestwardly flow to the coastal mangrove zone. The easternmost flow way is Lostmans Slough along the border between the Big Cypress and the Everglades. Historically — prior to man-made flow obstructions including the Tamiami Trail, Loop Road, and especially the southern leg of the L-28 dike — some water from Mullet Slough freely joined flows from rainfall in the Big Cypress interior and moved through surface routes such as Dayhoff Slough, Gum Slough, Roberts Lakes Strand, Sweetwater Strand, and Sig Walker Strand to the Chatham River in the mangrove zone (Color Figures 2, 3, and 5).[456]

The west side of the Big Cypress has no agreed boundary. It merely grades into sandy uplands near Naples, Florida. The original terrain there — now heavily developed — was dominated by slash pine forests, but often with intermixed cypress and areas of deeper cypress swamp. The southern edge of the Big Cypress slopes into the coastal saline communities, first into coastal marshes and then into mangrove swamps. Continued rising sea level coupled with canal diversions of freshwater, which have reduced water levels and flows from the Big Cypress, have induced mangroves and other salt tolerant plants to encroach northward into the Big Cypress Swamp.

GEOLOGY AND SOILS

While the surface terrain of the Big Cypress Swamp is relatively flat, the underlying bedrock is irregular and often exposed at the surface, unlike most of the Everglades where exposed rock is the exception. Despite the apparent complexity, the bedrock and soils are simplified when approached from the prehistoric view of successive layers deposited by high stands of sea level, and intervening low sea level where weathering, erosion, and dissolution dissected or even erased portions of earlier deposits.

The most widespread rock at the surface, and reaching to depths of 150 feet, is the Tamiami Formation, a marine limestone deposited from early Pliocene until about 4 million years ago. The Tamiami Formation also lies beneath the Everglades where it is covered by more recent limestones (see Figure 7.2). Following its deposition, the sea level receded, leaving the then-unconsolidated Tamiami Formation exposed to weathering and recrystallization from seasonal wetting and drying, which solidified the upper foot or two into a very hard, dense limestone called "cap rock." However, during cap-rock formation, solution holes also formed, often where tree roots provided an initial opening. Subsequent percolation of rainwater then dissolved the softer, underlying limestone, forming shallow subterranean cavities. In time, many of these solution cavities collapsed, making shallow depressions that usually contain fragments from the overlying cap rock.

Subsequently, sea level rose and fell at least five times, successively inundating and exposing southern Florida. With each inundation, new deposits occurred, mostly marine and freshwater marls that would ultimately produce limestone, but also silica sand from the north, carried by wind or by waves along shorelines. The sand deposits, such as the "Pinecrest beds," were thicker in the northern and western coastal areas, but sand pockets occur in many places throughout the Big Cypress. These various sediments filled the weathered surface of the older Tamiami Formation, including the solution holes and cavities, and produced relatively thinner veneers of new sediments that, in turn, consolidated and produced sandstone or additional cap rock in some locations. The last flooding — a relatively short stand of high sea level that formed the Miami Limestone and deposited the last seashore and dune sands mostly in the northern and western Big Cypress — ended about 100,000 years ago. It had flooded South Florida to 25 feet above today's sea level. It left the topographic shoreline feature called the Pamlico Terrace, locally identifiable around the Immokalee Rise, then an island (see Figure 1.4). The Miami Limestone forms the uppermost bedrock of the southern Everglades but remains only as patches in the easternmost Big Cypress, lying over the Tamiami Formation and younger layers such as the Pinecrest sands. Thus, at any one location in the Big Cypress, the end result may look very complicated, but, with data from many excavations and well cores, a story of South Florida geologic history has emerged that is now reasonably well understood.

Most of the Big Cypress is now covered by shallow, recently developed soils, deeper only in depressions. The soils include sands, most prevalent in the northern and western Big Cypress, marls that have developed in freshwater marshes, and peat soils that have filled depressions that are flooded most of the time. The peat soils are in everything from small pits in the rock to extensive troughs in the rock surface where sloughs have developed in major flow ways.

VEGETATION

The vegetation of the Big Cypress developed in response to hydroperiod and water flow, which in turn responded to the topography and underlying geology. While knowledge of Everglades vegetation and plant communities makes most of the Big Cypress Swamp familiar, there are interesting differences. The major plant communities include the following:

- Hardwood hammocks
- Pinelands (flatwoods)
- Cypress forests and domes
- Mixed pine and cypress forests
- Mixed swamp forests
- Freshwater marl prairies
- Freshwater sloughs

Mapping the plant communities in the Big Cypress has been a challenge relative to the Everglades because many communities exist as small and/or irregularly shaped patches. A representative area of the Big Cypress Swamp is presented in Color Figure 5. If the whole region were mapped the size of one page, it would be hopelessly complex and unreadable.

To generalize, most of the plant communities of the Big Cypress are floristically more diverse than those of the Everglades, having more temperate species especially toward the north, and more endemic species with very limited ranges. Emphasizing the unique features that set the Big Cypress apart from the Everglades, each plant community is described as follows:

Hammocks

Just as in the Everglades, these upland forested habitats are the final stage of succession where there is protection from fire, and the elevation is above regular annual flooding. They are most abundant in the eastern Big Cypress where there are isolated, elevated "islands" of Miami Limestone lying on top of the older Ft. Thompson and/or Tamiami formations (see Figure 7.2). A few small hammocks are seen in Color Figure 5.

Only the hammocks in the southern Big Cypress can be called tropical. The interior and northern areas are more conducive to reproduction of temperate species and more sensitive tropical species are restricted by cold. Examples of temperate trees are the common persimmon (*Diospyros virginiana*) and oaks including laurel oak (*Quercus laurifolia*) and water oak (*Quercus nigra*) in addition to the ubiquitous live oak, which is also abundant in the hammocks of the Everglades and Florida Keys. A tropical tree that persists throughout the Big Cypress is the strangler fig, although northward this species is usually pruned by cold weather to a dwarf stature except where protected by moist environments such as in Corkscrew Swamp.

Pinelands

The pinelands in the Big Cypress are like islands, usually surrounded by shallow marsh communities. They range from areas as large as 15 square miles in the center of the Big Cypress National Preserve, to thousands of patches measuring only a few acres (Figure 7.3). Accessible pines were logged at the same time as cypress.

The region's pinelands are of two types, depending on elevation: high and low pinelands. Both kinds can occur on rock or sandy soil but not on peat, and are very tolerant of fire. Wax myrtle is a ubiquitous shrub throughout. The "high pine" is seldom flooded, has upland soils, and has an understory dominated by saw palmetto but with smaller numbers of cabbage palms. Unlike the

FIGURE 7.3 The Loop Road area of the Big Cypress National Preserve, showing a small pineland (center) amid marsh and cypress domes, in early dry season defoliated condition. The location is near the east edge of Color Figure 5, looking north (the pineland is too small to appear on the color figure). (Photo by Clyde Butcher, with permission.)

pinelands of the eastern and southern Everglades, several temperate shrubs are common in the Big Cypress, including gallberry (*Ilex glabra*), staggerbush (*Lyonia fruticosa*), and fetterbush (*Lyonia lucida*). Upland tropical species such as West Indian lilac, common in the rocky pinelands of the southeast Everglades, are few or absent in the Big Cypress, where only flooded swamp habitat provides sufficient protection from cold.

The areas of lower elevation have seasonal shallow flooding, hydric soils, and few shrubs except that wax myrtle may be abundant. The hydroperiod is short, up to about two months. The ground cover is of grasses and herbs, many of which occur also in the marl prairies. Cabbage palms, more tolerant of flooding than saw palmetto, may be common and may be dominant in areas where pines were removed by logging. Of botanical interest, the wet pineland is the most diverse South Florida community with 361 species known.[365]

Cypress Forests and Domes

In aerial coverage, cypress forests — including domes, cypress strands, and dwarf cypress prairies — dominate the landscape, especially in the Big Cypress National Preserve. In areas of peat soils, bald cypress dominates, and in marl soils, pond cypress dominates. Their growth forms are distinctive — bald cypress limbs extending horizontally and pond cypress extending at a uniform upward-reaching angle.

Cypress domes are perhaps the most structurally unique feature of the Big Cypress. They are particularly common south of Alligator Alley (I-75) in the eastern Big Cypress, where, from the air, the terrain looks like thousands of hills (Figure 7.4). Cypress domes are a response to bedrock depressions, where dissolution of the softer rock beneath overlying cap rock caused the surface to collapse. The longer hydroperiod, sometimes with nearly perennial water in the center, allowed for accumulation of peat soils. The dome shape of the cypress forest in the depression is a graphic depiction of growth conditions, better in the longer hydroperiod, deeper peat soil toward the center (see Figure 4.3). Thus, height signifies vigor not age of the pond cypress species that dominates this community. In some cases, the central depression is too deep for seedling survival so that the center is an open pond, often filled with aquatic vegetation. The condition of separate domes may explain why soil fire does not eliminate cypress here as it does in the Everglades. Soil fires may start in domes during severe droughts, burning the peat soil sometimes to bedrock, and killing trees in the dome's central portion. However, fire is unlikely to spread between domes because peat soils are not continuous. Short-hydroperiod marl soils that do not support fire usually occur between domes. Even the outer rims of domes have marl soil, so that trees there survive and provide a seed source that leads to dome recovery.

Occasionally, cypress-dominated communities are in the form of a strand but lacking significant numbers of broad-leaf trees and shrubs that characterize the mixed swamp forest. Alternatively, there are vast areas of short-hydroperiod marsh on marl soils that support short-stature cypress, often called "hatrack" cypress, which also occurs in limited areas of the Everglades under the same soil conditions (see Figure 3.5).

Mixed Pine and Cypress Forest

This community is a variation of cypress swamp that occurs on nearly level, sandy soil where drainage is poor. Widespread areas occur west of the Fakahatchee extending nearly to Naples. The particular condition does not favor peat accumulation because fire is common, thus removing surface litter, and the hydroperiod is short. However, the wet sand, a hydric soil, restricts vegetation to wetland species and protects roots against fire. The overstory is almost completely dominated by slash pine and pond cypress. The understory commonly contains cabbage palms and various herbaceous species.

FIGURE 7.4 Aerial view of cypress domes of the Mullet Slough area just south of Alligator Alley, Big Cypress National Preserve. The large dome in the foreground contains a pond in the center. The cypress trees are in winter, defoliated condition. (Photo by T. Lodge.)

MIXED SWAMP FOREST

This wetland community is botanically related to the wetland tree islands of the Everglades (see Table 4.1). Examples of wetland trees that are abundant in Big Cypress mixed swamp forests but uncommon in the Everglades tree islands, especially southward, are pop ash and red maple. Prior to logging, these communities were dominated by bald cypress.

The most striking and unique plant community feature of the Big Cypress is the linear corridor of swamp vegetation called a "strand." Because the bedrock lies close to the surface, linear bedrock depressions are occupied by these swamp corridors, which were naturally dominated by taller cypress prior to logging, and now dominated by hardwood swamp trees. In the Everglades, such bedrock features would not be apparent because of the widespread deeper veneer of peat soils that obscure bedrock irregularities. While not quite as floristically diverse as the wet pinelands, these strands are of extreme interest to botanists because of their unusually rich tropical character, most pronounced in epiphytic plants, notably orchids, bromeliads, peperomias, and ferns.

Prominent examples of mixed swamp forest strands that are identified on many Big Cypress maps are Fakahatchee Strand, Deep Lake Strand, Gannet Strand, Gator Hook Strand, Picayune Strand, Sweetwater Strand, Roberts Lakes Strand, Skillet Strand, and Kissimmee Billy Strand. The Fakahatchee is by far the best known and is one of the most noted botanical areas in Florida (Figure 7.5). Its name came from the Seminole Indians and means "muddy creek," referring to its relatively deep muck/peat soils, not to the character of the water, which is normally clear. The Seminoles and earlier Indians used the area, evidenced by many habitation mounds along its course. The central part of the strand is a forested slough — often with pond apples having spectacular buttressed trunks (see Figure 4.7) — punctuated by small lakes. Prior to logging, tall bald cypress formed the canopy of the strands, and should again in the future, but based on over 50 years of regrowth to date, full recovery to former stature will take at least another century. Also, lining the edges of the slough communities, royal palms were very common on the damp peat soils. Most of these were

FIGURE 7.5 A profusion of airplants, mostly West Indian tufted airplant (*Guzmania monostachia*), growing on cypress, pond apples, and pop ash in a deep slough of the Fakahatchee Strand. (Photo by Jeff Ripple, with permission.)

removed by landscapers after the completion of the Tamiami Trail in 1928, but their numbers are now recovering.

Because of the protective humidity emanating from the open water in the central slough and the saturated soils, many sensitive tropical plants thrive in the Fakahatchee, some found nowhere else. An inventory lists 403 kinds of native plants in the Fakahatchee Strand State Preserve, 183 specifically in the swamp forest portion. Of special note among the preserve's native species are 35 ferns, 13 bromeliads (airplants), 45 orchids, and three peperomias. Like in other South Florida natural areas, many exotics — about 90 species in the Fakahatchee — have become established.[46]

The Fakahatchee Strand is accessible by hiking trails off of Janes Memorial Scenic Drive, a maintained dirt road that crosses the entire strand.[*] Many of these trails are on the remnant elevated railroad tramways used for logging. But to traverse areas between tramways, you must often walk through waist-deep sloughs, rewarded, however, by a profusion of epiphytes on pond apple, pop ash, red maple, and second-growth cypress, all growing in water. For the less adventuresome, there is Big Cypress Bend, a spectacular boardwalk off of the Tamiami Trail seven miles west from State Road 29. Big Cypress Bend and Corkscrew Swamp are the only areas where no logging occurred so that huge bald cypress still thrive there, resembling the 130-foot stature of the former Fakahatchee giants.

Deep Lake Strand lies just east of the Fakahatchee Strand. It also harbored large cypress, which were logged at the same time as the Fakahatchee. Deep Lake, the strand's namesake feature, is unusual in being a deep sinkhole, formed during low stands of sea level. Only five sinkhole lakes occur in southern Florida, the others farther north in southwest Florida. The closest is Rocky Lake,

[*] Few are aware of this rough but generally passable road and its connection to Immokalee by following it from State Road 29 to its end on Stewart Blvd., then west through Picayune Strand State Forest to Everglades Blvd., then north over I-75 (no access) to State Road 846. Most of this 37-mile connection between routes 29 and 846 is paved but not shown on Figure 7.1: consult detailed area maps.

located about 25 miles north by northeast of Deep Lake. Deep Lake is only 1.7 acres in size but is 95 to 98 feet deep (depending on seasonal water level) and exhibits interesting water quality features with depth.[349, 524] It is also intriguing that Deep Lake was an open sink hole, possibly with a pool of water at the bottom, when the first Native Americans probably entered the region over 11,000 years ago. Although no evidence yet exists of prehistoric use of Deep Lake as a water source, such is well documented for two other sinkhole lakes, namely Little Salt Spring and Warm Mineral Spring in Sarasota County (see Chapter 21, Paleo-Indians section).

Roberts Lakes Strand runs from the Tamiami Trail in the eastern Big Cypress southwestward toward the west segment of Loop Road, where its waters feed into Sweetwater Strand. It was also an area having large bald cypress that were logged using rail tramways that served a local sawmill on Loop Road. Roberts Lakes Strand is of interest because of its easy access from Loop Road, which itself provides excellent views of all Big Cypress plant communities at a leisurely pace, without having to worry about high-speed traffic.

Corkscrew Swamp, although not in the form of a strand, is the largest of the remaining mixed swamp forests that still shelter huge bald cypress. Maintained by the National Audubon Society, it is easily accessible via an interpretive boardwalk two miles long. It is famed for having the largest wood stork nesting colony in the United States.

FRESHWATER MARL PRAIRIES

These areas are virtually the same as those occurring in the southern and southeastern Everglades. One distinctive plant here, however, is corkwood (*Stillingia aquatica*), a low shrub that may resemble a sapling cypress to the novice. As in the Everglades marl prairies, muhly grass may be very abundant, and the Everglades or paurotis palm (*Acoelorrhaphe wrightii*) was formerly common here and in other moist plant communities. It is recovering from extensive collecting by landscapers.

FRESHWATER SLOUGHS

Unlike the sloughs in the Everglades, sloughs in the Big Cypress are not characterized by waterlilies. Big Cypress plant community maps use "freshwater slough" to mean marsh areas slightly deeper than marl prairies and dominated by cattail, sawgrass, arrowhead, pickerelweed, fireflag, narrowfruit horned beaksedge, spikerush, and bladderworts. By far the largest such slough community of the Big Cypress is Okaloacoochee Slough, which carries water southward from the lower Immokalee Rise into the Fakahatchee Strand. However, several flow-ways of the Big Cypress bear the name "slough" but are characterized by other plant communities. For examples, Mullet Slough and California Slough, both in the Big Cypress National Preserve (described below) are areas dominated by cypress. Similarly, the forested central flow-ways in strands are called sloughs.

INTEGRITY OF THE BIG CYPRESS

There have been many alterations of the Big Cypress Swamp that have impacted the natural environment. The main impacts have been the following:

- Timbering for cypress and pine in the 1940's and '50's. Recovery of tree stature may require another century or more.
- Plant collecting, still occurring by individuals, especially for orchids and bromeliads, and in the past by landscapers, particularly for paurotis (or Everglades) and royal palms.
- Off-road vehicles for hunting and outdoor recreation. Extensive rutted trails have created enormous impacts and controversy (Chapter 21).
- Hydrologic impacts due to: 1) drainage and residential development mostly west of the Fakahatchee Strand, 2) agricultural development mostly in the northern Big Cypress,

and 3) diversion of waters by Everglades works, primarily the L-28 levee and canal. Drainage began with the construction of the Tamiami Trail and canal in the 1920's.

- Loss of peat soils (for examples, 1.5 feet in Deep Lake Strand and Gum Slough) due to interior drainage canals and to reduced water from the Everglades.
- Change in fire seasonality from late spring and summer, set by natural lightning, to winter, due to controlled burns and arson.
- Road kills and barriers due to highways. Major highways include the Tamiami Trail, State Road 29, and especially I-75 (Alligator Alley). The development of Alligator Alley required the complete clearing of over 2000 acres (more than three square miles) of Big Cypress habitat and was a formidable physical barrier to movement of animal populations. However, road kills and barrier impacts were subsequently minimized by the highly successful installation of fencing and wildlife underpasses.[202]
- Oil and gas extraction. Exploration for oil and gas began in 1940, and has continued to the present, with several hundred wells drilled, but only a few have been successful. Understandably, conservationists have been alarmed, but impacts have been relatively small, especially in later phases where environmental controls have been strict. Compared to highway development and associated wildlife impacts and to recreational off-road vehicle use, the oil and gas impacts have been negligible.
- Airport construction. In the late 1960's, a major South Florida regional airport was planned for the southeastern Big Cypress. Development was halted due to envionmental concerns, but one runway was constructed. It has been subsequently used for pilot training, mostly in large commercial aircraft, and is called the Dade-Collier Training and Transition Airport.
- Invasive nonindigenous plants and insects, especially melaleuca, Brazilian pepper, the lobate lac scale insect, and the bromeliad beetle. Only melaleuca receives effective control, but the program is subject to funding problems. Of ominous implication, Old World climbing fern has recently become established (see Chapter 21). Example areas are in the Seminole Big Cypress Reservation.

In spite of all these impacts, much of the region is still wild. It resembles a pristine environment as much as can be found anywhere in Florida. Visitor use is enhanced by the Florida Natural Scenic Trail (a hiking trail that traverses the Big Cypress National Preserve from Loop Road on the south to Alligator Alley on the north). The Fakahatchee is accessible by rough foot trails off of Janes Scenic Drive (see Figure 7.1) and by the Big Cypress Bend boardwalk on the Tamiami Trail about 7 miles west of State Road 29. To the northwest of the Big Cypress, the boardwalk at Audubon's Corkscrew Swamp Sanctuary is exceptional.

8 Mangrove Swamps*

Throughout the world, wherever shorelines are protected from the direct action of waves, the land lying between high and low tides supports wetland vegetation. Where coastal lands are flat, this *intertidal* zone may be huge, permeated with rivers and tributary streams that carry the ebb and flow of tidal waters through the wetlands. In temperate climates, protected intertidal habitats are dominated by various grasses and related herbaceous (non-woody) plants. In northern Florida's gulf and Atlantic coasts, where sufficient protection exists, the intertidal areas are salt marshes, dominated by smooth cordgrass (*Spartina alterniflora*) and black rush (*Juncus roemerianus*). While both of these species are found in southern Florida, they are restricted to much smaller areas by a completely different intertidal community: the mangrove swamp.

The mangrove swamps in southern Florida lie primarily within Everglades National Park and cover more than 500 square miles, forming dense forests from the southernmost freshwater marshes of the Everglades and Big Cypress Swamp seaward to the open waters of Florida Bay and the Gulf of Mexico (Figure 8.1). This area amounts to about two-thirds of all mangrove forests in Florida, and is the largest contiguous mangrove swamp in the world. Its great size evolved because of coastal deposits of sand, marl, shells, and mangrove peat that extended intertidal and shallow-marine habitats seaward. Slowly rising sea level over the past three thousand years provided the condition for these thick accumulations (see Chapter 10).

Intertidal habitats present strong environmental stresses. They are alternately drained and flooded through tidal cycles. Salinity may vary seasonally from freshwater to hypersaline (sometimes more than twice the salinity of sea water due to evaporation in some locations), and can change radically within minutes if there is heavy rainfall at low tide. Temperatures may also vary greatly. A mangrove swamp in Florida may experience near freezing temperatures in winter, and the surface mud may be so hot during a summer low tide that it is uncomfortable for a barefooted person to walk through open, sunny areas. Thus, the plants and animals of mangrove swamps must be adapted to a wide range of conditions.[492]

TYPES OF MANGROVES [284, 598, 599]

Mangroves are tropical trees that are adapted to salt water and to the rigors of tides. Through the world's tropics 34 tree species are called mangroves. An additional 20 minor species that occur in mangrove swamps, such as the leather fern in South Florida, are called mangroves in some regions. Mangroves are ecologically related by their adaptation to the intertidal habitat. Although several are members of the same plant family, many are unrelated by evolution. In fact, nine different plant families account for the 34 main mangrove species. Even the palm family is represented by one species, which occurs in the Indo-Pacific, not South Florida.

Only three species are normally referred to as mangroves in Florida, the red mangrove (*Rhizophora mangle*), the black mangrove (*Avicennia germinans*), and the white mangrove (*Laguncularia racemosa*), all from different families. The common names of these species may be mystifying at first. The red mangrove receives its name from a red layer *under* the thin, gray bark. "Black" refers to the color of the bark of a mature black mangrove. The white mangrove has the lightest colored bark of the three species, but it is not really white. Another species, the buttonwood (*Conocarpus*

---

* The main sources for this chapter are references 144, 266, 284, 340, and 447.

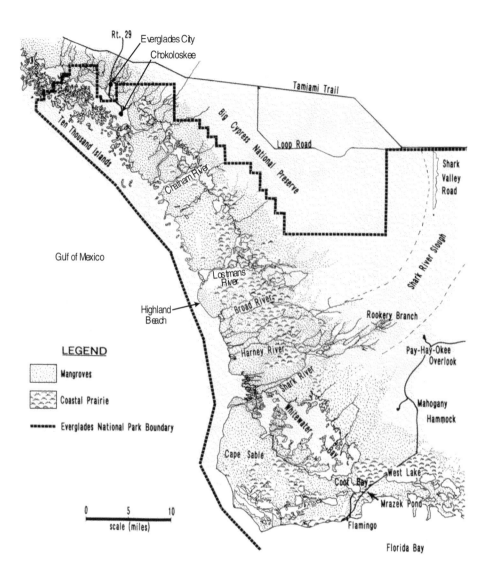

FIGURE 8.1 Map of the expanses of mangrove swamp and coastal prairie in Everglades National Park. (Redrawn from Ref. 428.)

erecta), is sometimes called a mangrove. It belongs to the white mangrove family and is best suited to slightly higher ground around the edges of mangrove swamps. A "silver-leaved" variety of buttonwood is used as a native landscape plant in southern Florida.

RED MANGROVE

The red mangrove (Figure 8.2) is adapted to its intertidal habitat by a most curious feature, which has rendered this species as a prime subject for photographers and landscape painters. It develops aerial adventitious roots, commonly called "prop" or "stilt" roots, from locations sometimes many feet high up on the trunk, or even from branches. These roots initially reach outward, then curve down to the soil. The roots themselves often develop prop roots, sometimes with a half dozen successive arching tiers reaching outward from the tree.

FIGURE 8.2 The author amid extensive prop roots of an old red mangrove along lower Oleta River, north-eastern Miami-Dade County (1972). (Photo by T. Lodge.)

The prop roots provide two adaptive functions. First, they form extra support for the tree, with prop roots reaching in all directions. Red mangroves normally occupy the most exposed, outer locations of mangrove swamps, where waves and tidal currents are the most rigorous and where survival requires a firm grip. Second, the prop roots are an adaptation to hydric soils. Because intertidal soils are continuously saturated with water, there is no opportunity for atmospheric oxygen to get into the soil. Root growth and function requires oxygen as do other plant tissues (as discussed earlier in the case of cypress trees in freshwater wetlands). Pores (white spots on the thick part of the exposed root just above the soil) exchange gasses with the atmosphere and supply oxygen to the subterranean root system. This adaptation also allows red mangroves to survive in the interior of mangrove swamps. Under certain conditions — usually relatively lower elevation — red mangroves may dominate the interior where black mangroves usually rule.

Red mangroves do not require salt water. As a notable example, Hurricanes Donna (1960) and Betsy (1965) pushed huge numbers of seedlings inland into freshwater marshes in Everglades National Park. Many of these are thriving today and can be seen along the main park road between Mahogany Hammock and Paurotis Pond. A group is even growing among some cypress trees, which are highly intolerant of salt. Red mangroves, however, do much better in estuarine habitats where sea water and freshwater mix.[128,144]

Another notable and important characteristic of the red mangrove — and of other mangroves to a lesser extent — is a poor ability to resprout from branches and trunks after breakage or pruning. Most trees and shrubs retain dormant buds under their bark. These buds are normally inhibited by the tips of the actively growing branches above them. However, when outer branches are broken or pruned, these buds begin sprouting. The trunk and branch sprouting process is important, for example, in fire-prone communities where many shrubs are killed back but then quickly recover with new foliage. Gardeners take advantage of this characteristic in selective pruning. The red mangrove initially produces such tissues near its growing tips, but the dormant buds die within two or three years so that older portions of branches and trunks have no ability to sprout new branches. Thus, fire and wind damage in mangrove swamps causes far more tree death than would be expected, especially where red mangroves predominate.[128, 598]

BLACK MANGROVE

The black mangrove is also adapted to hydric soils, but its approach is opposite that of the red mangrove. Black mangrove roots beneath the soil send small extensions upward, reaching as much as a foot above the soil surface. They also function to exchange gases, and they bear the long name *pneumatophore* (Figure 8.3). Pneumatophores are functionally comparable to cypress knees but are so numerous that the whole forest floor seems to be covered with them wherever black mangroves are found — mostly in the interior of mangrove swamps where tidal action is very sluggish but shallow flooding may be prolonged.

Another peculiarity of black mangroves is salt balance, which differs abruptly from other Florida mangroves. Salts are mostly excluded from entering the roots of red and white mangroves (and related buttonwood) but enter black mangrove roots. In compensation, black mangroves control internal salt concentration by excreting excess salt through tiny glands in their leaves. Subsequent evaporation produces visible salt crystals on the upper surface of the leaves if it has not rained for a few days. Thus, another identifying character of the black mangrove is the taste test: just lick a leaf.

WHITE MANGROVE, BUTTONWOOD, AND THE BUTTONWOOD EMBANKMENT

The white mangrove can occur almost anywhere in a mangrove swamp or protected shoreline, but it is most abundant in the higher elevations, such as the edges of mangrove swamps adjacent to uplands. Under stressed conditions, its roots may form short, branched pneumatophores and its lower trunk may produce small adventitious roots. Compared to the respective adaptations of the black and red mangroves, however, these growths are minor. White mangroves can colonize available upper intertidal areas rapidly, sometimes forming dense, nearly pure stands. A distinctive identifying character is the presence of two obvious gland-like openings on the petiole (leaf stem) where it meets the leaf.

FIGURE 8.3 Black mangrove pneumatophores along the boardwalk at West Lake, Everglades National Park. (Photo by T. Lodge.)

The buttonwood occupies still higher locations where normal tidal flooding is only an inch or two deep to uplands that only flood a few times a century. It lacks the special seedling and root adaptations of most mangroves, but it is tolerant of saline soils. Its sensitivity to frost restricts it to southern Florida, where it is primarily coastal in distribution but occasionally found inland in pinelands or the edges of tropical hardwood hammocks. Buttonwoods form forests along the inland edges of mangrove swamps, saline coastal lakes, and tidal creeks. Buttonwoods also cover curious, naturally occurring levees that stand about a foot and a half above the surrounding mangrove swamp elevation, leading to the name *buttonwood embankment*. (The processes that form these levees are covered in Chapter 10.) At the inland edge of the mangrove swamps in Everglades National Park, buttonwood embankments act as dams, impounding freshwater during the rainy season, sometimes creating an audible rush as high water from the Shark River Slough spills over. These areas are important to, and rich in, wildlife.[34, 128, 286]

MANGROVE SWAMPS AND EVERGLADES WILDLIFE

The mangrove forest development of South Florida is most spectacular along tidal rivers in Everglades National Park, including the Shark, Harney, Broad, and Lostmans Rivers in the mangrove zone of the park (see Figure 8.1 and cover). While mangroves visible to most people seldom exceed 50 feet in height, many specimens along these rivers reached nearly 100 feet before Hurricanes Donna and Andrew and severe cold fronts during the 1980's toppled and pruned them to a lower stature.[144, 190]

The transition from the freshwater marshes of the Everglades to the tidal mangrove swamps begins in a series of creeks, which reach like fingers from the mangroves into the marshes. The creeks, form the headwaters of the tidal rivers. The rivers and their tributary creeks, coupled with the buttonwood embankment and its associated freshwater pools, are of great importance to wildlife. Freshwater flow from the marshes keeps the tributary creeks fresh to lightly brackish most of the time, with saline conditions only reaching the creeks during times of extended drought or storm-driven tides.

Alligators play a prominent role in the mangrove swamp tributary creeks, their activities adding substantially to the ecological function of the area by allowing interplay between the upper tidal and freshwater marsh areas. Alligators keep the upper reaches of creeks open and provide numerous trails into the marsh; these trails are advantageous to aquatic life moving between the creeks and marsh.[128] The alligator's "enhancement" of the creeks provides wading birds with an increased availability of prey, much being juvenile fishes of estuarine and marine species.[360] This enhancement was likely a key reason why the historic wading bird rookeries of Everglades National Park were located in these areas, giving one tributary, connected to the Shark and Harney Rivers, the name Rookery Branch (see Figure 8.1).

MANGROVE SWAMPS AND MARINE FISHERIES

Mangrove swamps can be extremely productive. Given the right combination of tidal "flushing" (movement of water and suspended material in and out of the swamp) and freshwater runoff, mangrove swamps are among the most productive natural communities in the world. These conditions are present where the giant Shark River Slough and Big Cypress Swamp pour their freshwaters into the Gulf of Mexico, across the very gentle slope of the Florida Platform in the western half of Everglades National Park. The freshwater input dilutes the saltiness of marine waters, making life easier for the mangroves by requiring less effort to exclude or excrete salt. Therefore, more of their photosynthetic catch of sunlight energy is directed toward growth. The tidal flushing also helps to move nutrients around, making them more available, and to transport products (dead leaves, twigs, etc.) away. The combined interaction of the tides and freshwater can be compared to the work of a farmer in his fields, tilling the soil and fertilizing, irrigating, and harvesting the final product. Just as the farmer's labors greatly increase the production of his land and of the region, so do the tides and freshwater runoff enhance the mangrove ecosystem and its benefactors.

The ecological values of the mangrove swamp arise from the fate of its products: the leaves, twigs, bark, pollen and so forth. As these materials decay, they become food for marine life. Mangrove "detritus" (dead material) is eaten by many organisms at the base of an extensive food web, including shrimp, larval crabs, and small fish. In fact, the excellent fisheries of the Florida Keys, Florida Bay, and the eastern Gulf of Mexico owe much to the mangroves of Everglades National Park. Not only do the mangrove swamps provide a food source, but they and the associated tidal creeks and bays also act as a "nursery ground" for a great many marine species (see Chapter 20). It has recently been documented that phytoplankton and algae in tidal rivers and sunlit parts of mangrove swamps also contribute importantly to the overall productivity and its food chains.[218]

MANGROVE SWAMPS AND SOIL-BUILDING [126, 284, 447, 501, 630]

Where they are protected from erosion, mangrove swamps are often soil-builders. As individual trees die, their prolific root systems and varying loads of leaf and twig litter become peat, which can rise to the level of high tides. However, mangrove swamps also trap sediments, including sand, shells, and marl produced in shallow coastal waters (see Chapter 10) and carried inland by storm tides. Thus, sediments in mangrove swamps are a mix of mangrove peat and other sediments. When sea level rises slowly, mangrove swamps are elevated by these soil-building processes, at least until some major catastrophe (most often a hurricane) kills the trees (see Chapter 9). A common misconception is that mangrove swamps are always land builders; rather they merely keep pace with rising sea level, at least during times while sea level is rising slowly, as it has during the past 3000 years. Actual "land" building is an unusual event that results when a hurricane deposits enough sediment into a mangrove swamp to raise the elevation above high tide, and such cases of land formation have been documented.

MANGROVES AND HYPERSALINITY [626]

Red and black mangroves are well known for their means of surviving in high salinity, which, because of evaporation, may greatly exceed that of normal sea water during dry periods. These two species have opposing adaptations to this condition. The red mangrove excludes nearly all salt from entering its roots. The black mangrove allows significant salt to enter but then excretes salt from its leaves (as mentioned above). If it has not rained within the past few days, the surfaces of the leaves will have a crust of salt.

MANGROVE REPRODUCTION AND DISPERSAL [598, 599]

Dispersal of mangroves is by floating seeds, as might be expected for an intertidal plant. Floating seeds explain how mangroves got to southern Florida and explain how occasional mangroves take root in northern Florida, only to be killed by cold weather. Of the three species, the black mangrove is the most tolerant of cold and survives farther north, with some individuals found in the Mississippi delta.

The seeds of the red mangrove are a special case. Called propagules, they actually mature and begin growing while still on the tree. Each propagule develops an elongated body that becomes the trunk of the seedling. The lower, brown-colored end is pointed like a missile, and this end develops roots if it lodges in soil (enough so that the tip is in the dark). When the propagule is mature, it drops from the tree and may lodge in the soil directly, where it will begin growing.

However, if it does not stick, it will float, on its side at first. After several days, it begins to change buoyancy and floats vertically. Finally, after several weeks, it sinks. If, by chance during the course of its tidal travels it becomes lodged in a calm area shallow enough for growth, it rapidly sprouts roots from the bottom and an initial two leaves from the top.

LEGAL PROTECTION OF MANGROVES

The story of mangroves and mangrove swamps does not end with mere descriptions of these interesting trees and how they live. Mangroves have been a focus of federal, state, and county legislation in Florida, most of which emanated from studies that began in the late 1960's and pointed out the great ecological benefits that mangrove swamps provide man and wildlife alike.[266] Mangrove swamps had long been recognized for their role in shoreline stabilization, particularly during hurricanes (see Chapter 9). Mangroves take the initial rage of storm-tossed seas and thus protect man's interests further inland. The vulnerability of mangroves to pruning has more recently been the basis of local regulations that restrict trimming mangroves, formerly a common practice for enhancing the view from private property.[*]

MANGROVES AND MOSQUITOS [290]

Mosquitos of the mangrove swamp are normally abundant and aggressive, and a smaller biting insect, the sand fly or "no-see-um" (technically a biting midge, or ceratopogonid), can be even worse. It is a popular misconception that mosquitos do not breed in salt water. The proper statement is that freshwater mosquitos do not breed in salt water. The saline waters of mangrove swamps throughout the tropics of the world, however, are breeding grounds for many species of mosquitos. In coastal South Florida, the black salt-marsh mosquito (*Ochlerotatus* [formerly *Aedes*] *taeniorhynchus*) is abundant. This small species has distinctive black and white striped legs, abdomen, and proboscis. Winds occasionally blow hordes of them into residential areas. However, in a mangrove swamp during any warm month of the year, the combination of mosquitos and no-see-ums can defy description — like trying to explain the idea of infinity to a child. The mosquito "problem" is also described in Chapter 5 on tropical hardwood hammocks.

VISITING A MANGROVE SWAMP

You can enter a mangrove swamp to see all this for yourself, but stepping over thigh-level prop roots and swatting mosquitos at the same time requires athletic ability, patience, and a touch of masochism. Some think a machete is required to get through a mangrove forest. My own experience is that using a machete results in blisters at the least, and possibly a major injury, in addition to falling far behind those who merely take the time to step over and walk around obstacles. The best way is to use the boardwalk[6] such as the one at West Lake in Everglades National Park, with a fresh application of mosquito repellent unless it is a cool, breezy day during the dry season when repellent is not needed.

---

[*] The perceived value of mangroves is often confused by assuming that a mangrove tree itself is the ecosystem value. But there have been many demonstrations that mangroves can grow even in residential lawns, given care. It should be recognized that the value of a mangrove comes from its living in a viable tidal swamp or mangrove island where ecosystem "goods and services" are produced and exchanged. Outside of that context, a mangrove tree does not have a special environmental value.

9 Coastal Lowland Vegetation...and Hurricanes!

The awesome power of hurricanes to reshape nature — as well as man's existence — is of obvious importance in southern Florida. Their ecological impact can be seen in many plant communities especially forested areas, but their effect is most prominent in the lower elevation, coastal communities where raging winds and rapidly moving high water rearrange not only trees but also terrain. These effects combined with rising sea level over the past several thousand years have shaped our coastal features (see Chapter 10). Were it not for hurricanes, certain coastal habitats would not exist.

My own experience with hurricanes began on a normal, sunny summer morning, August 22, 1992. I had stopped at Phil's Camera shop at Suniland on South Dixie Highway, where the air was abuzz with concern. A hurricane had developed in the Atlantic and was then some 800 miles east of Miami and heading west. Its name was Andrew, and without a change in course, it would arrive within 48 hours. Little did I know that 48 hours later, it would all be over — except for the years of recovery, both for nature and for countless human misfortunes.

Inexperience with hurricanes shaped my indifference to the impending dangers of Hurricane Andrew. I moved to Miami in 1966, and in the ensuing years, there had been several close calls but no hits. My timing had been lucky: Hurricane Betsy had passed the previous year, and the ferocious Donna in 1960 (Figure 9.1). In October 1966, Inez threatened after having devastated Haiti with 160 mph winds. With the memory of Donna and Betsy still fresh, experienced residents prepared: houses were boarded-up everywhere. But Inez, after an all-night standoff, failed to come ashore, merely brushing the South Florida mainland before crossing the Florida Keys with then minimal hurricane winds. On September 2, 1979, David took direct aim at Miami on the same path as Andrew. My wife was into her ninth month of pregnancy, and despite our doctor's opinion, we worried about possibly having to bring our daughter into the world in the midst of a raging storm. Winds became alarming in the first morning hours of September 3, but Miami was spared as David veered north to Palm Beach and Stuart. Even as a minimal hurricane, it did substantial damage there, as related by a friend who experienced the passage of David's eye in Stuart. With some justification, I thought Andrew would do the same. Little did I know that within 40 hours of my visit to the camera shop, in the last hours of darkness on Monday morning, August 24, my family and some friends would be prisoners in our house, witnessing the terrifying rising and falling roar of hurricane winds — contemplating our situation with each surge.

By 7:00 a.m. that morning, the winds had dwindled enough, save for some frightening gusts, to survey Andrew's damage. Our house suffered little, but trees and power lines in our neighborhood were in ruins. Most of the coconut palms that had lined both sides of our street lay on the ground, snapped off at the base and pointing southwest. We quickly determined that we had taken the northeast quadrant of the storm and assumed that we must have been very close to the eye to have such damage. With no electricity and little means of getting reliable news, it took several days to understand how wrong we were. The northern edge of the eye had actually passed nearly ten miles south. Communities that had bordered it, from Kendall to Homestead and Florida City, lay in shambles. In all, 117,000 homes in a 25-mile-wide path were destroyed or had major damage.[654]

This personal experience is passed along to relay the message that individual people do not often endure hurricanes, the most powerful of tropical cyclonic storms, defined as having winds in

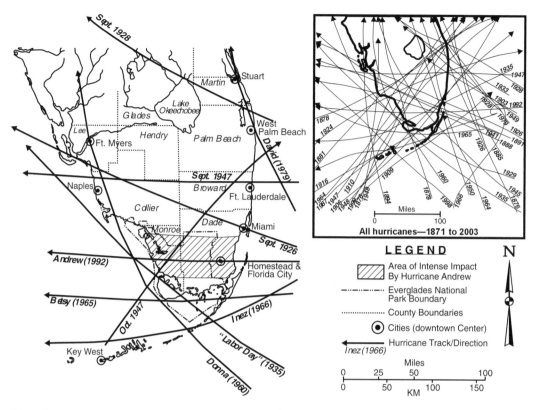

FIGURE 9.1 Map of selected South Florida hurricane tracks mentioned in the text (and all hurricanes 1871–2003). (Redrawn from Refs. 58, 226, and 654).

excess of 74 miles per hour. From the perspective of natural history, however, their common influence on coastal environments is an inescapable fact, worthy of considerable attention by ecologists and geologists.

IMPACTS OF HURRICANE ANDREW ON THE EVERGLADES
33, 54, 142, 154, 352, 367, 406, 443, 654

Small but intense, Andrew followed a westward track through the heart of the mainland area of Everglades National Park. (see Figure 9.1) Its path of heavy impact was about 30 miles wide. Maximum sustained winds were 145 miles per hour, with peak winds to 175. At its landfall along Biscayne Bay south of Miami, its storm surge — the rise in water above expected tidal level — was 16.9 feet. Andrew's fury crossed the park's pinelands, hammocks, and freshwater marshes before exiting through the expanses of mangrove forests where its winds had weakened to about 125 mph, but the storm surge tide was still very high — about 13 feet.[501]

Early evaluations indicated that less apparent destruction occurred in the second-growth pinelands in Everglades National Park, attesting to the strength and flexibility of the younger slash pine. Older, taller pines there and east of the park were commonly broken, but invariably high above the ground, the grip of their rock-anchored roots refusing to yield. However, within a few months, most of the pines in Andrew's path began to yellow and die as they succumbed to pine bark beetles and weevils because of their weakened condition.

There was extensive damage to virtually every tall hammock tree in Andrew's path, most easily observed at Royal Palm in Everglades National Park (see Figure 5.2). The trees were either stripped of leaves and branches or completely toppled. Within a few weeks following the hurricane, vines, many of which were introduced species, began overtaking the defoliated hammock trees, smothering the slow recovery of damaged but living trees.

Relatively little damage was observed in the vegetation of the freshwater marshes and where Andrew crossed the Shark River Slough and to cypress forests in the southern Big Cypress Swamp. There was massive defoliation of southwest coastal mangrove swamps and tall mangroves were uprooted or broken over a 110-square-mile area. In addition, huge sediment loads derived from the adjacent shallow marine waters were deposited in the mangroves, carried by the on-shore storm surge. In the vicinity of the Harney and Broad rivers, about one million cubic yards of sediments were deposited over about 40 square miles of mangrove swamp. Surprisingly, where Andrew exited over shallow marine habitats, mainly beds of turtle grass and shoalweed, there was little damage.

HURRICANE FREQUENCY AND ENVIRONMENTAL IMPACT IN SOUTHERN FLORIDA [58, 226, 443, 635, 636, 654]

Was Andrew an unusual hurricane? For the people who experienced its fury, the answer is emphatically, "yes!" During the prior 100 years, however, 15 major hurricanes passed directly through what is now Everglades National Park (established in 1947). Three were major hurricanes that wrought destruction specifically to the park's coastal area: the relatively small but intense Labor Day hurricane of 1935, which resembled Andrew, the much larger Hurricane Donna (1960), and Hurricane Betsy (1965). These storms altered the ecology of the coastal portions of the park to such an extent that hurricanes are an unavoidable topic in the ecology of southern Florida — as basic as alligators.

Based on records from 1871 through 2003, southern Florida has been struck by 40 hurricanes (see inset in Figure 9.1) for an average frequency of about once per three years, mostly confined to the peak season from mid-August through most of October. The coastal portions of Everglades National Park have been struck about once every ten years, and "great hurricanes" (with winds of 125 mph or more) strike the park's coast four times per century.

Hurricanes vary greatly in size as well as in the intensity of their winds. Despite their strong winds, Hurricane Andrew and the Labor Day Hurricane of 1935 were small, with zones of heavy impact less than 40 miles wide — approximately the area where sustained winds exceeded 100 mph. Using the 100 mph criterion, the giant September 1947 hurricane had a zone of intense impact 70 miles wide, and its hurricane force winds (75 mph or more) extended across 240 miles of Florida's coast.[226] Despite its huge size, the '47 hurricane's peak winds were 155 mph, significantly lower than Andrew (175 mph) and the '35 Labor Day Hurricane (forensically determined between 200 and 250 mph).[654]

Destruction by hurricanes is due to one or more of four factors: wind, waves, storm surge, and rain. For most people, storm surge is the least understood factor and, close to the coast, is potentially the most damaging aspect of a storm. Storm surges are a combination of the "mound" of water (the storm tide) induced by the hurricane's low barometric pressure and of the rough seas and current driven by the storm's winds. Storm surges of severe hurricanes reach more than ten feet above regular tidal levels. The relatively small, but deadly, Labor Day hurricane of September 1935, swamped Matecumbe Key in the upper Florida Keys with an 18-foot surge (its maximum surge was over 20 feet), and Andrew's surge neared 17 feet.

IMPACTS OF HURRICANE DONNA [128, 226, 654]

In addition to the effects of Hurricane Andrew, there were revealing studies of the impacts caused by Hurricane Donna, an immense storm that exceeded the stature of the September '47 hurricane in several statistics. On September 10, 1960, Donna's eye crossed the middle Florida Keys with sustained winds of 140 mph, peak winds between 180 and 200 mph, and a maximum storm surge over 13 feet. From the Keys it headed for the shoreline of Everglades National Park with a track that kept the northeastern quarter of the storm (usually the most powerful part) over the mangrove swamps and other coastal environments, as the eye moved to Naples and Ft. Myers on Florida's southwest coast. In crossing Florida Bay, it reduced the small U.S. population of the great white heron to about 600 survivors.

Some 120 square miles of mangrove swamp were devastated. It is estimated that half of the trees were killed. Many were sheared off by the high winds, and those left standing were stripped of most branches and leaves. These effects were noted immediately after the storm, but then a peculiar sequence of events began. Within a few weeks, as expected, intact trees began recovering, with new growth evident everywhere. Soon, however, the process stopped, leaving a complete kill of mangrove trees over extensive areas. Why did the recovery cease? The answer lay in the impact of the storm surge and the particular direction of Donna's approach. The ten-foot deep mass of moving water forced inland by Donna carried with it a tremendous load of lime mud (marl) from Florida Bay and the bays and lakes inside the coastline. The mangrove forests slowed the flow of the water, causing much of this load of marl to be deposited as a blanket up to about six inches deep over the forest floor. Examination proved that this was enough to interfere with the critical oxygen supply required by the mangrove roots, killing virtually all of the trees in low-lying areas where the sediment accumulation was greatest. Many trees on slightly higher elevations survived, and (oddly enough) even a large number of mangroves that had been completely uprooted, but with blocks of soil still clinging to the roots, also survived. An adequate oxygen supply continued to be available for the roots of these displaced trees.

Donna had another major effect. Prior to Donna, the mangrove forests had supported epiphytic plants in great abundance and variety, including airplants, orchids (see Figure 12.1), and ferns. In the damaged area, it was estimated that 90 percent of these perished, either due to outright removal by the high winds, or in subsequent exposure to direct sunlight in the dead, leafless forests. Perhaps a full century will be required for this profusion of epiphytes to become reestablished in similar abundance in the affected mangrove swamp areas. It is highly possible that some very rare, perhaps recently established species, were completely exterminated from Florida. Thus, while some hurricanes may serve to bring the seeds of tropical plants, such as epiphytes, to tropical southern Florida (see Chapter 12), others may destroy established populations.

THE HURRICANE LEGACY: COASTAL LOWLAND VEGETATION [128]

The patchwork of other kinds of lowland habitats in and around the great mangrove swamps are primarily the result of hurricanes. Even some areas of higher elevation that support species other than mangroves are the result of deposits that killed a previous forest. These non-mangrove coastal lowlands support interesting and often beautiful plant communities. If one species had to be identified as the most ubiquitous of these salty habitats, it would be the buttonwood (*Conocarpus erecta*). Buttonwoods may stand as scattered individuals, or may be closely spaced forming a buttonwood forest. Many dead snags that look like aged driftwood are buttonwood trunks, while others are mangrove "tombstones" left from Donna or even from the Labor Day hurricane of 1935 (Figure 9.2). The variability of these saline areas is great, and ecologists have named numerous plant associations including black rush marshes, saltwort marshes, coastal marl prairies, and scrub mangroves. Many such habitats are present near the town of Flamingo, at the end of the main road in Everglades National Park.*

---

* Many of the "higher" coastal storm levees (1.5 to 2 feet above sea level), especially near Flamingo, previously supported coastal hammocks, dominated by cabbage palms and salt tolerant tropical hardwoods such as West Indian mahogany, buttonwood, and Jamaica dogwood (*Piscidia piscipula*). Cleared for lumber and charcoal, these areas are now coastal prairies that were not born of hurricanes.

FIGURE 9.2 Dead mangrove and buttonwood snags are evidence of a former mangrove swamp decimated by Hurricane Donna or the Labor Day Hurricane (1935) and now a coastal prairie dominated by saltwort and glasswort. (Photo by T. Lodge.)

The most picturesque of these areas is the saltwort marsh, named for the abundant light, yellow-green ground cover of saltwort (*Batis maritima*). Prior to hurricane Donna, few areas were dominated by this shade-intolerant species, but Donna opened extensive habitat and spread its seeds. This decorative little plant has a woody stem that arches over and supports numerous short branches with fleshy, salty-tasting, inch-long leaves shaped like miniature cucumbers. Saltwort has a maximum height of three feet, and flats dominated by it are very open. With care not to trip over the stems, and with the decided advantage of long legs, you can hike through these areas. The soil is normally of marl and is firm, but usually wet, except in the height of the dry season.

In addition to buttonwood and saltwort, other plants in the areas opened by hurricanes (or sometimes by man) include: sea daisy (*Borrichia frutescens*); glasswort (*Salicornia virginica*); sea purslane (*Sesuvium portulacastrum*); and scattered black, white, and red mangroves. Seagrape (*Cocoloba uvifera*) — named for its edible, grape-like purple fruits — occurs above wetland elevations here as well as widely in other coastal communities. It forms a sprawling, low tree with large, stiff, round, prominently veined leaves. Dead snags and live buttonwoods support numerous epiphytes, making these open, sunny areas ideal for observation and photography. The epiphytes include several kinds of airplants (*Tillandsia* spp.) and some orchids, the most common being the butterfly orchid (*Encyclia tampensis*). However, in exploring any habitats near the coast, be prepared for mosquitos in all but cool, breezy, winter weather.

THE WHITE ZONE — A HURRICANE-PRONE LANDSCAPE [128, 182, 514]

A prominent coastal landscape of the southeastern Everglades, in an area known as the Southeastern Saline Everglades, is an extensive, sparsely vegetated feature called the *white zone*. Coastal white zones occur in many lowland areas inland from normal tidal influence where limited freshwater flow is present, restricted mostly to sheet flow. The drainage flow ways that do occur are limited

FIGURE 9.3   The white zone of the southeastern saline Everglades, looking northwest toward Card Sound Road. (Photo by T. Lodge.)

FIGURE 9.4   Dwarf red mangroves of the white zone near Paurotis Pond, Everglades National Park. (Photo by T. Lodge.)

and often discontinuous. The zone's name comes from its very light color as seen from the air or in aerial photographs (Figure 9.3). The light color is the result of the high reflectance — unobstructed by vegetation — of exposed marl soil during the dry season, or of calcium carbonate particles in the community's actively growing periphyton during the wet season. The nearly flat terrain and short hydroperiod have caused the development of a one-to-two-mile-wide white zone in this location. White zone plant communities can be seen in Everglades National Park between Paurotis Pond and Nine Mile Pond (Figure 9.4).

The white zone is a nutrient-impoverished environment. The vegetation is sparse and productivity of the area is notably low compared to other coastal communities, especially mangrove swamps. However, tides from tropical storms and hurricanes sporadically flood the white zone. With these floods come salt water and the seeds and propagules of mangroves. Thus, the soils harbor significant salt (the basis of somewhat misleading name, *saline Everglades*),[82] restricting many freshwater plants, such as sawgrass, to the inland edge of the zone. Scattered mangroves are widespread, but due to poor growth conditions, most are only two to four feet tall, although much taller mangroves occur in the normal tidal areas immediately coastward of the white zone.

Scattered through the white zone are short, depressed drainage ways that harbor lines of tree islands (see Chapter 4). The tree islands develop peat soils that accumulate to elevations above the white zone marl soils, but the drainage pathways around and between the tree islands are slightly depressed and host aquatic life that is an attractive food source for wading birds. Falling water levels concentrate small fish in these pathways, and roseate spoonbills, for example, can find important food sources there, although this species is highly vulnerable to interior water management that affects water delivery to the creeks.[368]

Because of the low relief and slow drainage of the white zone, it is easily affected by man-made obstructions such as berms, canals, and roads. A prominent example of such an impact occurred when Hurricane Betsy (1965) pushed salt water over U.S. Highway 1 (constructed on the original Flagler Overseas Railway embankment) into an area impounded by the berm of a then new major canal, C-111 (see Figure 21.5). Unable to drain, the saltwater killed several square miles of freshwater Everglades vegetation, principally sawgrass, in the upper white zone.[17] Obstruction of freshwater supply to the white zone and recently rising sea level have also had noticeable effects of extending the white zone inland.

HURRICANES AND GLACIAL CYCLES

Hurricanes and their progenitors — tropical storms and tropical depressions — are the result of warm sea surface temperatures and proper atmospheric conditions. During winter and spring months, these conditions do not exist in the tropical North Atlantic, Caribbean, or Gulf of Mexico. Beginning in June and extending into November each year (with a peak in September), the conditions are ripe to spawn and sustain hurricanes. Because hurricanes require warm, open water, they cannot develop over land the way that tornados do. Extrapolate these requirements back to the times of the interglacial sea that existed prior to 100,000 years ago. The warmth of those times and the expanded sea surface (the Gulf of Mexico was significantly larger then, in part at the expense of Florida) must have provided conditions that spawned hurricanes in great numbers. Thousands of hurricanes have probably revised southern Florida's coastal environments during the warmer parts of each sea level excursion through the glacial cycles.

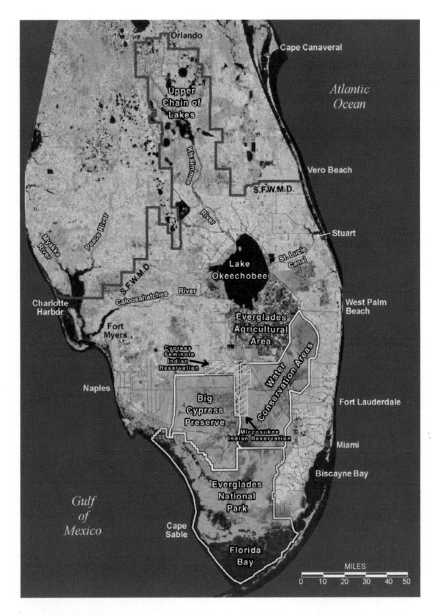

COLOR FIGURE 1 Satellite image of central and south Florida showing features of the entire Kissimmee-Lake Okeechobee-Everglades watershed (KLOE) and adjacent areas. The marked boundary of the South Florida Water Management District (SFWMD) generally includes this watershed, beginning with the "upper chain of lakes" just south of Orlando, but it also covers the Caloosahatchee River watershed, not historically part of KLOE, and lands northeast of Lake Okeechobee, parts of the coastal St. Lucie River watershed (lower river visible at Stuart) and the St. Johns River watershed. The St. Lucie River, connected to Lake Okeechobee since 1926 by the St. Lucie Canal, connects with the lower end of the Indian River, a coastal lagoon lying parallel to the coast northward past Cape Canaveral. The St. Johns River system (which flows northward beyond this image to the Atlantic at Jacksonville) can be seen as the series of lakes between the upper KLOE and the Indian River. Most of it lies in the St. Johns River Water Management District. Lands west of the SFWMD boundary are in the Southwest Florida Water Management District, which includes the Peace and Myakka rivers. Together with the Caloosahatchee, they flow into Charlotte Harbor, an important estuarine and marine complex bounded by the visible series of barrier islands. (Courtesy of the South Florida Water Management District, modified by the author.)

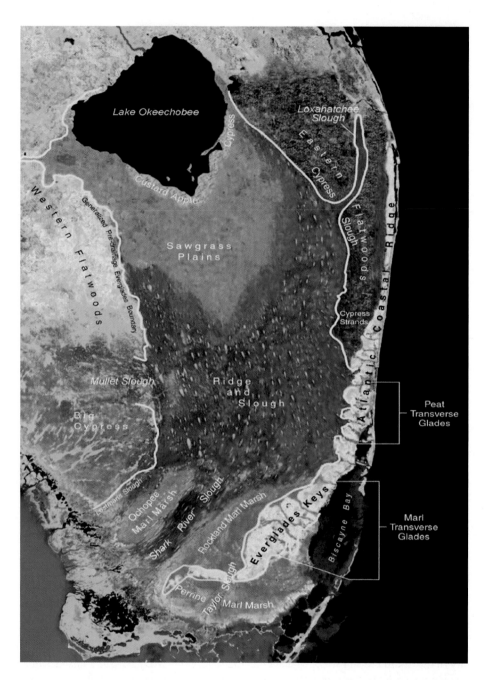

COLOR FIGURE 2 Simulated satellite image of original Everglades. The yellow line denotes where the vegetation changed from marsh to higher ground along the Everglades' east and northwest sides. Along the southwest border between the Big Cypress and the Everglades, south of Mullet Slough, the terrain merely changed from Everglades marsh to mixed communities where cypress and open glades were most abundant — there was no hydrologic boundary (see Chapter 7). (Courtesy of the South Florida Water Management District, see Ref. No. 402. Mullet Slough and Lostmans Slough {the latter shown larger than actual due to lettering size} were added by the author.)

COLOR FIGURE 3 Satellite image (circa 1995) showing the historic Everglades boundary, water conservation areas (WCAs), the Everglades Agricultural Area (EAA), and other landmarks. Compare to Color Figure 2 for additional labels. Note that Miami International Airport is entirely within the original Everglades. A notable feature is the green area in WCA-2A immediately south of WCA-1. The feature is primarily cattails, presumably induced by phosphorus-rich water flowing from the Hillsborough Canal inside WCA-1 through control gates located along the south half of WCA-1 into WCA-2A. "S-9" is the pump station shown in Figure 21.2. (Base image courtesy of the South Florida Water Management District with overlay features added by the author.)

COLOR FIGURE 4 Satellite image (enlargement of part of Color Figure 3) showing the remaining ridge-and-slough landscape in Water Conservation Area 3A and vicinity. The best-preserved ridge-and-slough pattern is in the center and upper center of the image, south of Alligator Alley. Tree islands are bright green, sawgrass ridges are the thin brown lines aligned with the southward flow pattern, and slough habitat is dark blue (exposed water). Note the complete absence (loss) of ridge-and-slough landscape east of the Miami Canal, and the loss of slough habitat in WCA-3B, now mostly covered by sawgrass but with tree islands still visible. The effect of Alligator Alley is apparent. Also, note the abnormal vegetation along the Miami Canal and at the release point for the L-28 and L-28 Intercepter canals at upper left. Those waters come from agricultural land north of the Big Cypress Swamp. Mullet Slough lies below that release point. Also notable is the effect of deeper water in the southern portion of WCA-3A near the Tamiami Trail, caused by impoundment of this southward sloping landscape.

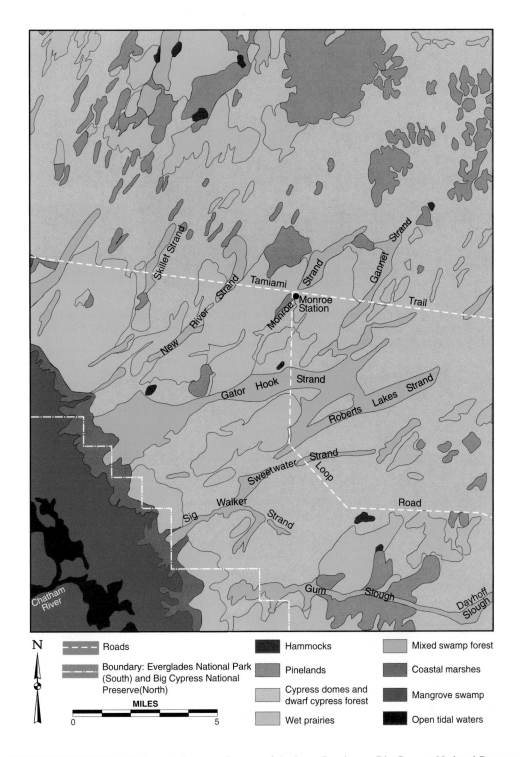

COLOR FIGURE 5 Detailed vegetation map for part of the Loop Road area, Big Cypress National Preserve and Everglades National Park. The location of this map is shown on Figure 7.1. Colors were selected in part so that comparisons can be made with Color Figure 3. The major strands and sloughs (green) and the edge of the mangrove zone can be located on the satellite image south of the western portion of the Tamiami Trail.

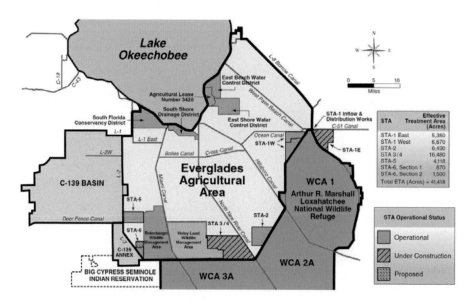

COLOR FIGURE 6 The Everglades Agricultural Area (EAA) and vicinity showing major features of the Central and Southern Florida Project and modern Stormwater Treatment Areas (STAs) of the Everglades Construction Project. (Courtesy South Florida Water Management District.)

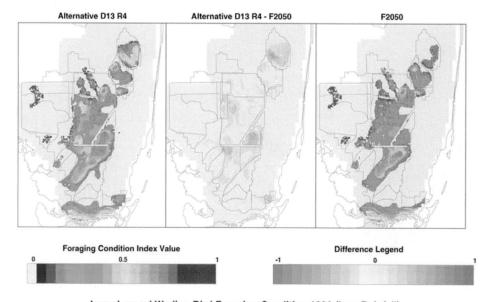

Long Legged Wading Bird Foraging Condition 1990 (Low Rainfall)

COLOR FIGURE 7 Sample comparison of the relative favorability of foraging habitat for long-legged wading birds for two possible hydrologic scenarios (the selected CERP alternative called "AltD13" in the left hand panel, and the year 2050 without CERP in the right hand panel) for the Everglades region, assuming the landscape hydrology pattern that occurred in 1990, a low rainfall year. The colors, ranging from dark blue to red, signify relative foraging quality during the breeding season on a scale from 0 to 1 (white signifies that the area is not potential foraging habitat). The middle panel represents the difference between the two scenarios, with yellows and browns indicating areas of better foraging under AltD13 and blue shades indicating better foraging under the 2050 scenario. Similar types of output were produced for all years from 1965 to 1975 using modeled hydrology for that period. These results are from one of a large number of ecological models developed by The Institute for Environmental Modeling of the University of Tennessee as part of the ATLSS (Across Trophic Level System Simulation) project sponsored by the US Geological Survey. More details are available at http://atlss.org (Courtesy of Donald De Angelis)

COLOR FIGURE 8 Alligator hole in the edge of a large bayhead, Everglades National Park. (Photo by Thomas E. Lodge)

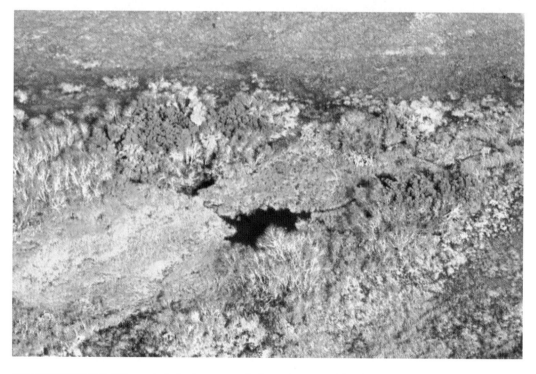

COLOR FIGURE 9 Alligator hole in the center of a burned-out tree island, Everglades National Park. (Photo by Thomas E. Lodge)

COLOR FIGURE 10 Florida treesnails aestivating on a wild tamarind, on Loop Road, Big Cypress National Preserve. (Photo by Thomas E. Lodge)

COLOR FIGURE 11 Marjory Stoneman Douglas and Tom Lodge, on her patio, March 1990. See "Introduction to the First Edition". (Photo by Masud Quraishy.)

10 Coastal Estuarine and Marine Waters[*]

The coastal receiving waters of the Everglades watershed are Florida Bay to the south and the Gulf of Mexico to the southwest (Figures 10.1 and 10.2). However, the interconnecting labyrinth of bays and tidal creeks lying within the coastal outline confuse the land/sea border. Furthermore, the great expanses of mangrove swamp grading inland into freshwater marsh make the distinction between land and water even more difficult. Excellent fishing and bird watching are available in these waters, but an experienced guide is all but required for exploration (see Figure 10.3).

FLORIDA BAY: A GEOLOGIST'S CLASSROOM

In addition to its great interest to fishermen, Florida Bay is famous among geologists. Ongoing sedimentation processes in Florida Bay have answered geologists' questions concerning the formation of certain fine-grained limestones. Most limestones have an obvious origin because they are made of identifiable inclusions such as fossils. In the limestones of southeastern Florida, there are corals in some (e.g., the Key Largo Limestone of the upper and middle keys), bryozoans in others (e.g., the Miami Limestone beneath the southern Everglades and Florida Bay), and *ooids* in still others (e.g., the Miami Limestone of the Atlantic Coastal Ridge). The last of which are small, egg-shaped grains visible to the naked eye and have a physical-chemical origin. The limestone of the Atlantic Coastal Ridge in the Miami area is called *oolite* (sometimes the "Miami oolite") and is commonly used as a building material for rock walls and even house façades (for older homes) from Miami to Homestead where locals commonly (and incorrectly) call it "coral rock." The Miami oolite was deposited before the last glaciation when sea level was substantially higher in a shallow marine environment similar to areas of the Bahamas banks where the same process is occurring today.

Studies conducted in Florida Bay have resolved the origin question of fine-grained limestones, in which the grains are far smaller than ooids. The puzzle in Florida Bay that attracted scientific interest was the vast amount of lime mud (marine marl) on the bottom. Grains of this material are similar to those of fine-grained limestones and of the filler between fossils in other limestones. The limestone bedrock under the sediments of Florida Bay dates back about 100,000 years, to the time when sea level started to recede at the inception of the last glacial cycle. Lying over the bedrock floor are patches of peat soil derived from freshwater Everglades plants. Above those patches and the bedrock is an accumulation of marine marl that averages about six feet thick, yet it has all accumulated only in about the last 4000 years, as determined by radiocarbon dating. That date is the time when rising sea level first overtook the southernmost extension of the historic Everglades and flooded Florida Bay with marine waters.

The search for the origin of Florida Bay's lime mud did not have to go far. The bottom is nearly covered with tropical marine vegetation, some of which is turtle grass[**] (*Thalassia testudinum*), a "higher" plant with true roots, leaves, and flowers (although they are too small to be noticed). The

---

[*] The main sources for this chapter are references 280 and 635.
[**] The leaves of turtle grass are often heavily colonized by bryozoans, which are commonly found as fossils in the Miami Limestone that underlies Florida Bay and the southern Everglades. Thus, it is probable that turtle grass was abundant in the shallow marine habitat that produced those sediments, just as it is today in parts of Florida Bay.

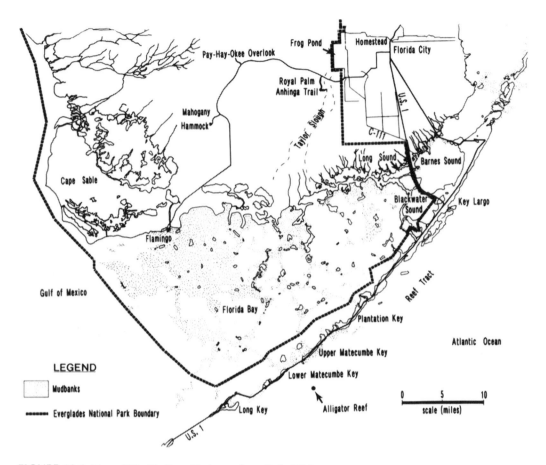

FIGURE 10.1 Map of Florida Bay. (Redrawn from Ref. 428.)

bulk of the vegetation, however, is many species of algae. Numerous among these are species that have a root-like "holdfast" and other parts that superficially resemble stalks and leaves of flowering plants. These types of algae have stiffened bodies, which aid in maintaining upright posture. Species of *Penicillus* (the four-inch-tall "shaving brush" alga) (Figure 10.4), *Halimeda*, and others are abundant in the area. The stiffening in these species results from calcium carbonate particles produced within their tissues, and these particles appear identical to those that compose the bulk of the sediments. Furthermore, when these short-lived algae die, they disintegrate so completely that all that is left is their deposit of calcium carbonate particles. This process is comparable to the production of marl by periphyton in freshwater marshes (see Chapter 3).

Studies on the rates of growth of these bay-bottom algae have confirmed that they could easily be responsible for a large proportion of the lime mud that has accumulated during the past 4,000 years. In some locations, the mud is ten feet deep. The nature of the sediments, however, is much more complex than just fine-grained carbonate material. The sediments are reworked by worms and other burrowing organisms to make pellets of carbonate. The shells of tiny foraminifera, bryozoans, and molluscs become incorporated, and physical sedimentation adds precipitated carbonates.[69, 634] Given the proper circumstances, such as another reduction in sea level, thus exposing this lime mud and other carbonate sediments to atmospheric weathering, these materials would consolidate and recrystallize into a rock layer. Much of it would be fine-grained limestone.

FIGURE 10.2 Northeastern Florida Bay, looking westward over Little Blackwater Sound (left) and Long Sound (center). Note the series of coastal lakes along the north side of Long Sound. (Photo by T. Lodge.)

FIGURE 10.3 Florida Bay channels and islands. Erosion by tides, and deposition in marine shallows constantly change conditions making navigation treacherous. A power boat could easily pass through the foreground channel from left to right, but would then run aground where the divided channel ends in shallows. (Photo by T. Lodge.)

FIGURE 10.4 Shaving-brush alga amid turtlegrass. The white material is flocculent lime mud (marl). Sediment produced by this alga and similar species add to the marl deposit after their death. (Photo by T. Lodge.)

SEA-LEVEL RISE AND COASTAL ENVIRONMENTS

Studies of the sediments underlying Florida Bay and the adjacent coastal environments have been instrumental in understanding the ecological effects of sea level rise. A graph of sea level changes since the end of the last glacial cycle is presented in Figure 10.5. By about 15,000 years ago, melting of the polar ice caps was in full progress. Melt water was rapidly entering the oceans, increasing sea level and radically changing shorelines not only by rising water but also by altered erosion and sedimentation processes.

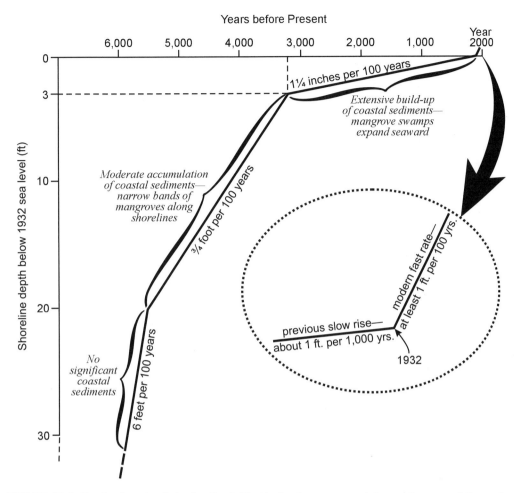

FIGURE 10.5 Graph of sea-level rise for South Florida for the past several thousand years and the modern rise since 1932, indicating a rate of about one foot per 100 years. (Modified from Ref. 635.)

The rate of sea-level rise evidenced at various locations around the world is not consistent because of regional instabilities of landmasses. In regions that were burdened under the massive load of glacial ice, land surfaces actually rebounded following the melting of the glaciers because of the lost weight of the ice itself. "Glacial rebound," as it is called — floating-up of the unburdened continents and islands resting on the earth's plastic crust — has been documented in the northeastern United States, Canada, Alaska, northern Europe, and southern South America. Accordingly, sea level rise in these areas does not appear to have been so great. In other locations, the sinking of land has confused the record. Coastal Louisiana is a prominent example, where subsidence from erosion and compaction of Mississippi sediments is occurring. The land surface there is receding where new sediments are not keeping pace. In such locations, sea level rise appears to have been excessive. The Florida peninsula, however, has long been geologically stable and was remote from glacial ice. Thus, Florida's record of sea-level change is relatively reliable. Land areas representing former sea levels have been accurately mapped for the Florida Keys, for example, showing how the former extensive and contiguous land of the lower peninsula finally diminished to the narrow string of islands we know today.[342]

In South Florida from 15,000 to about 9000 years ago, sea level rose at six to seven feet per century. From its lowest point (about 400 feet below today's level) it continued that rate until reaching about 60 feet below today's level then began slowing. By about 5500 years ago, the rate decreased to less than one foot per century. That slower rate continued until about 3200 years ago, at which time sea level was only about three feet below the present level. Sea level nearly stabilized then, slowing to a rate of about one foot per thousand years — one tenth of the previous interval — thus rising only about three feet until the year 1932. Since 1932, the rate of sea level rise has accelerated abruptly, returning to a rapid rate of nearly a foot per century (1932 to the present) — ten times faster than the rate of the previous 3000 years (Figure 10.6). Maps of the Florida Keys, for example, have been extrapolated for future sea levels, showing a nearly complete loss of land within 500 years at the present rate.[342, 633]

Coastal sedimentation processes in South Florida are most active in shallow water less than about ten feet deep. Waves and shallow-water coastal currents transport materials along shorelines. Mangroves lay down peat soils and trap storm-driven sediments in the intertidal environment, and algae, corals, and other organisms produce carbonate particles and shells that become sediments and structures in warm, shallow, sunlit marine environments. All of these processes have rates that are impressive so that a few hundred years produce results that are obvious even to casual, untrained observers. However, with sea level rising rapidly, the resulting deposits are only thinly spread as shorelines and adjacent shallow waters quickly advance landward. Such was the case during the early phases of sea-level rise following the last glacial cycle. At the rate of six feet per century, scarcely any sediment was deposited over the shallow marine environment. However, sediments dating between 5500 and 3200 years before present are substantially thicker. But, the most impressive record is derived from the newest sediments, those younger than 3200 years. They are not

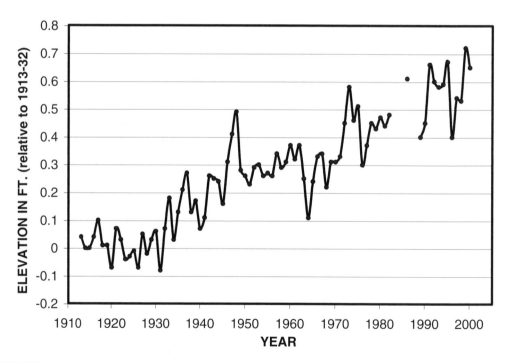

FIGURE 10.6 Graph of mean annual sea level at Key West, Florida. The general rising trend is obvious. Deviations in the trend represent ocean-wide anomalies. For example, reduced trade winds during the 1995 El Niño induced the lower tide levels around Florida and the Caribbean, with corresponding higher levels on the coast of Africa. (NOAA data plot courtesy of Hal Wanless.[632]) Data points are missing for 1983-85 and 1987-8. Station datum was converted by the author to set approximate mean sea level for period 1912-1932 at zero to simplify presentation.

only thicker, but they show an actual seaward advance of shoreline as deposits filled shallow waters more rapidly than sea level was rising.

Because of storms, the sediment accumulations in South Florida's shallow-waters are far from uniformly spread. Most prominent has been the formation of relatively deeper deposits on shore-lines, the result of tropical storms and hurricanes. Storms strip the growing shallow sediments of the adjacent marine waters — marl, sands, and shells — and deposit them where moving, storm-driven waters meet the friction of land and intertidal vegetation. Through successive storms, these deposits become elevated coastal levees, derived from sediments in the adjacent marine shallows. The campground at Flamingo in Everglades National Park is an example — land made from marl and shells carried from Florida Bay. Cape Sable is another very recent accumulation of storm-driven sands and shells from the adjacent shallows of the Gulf of Mexico.

These examples are the most recent storm levees. Moving landward into the Everglades there is a series of storm levees, now covered by mangroves and buttonwood and by salt-tolerant upland trees, such as West Indian mahogany and Florida fishpoison tree (*Piscidia piscipula*) where the elevation is sufficient. Between the levees are deeper areas of mangrove swamp, mostly growing over deep mangrove peat of recent origin, or even open water, such as Whitewater Bay, Coot Bay, and West Lake.[*] The most interior of these levees, where the mangrove-dominated coastal envi-ronment meets the freshwater Everglades, is a 3-foot-high levee (about 1.5 feet above the surround-ing swamp), the buttonwood embankment (see Chapter 8).[82, 286] All of this intertidal mangrove environment and successive levees have evolved in the past 3000 years as a seaward extension of "land" made possible by slowly rising sea level. It would not have developed if sea level had been stable, or had been rising rapidly.[**]

The seaward growth of intertidal habitats has been most impressive where adjacent shallow-marine sediments have accumulated most rapidly, supplying the material required for growth. In Everglades National Park, the more western areas have grown most, namely from Taylor Slough westward past the Ten Thousand Islands, including all of the major mangrove rivers that drain from the Everglades and Big Cypress Swamp. The freshwater outflows from these areas have contributed to the estuarine and shallow-marine productivity that has enhanced development of mangrove peat soils and coastal sediments. The northeastern portion of Florida Bay, east of Taylor Slough, has received far less freshwater outflow and has produced far less sediment. Accordingly there has been little seaward extension of land or intertidal habitat there.

Storms and tidal currents have also rearranged sediments in the open waters of Florida Bay. Florida Bay is anything but a uniformly shallow marine habitat. It is a patchwork of deeper areas, called "lakes," separated by shallow mudbanks. There are a few high islands, such as Carl Ross Key in western Florida Bay (Figure 10.7) and North Nest Key in the eastern part where enough land exists for camping. The islands and many intertidal mangrove islands on mudbanks are important for roosting and nesting birds (Figure 10.8). The lakes are about four to six feet deep, while the mudbanks are generally exposed at low tide where they are used extensively by a great variety of fish-eating birds, including the great white heron, reddish egret, roseate spoonbill, and, occasionally, the greater flamingo (a native Caribbean species but often thought to be represented by individuals having escaped from captivity). The lakes and mud banks make navigation a considerable challenge, and low tide can leave an inexperienced fisherman stranded many miles from land. As an additional challenge, storms regularly rearrange channels so that navigation charts are constantly changing (See Figure 10.3). [286, 473, 634]

---

[*] This "West Lake" is in Everglades National Park near Flamingo and should not be confused with another West Lake in Broward County near Fort Lauderdale.

[**] One should be cautioned that not all scientists agree on details of the sea-level rise that caused these coastal features, although generalizations are in agreement. In particular, there is evidence of a relatively rapid but brief rise in sea level from 800 to 1300 AD, the well documented Medieval Warm Period, and that resulting South Florida sea level may have temporarily exceeded the recent level by a foot and a half. The buttonwood embankment may have resulted from that interval, which is much more recent than the gradual sea-level-rise scenario would predict.[286]

FIGURE 10.7 Carl Ross Key in western Florida Bay, looking south to Sandy Key, Everglades National Park. Note the appearance of the beach — made of locally produced carbonate sand. (Photo by T. Lodge.)

FIGURE 10.8 A single red mangrove constitutes an island suitable for nesting ospreys in Florida Bay at Flamingo, Everglades National Park. A much larger mangrove island is in the distance. (Photo by T. Lodge.)

Interesting evidence of the sequential origin of Florida Bay lies in buried sediments, which verify that the freshwater Everglades once extended far out into Florida Bay. The lower portion of sediment cores taken throughout the bay reveals the presence of peat that contains recognizable remains of freshwater Everglades plants. Above this layer, many core samples contain peat derived from mangrove roots, but it is apparent that Florida Bay never had extensive mangrove swamps like those found in the western portion of the park today. When Florida Bay first began flooding with salt water, about 5000 years ago, sea level was rising a little too fast (about a foot per century). Any mangroves that could establish were soon overtaken by open marine waters, where marine sediments were laid down. Thus, the extensive upper sediment layers (mostly marl) are the marine deposits. As sea level rose, a forced succession of habitats occurred, from freshwater marsh, to a scattering of mangrove swamp areas, and finally to the open, shallow marine habitat that has remained for some 4000 years.[141]

The configuration of Florida Bay, with its pattern of mud banks and shallow lakes and its location pinched between the Keys and the mainland, restricts circulation, making the bay prone to extremes of salinity, temperature, and reconfiguration by storms, especially hurricanes. Hypersalinity has received much attention from ecologists. The main avenue of freshwater flow from the Everglades is, and historically was, the Shark River Slough, which releases its water to the west of Cape Sable into the Gulf of Mexico. Only the relatively small eastern area of the Everglades has provided freshwater directly to Florida Bay, principally through Taylor Slough. Thus, in dry seasons, evaporation may exceed freshwater input, and salts are concentrated. While historic data are few, this condition undoubtedly recurred naturally but has been aggravated by hydrologic modifications of the Everglades (see Chapter 21).[404, 587, 634]

OYSTERS AND MANGROVE SWAMPS [345]

The western-most coastal waters of Everglades National Park and tropical southern Florida lie in the Gulf of Mexico. The boundary between Florida Bay and the Gulf of Mexico is indefinite, but the interesting mudbanks of Florida Bay end on an approximate northwest-southeast line from Cape Sable, passing just west of Carl Ross Key and on to Long Key in the chain of the Florida Keys. The shoreline of the Gulf of Mexico from Cape Sable and northwestward looks very different from Florida Bay (see Figure 8.1). Along Florida Bay, much of the shoreline is a low coastal levee of marl. Cape Sable, however, is a high sandy beach. Stretching to the northwest are mangrove swamps and a labyrinth of tidal rivers, with a few curious sandy beaches, most notably the three-mile-long Highlands Beach between the Broad and Lostmans rivers. At the westernmost part of Everglades National Park, near the town of Everglades City, the apparent contiguous shoreline of mangrove swamps is preceded by a myriad of small mangrove islands. This region is called the Ten Thousand Islands.

Throughout this area, currents along the shore of the Gulf of Mexico have carried silica sand southward. This type of sand is familiar on the beaches of the Atlantic and Gulf coastal states. Here it mixes with the carbonate mud and abundant shell fragments making a firmer bottom than the carbonate mud of Florida Bay. The combination of the firmer bottom and the freshwater outflows from the Everglades and Big Cypress Swamp make an ideal habitat for the oyster, a species that has a very large ecological impact on shallow tidal habitats wherever it is abundant.

Oysters are related to clams, and like most clams, they are "filter feeders." Oysters strain water through a fine network in their gills, trapping minute particles including bacteria, plankton, and non-living material, all of which are the oyster's food. Unlike free-moving types of clams, oysters grow attached to a surface and are not free to move around once their planktonic larval form has settled down. They do best in areas where tidal currents transport food to them, remove wastes from them, and prevent the accumulation of smothering sediments. However, strong currents make it difficult for the attachment of larval oysters because of shifting sediments. Thus, a good site for a larval oyster's attachment is the shell of an existing, solidly anchored oyster shell, whether dead or alive.

Oysters live best in brackish water and at depths where they are periodically exposed by low tide. These conditions are unfavorable for some major oyster predators including snails called oyster drills and stone crabs. The result of all these factors is that oysters tend to form large colonies in shallow, tidal areas in the mouths of creeks and rivers. Because the colonies grow fastest where the current is swift, they tend to elongate across the path of tidal currents, forming long structures called oyster bars which tend to block tidal channels. As might be imagined, an oyster bar can be a rude awakening for a naive boatman! The tough, four-inch shells "cemented" together have cut right through many a hull and have ruined countless props.

In southern Florida, oyster bars have accelerated conditions conducive to the establishment of mangrove seedlings, especially the red mangrove. Under favorable conditions, a mangrove forest can overtake an oyster bar. This encroachment, together with sedimentation processes along the Gulf of Mexico shoreline of Everglades National Park, has lead to an especially rapid extension of mangrove swamps. Longshore currents have supplied the sand and shells that have added to local marl production, rapidly making shallower water. Oysters became established and then mangroves. This process has helped to create one of the world's largest mangrove swamps through the past 3000 years while sea level has been rising slowly.

SEA-LEVEL RISE AND THE FUTURE OF MANGROVE SWAMPS

The current, rapid rise in sea level may convert mangrove swamps to shallow marine habitats more rapidly than expected by the change in sea level alone, notably in areas that do not receive significant hurricane sediment loads. A large percentage of the total biomass of mangroves, particularly red mangroves, is the living roots below the soil surface. As a result, the substrate where mangroves are killed by hurricanes subsides several inches in the early years following the trees' deaths due to root decomposition. In combination with rising sea level, the loss of elevation increases marine tidal influence, and with this influence come marine burrowing organisms, such as several species of burrowing shrimps (*Callianassa* spp. and *Upogebia affinis*). Their excavations, largely the removal of soft, decayed mangrove roots, cause additional subsidence to the extent that the former mangrove habitat becomes far too deep for renewed colonization by mangroves. In some locations, the respective subsidence from the Labor Day Hurricane of 1935 has now approached two feet (sea level rise has only accounted for about 8 inches of that). The net result appears to be this: with rapidly rising sea level, more mangrove swamp is lost at the seaward edge than is newly created at the landward edge.[632]

11 Lake Okeechobee and the Everglades Headwaters*

In the forgoing chapters, discussion of modern human environmental impacts has been largely avoided in an effort to describe the historic, natural ecosystem. This chapter is a departure from that approach. It is not feasible or useful to describe the historic condition of Lake Okeechobee and its headwaters separately from the modern condition because so little remains and, in fact, so little detail is known. Thus, the modern condition is interwoven with what is known of the historic condition. For most locations, only general historic conditions were documented. Many details of relevance to restoration can only be estimated forensically. Historic water quality is the most important example, but there are also few locations that now resemble the historic condition of vegetation and scenic character.

With a modern area of 730 square miles and an average diameter of 30 miles, Lake Okeechobee (usually pronounced $ōk$-$ă$-$chō'$-$bē$) is by far the largest lake in the southeastern United States. Despite this regional acclaim, it is exceeded in North America by Utah's Great Salt Lake, by a lake in Alaska, by many lakes in the Canadian interior, and it is dwarfed by all five Great Lakes. To emphasize its size, some use the statistic that Lake Okeechobee is the second largest freshwater lake within the lower 48 states. The justification: of the Great Lakes, only Lake Michigan (about 22,000 square miles) is completely in the United States and the Great Salt Lake (varying between 1000 and 2500 square miles, depending on rainfall) is saline, not freshwater. In using whatever statistic serves a purpose, be aware that Lake Okeechobee is very shallow so its water volume is relatively small for its size, which is typical of lakes in flat, lowland regions.

Lake Okeechobee was historically the hydrologic hub of the greater Everglades ecosystem. Its role today is greatly modified — contained by the massive Herbert Hoover Dike, and regulated by controlled inlets and outlets, described under "The Modern Lake" below. Historically, southward seasonal overflows contributed to the Everglades, which would otherwise have prospered only from direct rainfall. The lake receives water from a 4000-square-mile watershed that begins in southern Orlando (Figure 11.1). In years such as 1947, which was unusually wet and exacerbated by two rainy hurricanes, the Okeechobee watershed contributed deep flooding to the Everglades.** Such periodic massive flows — certainly repeated more than a hundred times through the several thousand years of the lake's life — probably contributed to shaping the flow pattern of the Everglades landscape.

THE LAKE OKEECHOBEE WATERSHED

Historically and today, the water entering Lake Okeechobee has come from direct rainfall on the lake plus discharges from five main headwater sources — Fisheating Creek, the Lake Istokpoga drainage, the Kissimmee River (pronounced $kĭ$-$sĭ'$-$mē$), Taylor Creek, and Nubbin Slough.

---

* The main sources for this chapter are references 42, 80, 257, 262, 305, 574, 591, and 650.

** With the Hoover Dike completed around the southern half of the lake by 1937, replacing a former muck dike that had been breached in 1926 and '28 by hurricane floods, gated outlets regulated the lake's water level. Thus, overflows to the Everglades had been greatly modified from the historic condition prior to 1947. That year is merely used as an example of extreme flows (see "Recollections" by Taylor Alexander and Figure 9.1).

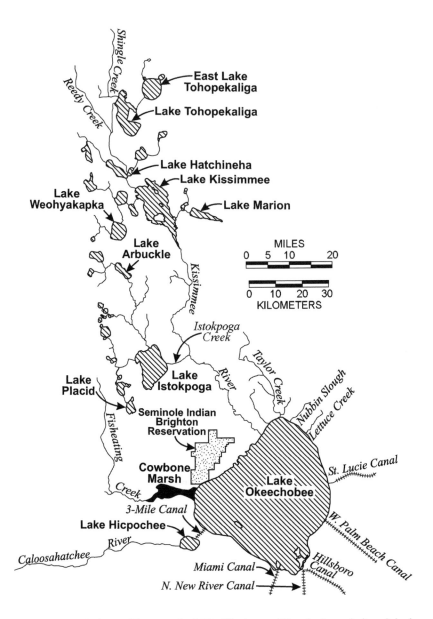

FIGURE 11.1 Lake Okeechobee and its watershed. The Kissimmee River is shown in its original meandering condition prior to canalization, but some connections in the watershed are shown as modified from the historical condition. See Figure 11.3 for historic character of the Lake Istokpoga drainage to Lake Okeechobee. (Selectively redrawn from USGS State of Florida map, scale 1:500,000, compiled 1966, 1967 edition.)

Together, the headwaters accounted for nearly half of the historic lake's water. Rainfall directly on the lake contributed the rest.[*] By far the largest tributary was the Kissimmee River, which alone accounted for over 60% of the inflow from tributaries. Groundwater inflows and outflows were

---

[*] Direct rainfall on the lake contributes just over half the water, which may seem disproportionate because the watershed is 4000 square miles and the lake is only 730 square miles. However, the inflows from tributary streams are what remain after evapotranspiration and percolation into ground water have removed about 70% of the watershed's rainfall. Evaporation from the lake's open water and evapotranspiration from its littoral zone are excessive, substantially exceeding direct rainfall, so that "atmospheric input" to the lake is negative (evapotranspiration exceeds rainfall), making the lake dependent on the watershed — i.e., it would often be empty without the watershed input.

probably small in comparison to surface water flows. Recently, the lake's hydrology has been complicated by periodic backpumping into the lake from the Everglades Agricultural Area (EAA) to relieve flooding. The EAA canals are normally used for drainage discharges from the lake and as a source of water for EAA irrigation (see Chapter 21).

The Kissimmee River's source is a network of lakes and wetlands just south of Orlando. Many of the lakes have Indian names, such as Lake Tohopekaliga (pronounced tă-hōp'-ă-kă-li'-gă but often shortened by locals to "Lake Toho"). Prior to their interconnection by canals, they overflowed slowly through wetlands and small creeks southward to the larger Lake Kissimmee. The Kissimmee River emerged out of the south end of Lake Kissimmee, meandering 100 miles through its one- to two-mile-wide floodplain wetlands over a straight-line distance of 50 miles to Lake Okeechobee. The trip by boat was described by early navigators as lacking scenery except where the river passed a few forested pinelands or oak hammocks at the edge of the floodplain. The Kissimmee regularly overflowed into its floodplain, typically on the same summer wet-season schedule as the Everglades. This schedule contrasts with rivers of northern Florida and the southeast where flows are swelled by late winter and spring rains and normally recede with summer dry spells.

With the Kissimmee's canalization* and containment by levees as a flood-control project from 1962 to 1971, much of its floodplain was drained, which benefited cattle farmers. The once meandering river became a straight, 56-mile-long canal, 30 feet deep and over 300 feet wide, designated "C-38" (Figure 11.2). It routed nutrient-enriched waters directly to Lake Okeechobee without the benefits of natural wetland treatment, water storage, and wildlife habitat previously provided by its bordering floodplain wetlands. Phosphorus inputs included natural, headwater geologic sources and drainage from steadily increasing cattle ranching that began in the late 1800's but greatly intensified after levees prevented flooding. This trend is now being reversed by land-management practices and by highly successful reconnections of some of the river's oxbows — cut-off meanders of the river's earlier configuration — thus rejuvenating the adjacent wetlands' original functions of nutrient uptake and wildlife habitat (see Introduction). The Kissimmee restoration is the largest river restoration project in the world. Some of the C-38 has been completely back-filled as part of the plan to eliminate 22 miles of the canal.[315]

Fisheating Creek, in contrast to the Kissimmee, largely resembles its original, wild character as first described in an 1842 military expedition.[160] Its upper reaches have been cleared and ditched where they drain ranchlands to the west and northwest of the lake, but its lower 30 miles of easterly flow to the west side of Lake Okeechobee, sometimes in multiple channels, still passes through majestic cypress swamps, then disperses into Cowbone Marsh for the final eight miles to the lake. With lower lake levels draining much of Cowbone Marsh, it has been partially tilled and channelized for agriculture, but traversing it with large canoes in 1842 was a considerable challenge. Fisheating Creek is the only uncontrolled lake tributary now remaining. Camping and canoeing near Palmdale, where U.S. 27 crosses Fisheating Creek, are exceptional among South Florida's outdoor experiences.

Lake Istokpoga and its headwaters lie between Fisheating Creek's watershed and the Kissimmee River. Of the many lakes in Okeechobee's watershed, Istokpoga is exceeded in size only by Lake Kissimmee. It originally connected eastward to the Kissimmee River by a small creek — Istokpoga Creek[548] — but most of its water probably overflowed historically into wetlands in an area known as the Indian Prairie, where the sheet flow was southeastward to Lake Okeechobee.[322] Vividly apparent in a 1943 vegetation map,[145] remnants of the former wetland flow-pattern are still visible on aerial photographs. The Indian Prairie superficially resembled a miniature Everglades complete with tree islands, except that the major vegetation type around the tree islands was "palm savannah," a short-hydroperiod, seasonally wet grassland with

---

* Most readers might prefer the more common word, "channelization." In the case of the Kissimmee, the distinction is important. The river was not simply modified for a navigation channel, as in "channelization." The river was literally obliterated and replaced with a canal, a distinction also relevant to parts of the Caloosahatchee (see Introduction).

5.5 miles to upper arrow on C-38

FIGURE 11.2 The Kissimmee Canal (C-38) and River. This view is looking south, the foreground 5 miles south of Lake Kissimmee, and the distant point where the canal bends right and disappears is 5.5 miles farther south. The original river can be seen winding along in the right foreground, together with old, filled-in channels from natural shifts in the original river's course. Farther in the background, the original channel winds along to the left of the canal. Vegetation along the original river's course can be seen encroaching on the river at lower right, a response to reduced flow. For restoration, sections of the canal have been filled in so that flow is directed again through the winding river and forced to spread out over portions of the original floodplain wetlands. The wildlife response has been excellent. (Courtesy South Florida Water Management District.)

abundant cabbage palms, developed on sandy soils — insufficiently wet for peat development. The Indian Prairie drained into Lake Okeechobee across a 25-mile-wide area that merged into the lake's peripheral wetlands (Figure 11.3).

Lake Istokpoga's original drainage is now completely regulated by a canal network that carries most of its outflow through the Indian Prairie, now mostly drained, to Okeechobee, and a canal link to the Kissimmee parallels Istokpoga Creek, now just an overgrown remnant. Lake Istokpoga is experiencing problems of eutrophication (nutrient-enrichment) though less severe than Okeechobee's condition, described below.

Taylor Creek, Nubbin Slough, and a few smaller drainages enter Okeechobee to the east of the Kissimmee. Although relatively short, Taylor Creek was once a beautiful stream, passing through

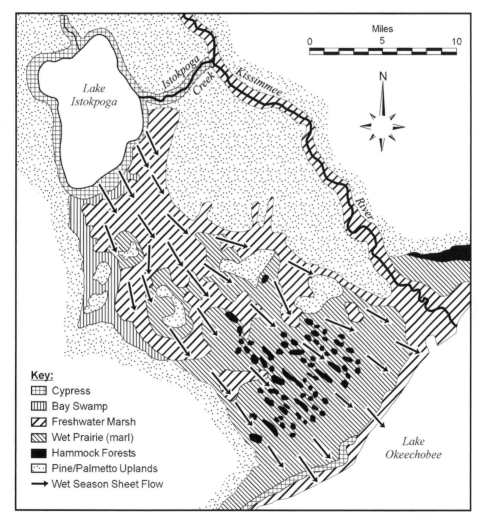

FIGURE 11.3 Lake Istokpoga and its historic "mini-Everglades" flow-way to Lake Okeechobee. The vegetation map is mostly based on Davis' work in the early 1940's[145] with modifications following Harshberger.[261] Today, most of the flow-way is developed for agriculture, and the south end is crossed by Route 78, which is parallel to the Hoover Dike at the edge of Lake Okeechobee. The Brighton Indian Reservation of the Seminole Tribe (see Figure 11.1) covers some of the southern portion, and three major canals (see Figure 11.4) drain the region.

cypress swamps and bordered here and there by upland oak forests. It entered the lake's northern-most tip. After the railroad was extended to the town of Okeechobee in 1915, Taylor Creek became crammed with fish houses and docks for the lake's commercial catfish industry. With considerable dry land in the vicinity, the cattle industry subsequently developed and later intensified with dairy feed lots, which contributed excessive phosphorus through Taylor Creek and nearby Nubbin Slough. Like other smaller tributaries, they now connect to a local rim canal outside of the Hoover Dike. In turn, the rim canal is connected through the dike to Lake Okeechobee through spillways, pump stations, and occasional boat locks.

LAKE OKEECHOBEE'S DISCOVERY AND NAME

Lake Okeechobee was used by Native Americans long before the relatively recent arrival of the Seminoles and Miccosukees, evidenced by Indian mounds and relics. Early accounts of the lake by Europeans — using names such as "Sarrope," "Mayaimi," (Calusa Indian name meaning "big water") "Lake Mayaco," or the Spanish "Laguna del Esperito Santo" — were highly varied and often fanciful. The lake's inaccessibility — nearly surrounded by wetlands and without navigable outlets — obstructed explorers, and the few who got there only saw it from the surrounding low, flat terrain, devoid of elevated overlooks. An 1822 military account flatly denied the presence of a large inland lake[334] and an 1823 report by Charles Vignoles, an English engineer, who had explored some of Florida's interior and had described the Everglades, regarded the lake as possibly fictional.[622, 650] Not until the "Seminole Wars" of the 1800's did the lake and its vicinity become known. The Seminole Indian's Hitichiti language name "Okeechobee" — also meaning "big water" as did the earlier Calusa name — became established in the 1830's during the Second Seminole War. The U.S. military pursued the Seminole Indians, who had earlier fled their homeland in Alabama and Georgia, into southern Florida, where — in the vicinity of the lake — sporadic confrontations raged for several decades. Near the end of the third and final Seminole War, the first accurate map of the lake and its vicinity was completed — the Lieutenant Ives' Military Map of 1856.[294]

ORIGIN OF LAKE OKEECHOBEE AND ITS BASIN

Lake Okeechobee's roughly circular shape spawned the hypothesis that it originated as a meteor crater, and a few speculative accounts promoted the idea. Evidence of a crater origin stops with the lake's outline — nothing in the underlying sedimentary sequences supports that idea. Sedimentary evidence indicates that a bedrock trough formed in South Florida's interior by a relatively greater subsidence of thick underlying clays (which shrink as water is slowly squeezed out) contrasted to the more substantial limestone and sand deposits that developed closer to both South Florida coasts. The deepest spots of the bedrock trough — about a foot below sea level — are east of the lake's center. The trough shallows and extends southeastward into what is now Water Conservation Area 1. To keep dimensions in perspective, if the area were still all dry land as it was more than 6000 years ago, you would scarcely notice the depression; it would appear nearly flat.

An increasingly wet climate and rising sea level led to Okeechobee's initial filling. Based on carbon-14 dating, the lake's birth occurred just over 6000 years ago. Peat deposits show that wetlands occupied parts of the youthful lake's basin for about 2000 years. Rising water eventually overtook most of the wetlands, leaving their peat soils in locations that are now in relatively deep, open water,[311] except at the lake's southern rim where the formation of peat soils kept pace with the lake's development. More rainfall also meant greater flows from northern tributary rivers and streams, and with those flows came sand from the ancient beach ridges of central Florida. Sands then came to cover much of the bottom, burying some of the wetland peat deposits.

As the lake's level increased, the Everglades rim to the south and southeast developed as a dam, retarding Okeechobee's southward elongation in the trough, although peat underlies much of the adjacent open water. The thickness of peat soils grew in response to the hydration provided by the lake's level, reaching a final thickness above the 7-foot-high bedrock of 13 to 14 feet just south of the lake, in the northernmost Everglades. Thus, Lake Okeechobee's "biological dam" grew to a final elevation of 20 to 21 feet above sea level.[237] Through the years, soft, black, organic-rich sediments up to 30 inches thick have covered the lake's deeper areas.[311]

THE PRE-DRAINAGE LAKE

Prior to perimeter diking and engineered water-level control, Lake Okeechobee was more variable in size. The only stable shoreline was a sandy beach ridge extending from the northern tip on around the lake's northeastern quarter. This narrow ridge was forested by oaks and palmettos but was mostly followed behind by sawgrass wetlands. It was probably the only "walkable" shoreline around the entire lake, but for a sandy beach on part of Observation Island in the southwest area of the lake. The low, flat, terrain with soft peat and muck soils on the rest of the lake was seasonally flooded. Lawrence Will, a local resident recalling the early 1900's, said that normal high water extended the lake five or six miles on the west side.[650] The southeastern and southern shore was bordered by a wetland forest dominated by pond apple trees, which are tolerant of deep flooding and locally called "custard apple."* During rainy-season high water, the lake flowed into this forest, which extended southward two or three miles, then through a zone of willows and elderberry bushes and into the sawgrass Everglades.**

Numerous explorers tried to find navigable outlets from the lake. They found no southwestern outlet to the Caloosahatchee River. The only recognizable outflow was seasonally over a wide area of the southern shore. Interestingly, the outflow there was not uniform sheet flow, but was mostly channeled through what were called "dead rivers." One account in the early 1880's recorded 17 such rivers[650] leading out of the southern half of the lake. Eight of them were large enough to be investigated as possible navigable outlets, the largest described as nearly 300 feet wide and up to ten feet deep near the lake. All of them became gradually shallower southward, and narrowed or divided into smaller streams that ended within a few miles, their flows spreading into pond apple, cypress, willow thickets, or sawgrass. Anything but biologically dead, first-hand accounts spoke to the excessive numbers of alligators and profusion of other wildlife, and the abundance of catfish in the lake has been ascribed to an important spawning and nursery function of these "dead-end" rivers and their adjacent wetlands.[650]

The largest of the dead rivers became known as the "Democrat River," the name derived from the second exploration party of the New Orleans Times-Democrat newspaper in late 1883. The river originated near the southernmost bay of Lake Okeechobee and divided into branches that passed on the northern and eastern sides of where the town of Belle Glade is today. The north branch flowed east several miles before dispersing into sawgrass. The other branch flowed south through the pond apple forest and finally into sawgrass. Using skiffs, the Times-Democrat expedition apparently followed the south branch to its end. From there, they pushed and burned their way through the dense sawgrass of the northern Everglades, then even used sails through some of the deep sloughs of the central and southern Everglades, finally navigating through the Shark River to Whitewater Bay and the Gulf of Mexico. Their journey went a straight-line distance of 90 miles from the lake and took almost a month.[650]

---

* See footnote regarding name confusion in Table 4.1.
** The presence of the pond apple forest was well documented by accounts beginning with the Times-Democrat expeditions in 1882 and '83, but earlier maps and written accounts fail to show or mention such an extensive forest.[80] Thus, it is possible that the pond apple forest was a new feature, a response to changing conditions for pond apple seed germination and growth that occurred in the second half of the 19th century.

The historic condition of Lake Okeechobee allowed its potential sediments to settle in the adjacent wetlands. Hurricanes and perhaps lesser storms would entrain lake sediments accumulated during calm periods and carry them into the wetlands where slower flow and the lack of waves allowed settling among plant stems. This process did two important things. First, it prevented the accumulation of fine sediments in the lake, a condition that would eventually fill-in the lake, reducing its open-water functions for fish and wildlife. Second, the fine sediments enriched the adjacent wetlands. It is known, for example that the northern Everglades soils were substantially more desirable for agriculture than more distant soils — the result of inputs from the lake.[575]

THE MODERN LAKE

The Caloosa Indians, and possibly earlier Native American tribes, constructed small channels and numerous mounds in the vicinity of Lake Okeechobee. But the first modification of the lake that significantly altered its hydrology was the work of Hamilton Disston beginning in 1881. The State of Florida offered him extensive land in exchange for drainage "improvements" through the Kissimmee River region and Lake Okeechobee. Disston realized that the only way to drain lands in the Kissimmee region from Orlando southward was to reduce the level of Lake Okeechobee. He determined that the easiest route was to excavate a channel to connect the lake to the Caloosa-hatchee River, a distance of only three miles from the lake* to the river's headwater lake, Lake Hicpochee (pronounced *hĭk'-po-chē*). Probably his farthest reaching alteration, however, was the dynamiting of a rocky obstruction in the Caloosahatchee near the present town of LaBelle. The rock sill acted as a dam so that the original upper river had little flow and was described as a series of shallow lakes, including Hicpochee. The largest was Lake Flirt, and its lake-bottom lands just east of LaBelle are still visible on aerial photographs. The removal of the rock sill drained Lake Flirt. With a channel dredged through the old lake bottom and boat locks to control flow, Lake Okeechobee was connected to the Gulf of Mexico, a navigable route of 70 miles. The first steamboat made the trip from Ft. Myers through the lake and up to Lake Kissimmee in 1882.[65]

Disston also excavated an 11-mile canal from the south shore of Okeechobee into the Everglades with the intent of continuation. But unexpected harder rock and financial misfortune stopped his work in 1893. Despite some success — drained lands in the Kissimmee headwaters, improved navigation on the Kissimmee, and, of course, the navigable connection to the Gulf of Mexico via the Caloosahatchee — Lake Okeechobee itself was still largely "untamed." The wet-season waters draining through the Kissimmee-Okeechobee system were so massive that seasonal flooding was extensive. In the vicinity of the lake, flooding may have been exacerbated by Disston's work because the drainage connections in the upper river system would have shed water more rapidly into the Kissimmee, which, with removal of numerous navigational obstructions, would have directed water to the lake faster.

In an effort to expand agriculture and further land development, the State of Florida undertook new flood-control and navigation projects, mostly around the southern half of Lake Okeechobee. A dike of local muck soil was constructed there to prevent flooding of new agricultural fields and small urban areas. By about 1917, four canals had been constructed to drain Okeechobee southward — the Miami Canal (extending Disston's earlier effort), the North New River Canal to Fort Lauderdale, the Hillsboro Canal to Deerfield Beach, and the West Palm Beach Canal. In 1926, the St. Lucie Canal was completed, although initially of minimal configuration, connecting Lake Okeechobee eastward to the Atlantic at Stuart, the shortest route for drainage waters. Reduced

---

* The canal connected the lake next to a large cypress tree that could be seen from a mile or so out in the lake. Descriptively called the "flat-top cypress," and named "Lone Cypress" and "Sentinel Cypress," it was used as marker for navigators on the lake to find the Caloosahatchee Canal. This tree still lives today, but it is on the canal's edge in downtown Moore Haven. Lake Okeechobee's open waters are now some 7 miles distant. A plaque commemorates the tree's historic significance.[65, 257, 300]

water levels and transportation provided by the canals spurred population growth until two major hurricanes — 1926 and 1928 — overwhelmed the canals and breached the soft, inadequate dike, causing extensive flooding both south and north of Okeechobee. The 1928 hurricane killed nearly 2000 people south of the lake, attracting federal attention to the situation.[58, 390, 402]

Using navigation as a means of securing federal funding,[*] control of Lake Okeechobee's waters was greatly expanded in the 1930's by the U.S. Army Corps of Engineers. Among the improvements was construction of the huge limerock levee — the Hoover Dike — eventually encircling the entire lake except where Fisheating Creek enters (where the dike turns and flanks the sides of the creek's floodplain westward for several miles). Inside the dike in the southern half of the lake, the Corps dredged a deep, navigation channel. It connected to enlarged navigation channels and boat locks in both the St. Lucie Canal and Calooshatchee River, giving birth to the official "Okeechobee Waterway" from Stuart, Florida on the Atlantic to Ft. Myers on the Gulf of Mexico.[611] Small ships and barges could take the route directly across the open lake, or follow the more protected route through the south perimeter channel, sometimes dubbed the "rim ditch."

Severe weather in 1947 provided a major test of Lake Okeechobee's modifications and lead to the water-control system that is in place today. Following an unusually rainy wet season that began early, there was a major hurricane on September 17 that drenched the northern Everglades and caused extensive flooding there and in Fort Lauderdale, where floodwaters were carried by the North New River Canal. Then, on October 11, a second hurricane progressed northward from the south end of the peninsula and exited to the Atlantic near Fort Lauderdale, extending the flooding (see Figure 9.1). The entire greater Everglades area was inundated, Kissimmee valley to the southern Everglades. Downtown Fort Lauderdale's streets and businesses were in deep water for weeks. Lake Okeechobee did not crest until November 2, 1947, because of delays in water flows through the Kissimmee River. The Hoover Dike, however, survived the waves and high water with little damage.

The U.S. Congress, already immersed in South Florida's flood control, responded in 1948 by authorizing a massive plan, the Central and Southern Florida Project for Flood Control and Other Purposes (the C&SF Project, see Chapter 21) to provide new flood-control infrastructure and to greatly modify the existing features. The modifications to Lake Okeechobee were: 1) an increase in the integrity and size of the Hoover Dike; 2) channelization of the Kissimmee River, with the addition of levees, water-control structures, and navigation locks; and 3) enlargement of the Caloosa-hatchee River/Canal and St. Lucie Canal for flood releases and improved navigation. The entire C&SF Project took several decades to complete, with Okeechobee's modifications done by 1966. To say the least, the Lake Okeechobee that emerged bore little resemblance to its original ecological role and configuration as the hydrologic hub of the greater Everglades ecosystem (Figures 11.4 and 11.5).

Today, water levels are regulated primarily for flood control and for agricultural and urban water supply. In anticipation of heavy rainfall, water is released to provide capacity for flood-control, and in anticipation of dry conditions, water is retained to provide for water supply. The present regulation schedule is complex, with water release decisions made according to tributary flows and climate outlook. Generally, the lake can be lowered to 13.5 feet, pending outlook, at the beginning of the wet season, and maintaining it at 15.5 feet from the end of the wet season through the dry season. A much simpler former schedule used 15.5 and 17.5 feet respectively, which was shown to impair the fish and wildlife functions with water too deep for the lake's fringing littoral zone, confined within the lake's dike since the 1930's, thus unable to shift location as it did historically. And with all the water controls, the important historic lake's function as a wet-season

---

[*] Up to that time, flooding and drainage were considered problems to be solved by state governments. Navigation, however, being in the national interest, could be funded by Washington. Because navigation from the coasts through Lake Okeechobee required control of water levels, federal funding became linked to water control, and thus to drainage and flood control in southern Florida.[300, 611]

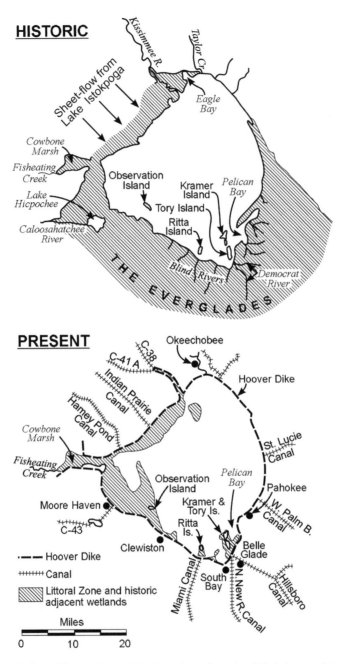

FIGURE 11.4 Historic Lake Okeechobee and its incoming tributaries, blind-river outlets to the Everglades, and littoral zone contrasted to the modern lake with its new littoral zone within the historic lake's pelagic zone. (Primarily from Refs. 145 and 650, combined with USGS State of Florida map.)

source of evenly distributed sheet flow to the Everglades is gone, especially in providing the occasional enormous flows that appear to have been important in developing the Everglades flow pattern and in providing soil enrichment to the northern Everglades. In turn, ecologically damaging freshwater discharges to the Caloosahatchee and St. Lucie estuaries have been another adverse outcome of the lake's modifications.

FIGURE 11.5 View of the northeastern portion of Lake Okeechobee showing the Hoover Dike and parallel canal outside. Portions of littoral zone can be seen in the upper center and upper left of the lake. (Photo by T. Lodge.)

The water-level history, or hydrograph, from March 1912 through December 2002 is shown in Figure 11.6. This record also shows the estimated, much higher historic range from the "Natural Systems Model" (NSM). In general agreement with the NSM, a Corps of Engineers' report in 1913 established the wet-season level at 20.6 feet NGVD (see inset box: Elevation Surveys, NGVD, and NAVD), where flows to the Everglades occurred. Seasonal low stage was described as 19.2 feet. [574]

VEGETATION AND WILDLIFE

Our understanding of Lake Okeechobee's ecology is hampered by the lack of studies of its historic condition — surprising for such an important lake. There are few published studies having to do with the lake's biota or water quality prior to the 1980's.

The native plants and animals of Lake Okeechobee are mostly the same species as found in the Everglades. The differences come from those species that thrive in the lake because of the expanse of deeper, permanent water. Submerged aquatic plants, migratory ducks, and some larger fishes are prominent examples.

Limnologists — scientists who study lakes — divide lakes into zones that have different functions. The shallow, near-shore environment, where rooted plants grow to or above the surface, is called the littoral zone (preferred pronunciation is *lĭ′-t ĕ-räl*, identical to the word "literal"). Littoral zones are essentially wetlands in appearance and function. The open water area is the pelagic or limnetic zone, where water is deep enough that the bottom is rarely if ever exposed, and no rooted vegetation is present.

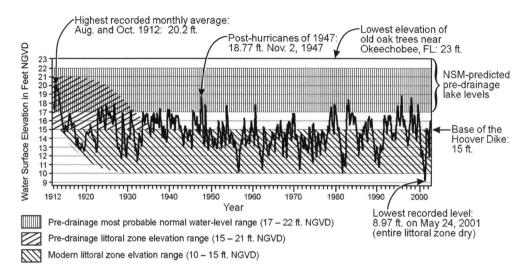

FIGURE 11.6 Lake Okeechobee hydrograph, March 1912 to December 2002, annotated with Natural Systems Model (NSM) predicted historic lake levels, and ranges of historic and modern littoral zones. From FFWCC (Mar. 1912 to Oct. 1931 data); SFWMD Hydrology and Hydraulics Division (Oct. 1931 to Dec. 2002 plot); Dr. Paul Gray of National Audubon Society (oak tree elevations); and Refs. 574 and 612.

LITTORAL ZONE

Almost the entire modern littoral zone is new in Lake Okeechobee's history. The original littoral zone probably occupied all areas around the lake that were lower than 21 feet in elevation, based on what is known about pre-drainage water levels. Most extensive around the southern and western sides, the historic littoral zone was almost entirely excised by the construction of the Hoover Dike, which was located on or close to open water, at the 15-foot elevation — close to the open-water edge of the pre-drainage littoral zone. Subsequent water control reduced the lake's level, and the modern littoral zone developed, mostly in the shallow western portion of the lake that had previously been open water. The original littoral zone, now outside of the dike, became drained farmland.

The depth to which the littoral zone may extend depends on light penetration so that, in clearer water, rooted plants may prosper in deeper water. Because water clarity is reduced in the modern, nutrient-enriched conditions, the littoral zone is undoubtedly restricted to shallower water. The littoral zone of the modern lake occupies just over a quarter of the surface area (almost 200 square miles) and is almost entirely confined to areas where the bottom is at or above 9 feet NGVD, so when the lake stage is at 15 feet, the littoral zone's deepest edge is in six feet of water, and most of littoral habitat would be less than 4 feet deep. The bottom in shallower parts of the littoral zone should be exposed annually due to seasonal variations in lake level (although lake regulation has seldom allowed for so shallow water in recent years). Deeper parts are exposed only in severe droughts, such as in 1981, 1989, and particularly 2001 when the lake level reached a record low just under 9 feet.

Common plants of the lake's littoral zone are listed in Table 11.1. These species are not randomly distributed, but usually form distinct communities, named for their dominant plant species. Thirty different littoral zone communities have been named. Moving in a successional sequence from shallow, seasonally exposed lake bottom, to deeper perennially flooded areas, examples are the following (with approximate square-mile extent as mapped in 1992):

TABLE 11.1
Common littoral zone plants of Lake Okeechobee[a]

Common name[b]	Scientific name[b]	Origin/Comments
Emergent Species - Herbaceous		
giant bulrush	*Scirpus californicus*	native/important gamefish habitat
gulf coast spikerush	*Eleocharis cellulosa*	native
Tracy's beaksedge	*Rhynchospora tracyi*	native
sawgrass	*Cladium jamaicensis*	native
bluejoint panicum	*Panicum tenerum*	native
sand cordgrass	*Spartina bakeri*	native
bulltongue arrowhead	*Sagittaria lancifolia*	native
pickerelweed	*Pontedaria cordata*	native
dotted smartweed	*Polygonum punctatum*	native
southern cattail	*Typha domingensis*	native
broadleaf cattail	*Typha latifolia*	native
sweetscent (camphorweed)	*Pluchea odorata*	native
Shrubs and trees		
coastalplain willow	*Salix caroliniana*	native/important nesting tree
buttonbush	*Cephalanthus occidentalis*	native
punktree, melaleuca	*Melaleuca quinquenervia*	Invasive exotic, introduced in mid-1940's to control levee erosion
Floating and floating-leaved species		
American white waterlily	*Nymphaea odorata*	native, rooted
yellow waterlily	*Nymphaea mexicana*	native, rooted
American lotus	*Nelumbo lutea*	native, rooted
bladderworts	*Utricularia* spp.	native, free-floating, two or more species
common water-hyacinth	*Eichhornia crassipes*	invasive exotic, free-floating, appeared in Lake O. in early 1900's
torpedograss	*Panicum repens*	invasive exotic, may be rooted or floating
Submerged species		
Illinois pondweed	*Potamogeton illinoensis*	native
American eelgrass, tapegrass	*Vallisneria americana*	native
waterthyme, hydrilla	*Hydrilla verticillata*	invasive exotic, appeared in Lake O. in 1967

[a] Selection of species is from reference 499.
[b] Common and scientific names follow references 222 and 661. Alternate South Florida vernaculars are in parentheses.

- Willow community (16.5) — occupies shallow areas with peat soil and is dominated by coastalplain willow but commonly contains buttonbush. Willows are important for wading bird nests and are killed by prolonged high water.
- Beakrush community (0.0) — dominated by Tracy's beaksedge, occurring over sandy soils in shallow areas. Other plants included spikerush, camphorweed, and occasional clumps of sand cordgrass. An important community in the early 1970's (24.3 square miles), it was completely overtaken by other communities by 1991.
- Sawgrass community (4.6) — occurs over peat soils and usually also contains spikerush and pickerelweed interspersed with the dominant sawgrass. It is very similar to sawgrass marsh of the Everglades.

- Waterlily community (4.8) — occupies deeper water and is dominated by white and/or yellow waterlilies. It closely resembles the Everglades slough habitat.
- Bulrush community (4.6) — a community not found in the Everglades (Figure 11.7). It typically occurs in still deeper water, near the littoral zone's open-water edge where it is exposed to wave action. Species associated with bulrush are spikerush and submerged plants. This community is particularly important for gamefish.[219] Bulrushes nearly disappeared in the lake due to high water during the mid and late 1990s but rebounded quickly following low water in 2001.
- Submerged community (47.7) — found in the deepest parts of the littoral zone. It is dominated by submerged aquatic plants, primarily American eelgrass, Illinois pondweed, and hydrilla (an invasive exotic species that first appeared in the lake in 1967). Mirroring the changes in extent of the bulrushes, the submerged community (especially Illinois pondweed) responds to high and low water.

The littoral zone harbors some communities dominated by invasive exotic plants. Floating masses of water hyacinth have been common since the 1920's. Melaleuca encroached far out into the lake's west side, advancing every time low water exposed bottom habitat where its seeds germinated. In 1992, melaleuca covered about 16.5 square miles of the littoral zone, but is now all but eradicated there. Hydrilla, mentioned above, and torpedo grass are more recent invaders. The latter spreads quickly by runners over the top of other floating-leaf vegetation, damaging wildlife values and obstructing shallow-water access for small boats such as canoes.

FIGURE 11.7 Outer edge of the littoral zone of Lake Okeechobee's north shore. Bulrush, often at the edge of open water, is shown here growing in about three feet of water. (Photo by T. Lodge.)

Pelagic Zone

The open-water or pelagic (or limnetic) zone does not have rooted plants, and primary productivity there is by phytoplankton — microscopic plants suspended in the water. Despite their small size, they are grazed by several fishes and invertebrates, such as clams that can filter them from the water. Contrary to what might be casually expected, Lake Okeechobee's pelagic zone is not always homogeneous but develops as many as four ecological sub-zones recognized by water quality characteristics. They may be stable through several years when water levels are below normal but disappear by mixing during extended high water. High water extends the plankton-dominated system into the littoral zones, often killing submerged littoral vegetation because of reduced light.

Fishes

Today, the fishes of Lake Okeechobee are a mix of strictly freshwater species and species that live part of their lives at sea, such as tarpon, snook, striped mullet, and Atlantic needlefish. It is improbable that these marine-related peripheral species occurred in the lake prior to access provided by canals. The route through marine waters and the Everglades was too long and too obstructed by the dense sawgrass of the northern Everglades for those species to reach the lake. Temporary exceptions may have occurred when extraordinary floods allowed schools of mullet, for example, to ascend to the lake, where they would be trapped and never reproduce. A study of the lake's fishes conducted in the late 1960's[5] documented 40 strictly freshwater species and eight marine-related species. Today, about 15 exotic species, primarily from the aquarium trade, can be added. Most species occurring in the lake begin their lives in the protection of the littoral zone, which provides a highly important function to many gamefish. Some small species such as mosquitofish live their entire lives in ther littoral zone, while others such as the bluegill occupy both littoral and limnetic areas, and still others such as the channel catfish and threadfin shad become mainly pelagic.[107]

Lake Okeechobee has supported important fisheries, which continue today. Beginning in the first decade of the 1900's, after the railroad reached Ft. Myers, then abundant catfish were netted or caught on lines and shipped throughout the southeast. Of three catfish species taken commercially, the channel catfish and the white catfish were the larger and regularly reached 20 pounds and 5 pounds, respectively, with the largest channel cats weighing over 50 pounds. Sizes in today's catches are considerably smaller and dominated by the white catfish. Subsequent to development of its commercial fishery, Lake Okeechobee has become widely recognized for sport fishing. The largemouth bass is the most prized, but huge numbers of black crappie and bluegills are also taken. There has been considerable friction between commercial fishermen using large nets and sport fishermen, resulting in stiff regulation of commercial fishing. Despite Lake Okeechobee's deteriorating water quality, commercial fishery catches have increased, apparently a response to greater lake productivity from nutrient enrichment (discussed below).[204]

Other Wildlife

Other important wildlife species that utilize the lake include all of the wading birds that also use the Everglades, snail kites that find abundant apple snails in the lake's littoral zone and use willow trees for nesting, and wintering ducks such as huge numbers of lesser scaups that use the lake's pelagic zones. The largest wildlife species in the lake is the Florida manatee, which enters during warmer months of the year and even travels far up the Kissimmee River. Like the marine fishes mentioned above, it can reach the lake through canals, moving through boat locks and under opened flood-control gates. Manatees could not have occurred in the lake prior to the canal connections.

WATER QUALITY [78, 262, 567, 573]

As in most Florida lakes and the Everglades, Lake Okeechobee's limiting nutrient that controls plant growth was historically phosphorus but has now switched to nitrogen (discussed below). No phosphorus data exist for the lake prior to the first studies in the late 1960's, but, based on sediment data, it has been determined that the original waters contained much lower levels than today, with original concentrations estimated to range between 20 and 50 parts per billion (ppb) (compared to the historic Everglades condition of about 10 ppb or less). This level is described as moderately eutrophic — a healthful condition conducive to the historically profitable fishery while also exhibiting fairly clear water, unlike the turbid condition in more enriched lakes.

Phosphorus inputs into Lake Okeechobee increased greatly in the 1900's, primarily due to agriculture. Bottom core studies have shown that the rate of phosphorus accumulation into sediments has increased about four fold between pre-1910 and the 1980's. During the 1970's, averaged phosphorus concentrations in Lake Okeechobee's pelagic waters doubled to about 100 ppb. The trend then became erratic, but the longer term shows a continued increase (Figure 11.8). The ability of the lake to assimilate phosphorus was overwhelmed by the early 1990's. During normal water levels, the littoral zone water remains clear and low in nutrients, a result of restricted water movement between the pelagic and littoral zones due to the dense littoral vegetation and nutrient uptake by the littoral vegetation. However, when water levels are very high, the impaired pelagic

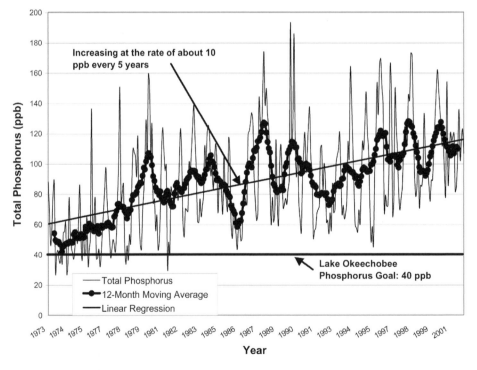

FIGURE 11.8 Lake Okeechobee's increasing phosphorus concentration: January 1973 to December 2001. The thin-line plot is the phosphorus data. The thick line with dots is a 12-month moving average, which helps smooth the highly variable data (much due to rainy versus dry periods) into a more understandable form. The straight line is linear regression of the entire phosphorus data set to give a best approximation of the trend for the entire period shown. However the mid-1970's to mid-1980's appears to have been steeper, leveling somewhat after that time. Turning this trend around to head back toward the 40 parts per billion goal is the focus of a huge ongoing restoration effort. (Data provided by SFWMD, plotted by EA Engineering, Science, and Technology.)

water quality enters the littoral zone. The lake's pelagic area is now described as being in an "advanced eutrophic" condition, described in the following paragraphs.

Increased nutrients have promoted the growth of phytoplankton — microscopic plants (many classified as "algae") that live suspended in the water. With more abundant phytoplankton, light is attenuated as it passes through the lake's water, obscuring vision commonly to less than two feet and sometimes to only a few inches — a condition called an "algal bloom." The enrichment has promoted a disproportionate increase in phytoplankton belonging to the group called cyanobacteria, or "blue-green algae" in older literature. Abundant cyanobacteria are often recognizable because of a thick surface scum caused by their production of gas. Between 1974 and 1989, the proportion of cyanobacteria more than doubled, from 28% to about 65% of the total phytoplankton. Accordingly, phytoplankton indicative of a more balanced ecosystem was reduced, including green algae and diatoms. The explanation is that phosphorus is no longer a limiting nutrient in Lake Okeechobee, and nitrogen has become the least available nutrient. Under nitrogen-limiting conditions, cyanobacteria have an advantage over other phytoplankton because they fix nitrogen from nitrogen gas dissolved in the water into combined nitrogen forms that can be used in their growth. Thus, the enriched phosphorus condition promotes a greater growth of cyanobacteria than would the same nutrient condition with the lake's normal phytoplankton because the latter would be limited by nitrogen.

Fishery production has not yet been reduced by the enriched condition although it has often damaged sport fishing by reduced water clarity and periods of extended high water have damaged fish populations by disrupting the littoral zone. It is predicted, however, that more enrichment would likely cause periodic reductions in dissolved oxygen, a condition caused by too much oxygen demand by phytoplankton — at night and on cloudy days when light-induced photosynthesis does not produce a net excess of oxygen. Depressed dissolved oxygen is the most common cause of fish kills, but Lake Okeechobee is so large and shallow that surface aeration and mixing will likely protect it from anoxic conditions. However, the lake will experience increased algal blooms and damage to the littoral zone and its important nursery functions for fish. Another large Florida lake — Lake Apopka — underwent such a fishery collapse much earlier as a result of similar circumstances.[520]

Other indications of the enriched condition are in the lake's bottom-dwelling or benthic invertebrates. In addition to clams and small crustaceans, important organisms that thrive in bottom sediments are small worms — oligochaetes — related to earthworms. There has been a significant increase in oligochaetes of the particular kinds known to tolerate polluted conditions where dissolved oxygen is low in the sediments and bottom waters. In fact, it is the accumulation of phosphorus in flocculent bottom sediments that is a major concern to our ability to reverse the lake's condition, described below.

There are extensive ongoing and planned improvements aimed at correcting Lake Okeechobee's condition. Best management practices (BMPs) have been promoted and used to clean up agricultural operations. The results have been very encouraging. A phosphorus "loading" target of 100 metric tons (2200 pounds is one metric ton) of phosphorus per year from tributary streams has been established for the lake. It is called a total maximum "daily" load (TMDL), yet is an annualized value. This loading is based on average tributary stream inflows achieving a phosphorus concentration of 50 ppb, and the lake's calculated ability to assimilate phosphorus.[613] The ultimate goal is to return the lake's average phosphorus concentration to 40 ppb (see Figure 11.8), which should preserve the ecological health of the lake. However, even if the inflow-loading goal is achieved, attaining the 40 ppb goal within the lake will require time — calculations range from 35 to over 200 years — primarily because abundant phosphorus now resides in a flocculent layer lying over the normal, denser bottom sediments. Every time wind-driven currents resuspended this layer (essentially a phosphorus reservoir), it contributes phosphorus to the overlying water, promoting objectionable phytoplankton. Sediment removal has been shown to be impractical. Also, for the past thirty years the inflow loading has averaged in the range of 400 to 500 tons per year, a condition that continues far above the 100-ton target. Nevertheless, extensive additional phosphorus controls are planned as part of the Comprehensive Everglades Restoration Plan (CERP, see Chapter 21) giving many people optimism that the deterioration can be stopped and the lake improved.

SUMMARY

Lake Okeechobee and its watershed lakes and wetlands historically stored vast amounts of water from rainfall in the south-central Florida interior south of Orlando. Most of this water was carried southward by the lake's largest tributary, the Kissimmee River. The lake annually spilled into the Everglades, adding substantial water flow and hydroperiod to the already flooded ecosystem. With this function, Lake Okeechobee was the hydrologic hub of the South Florida ecosystem. Today, that function has been curtailed by flood-control projects through the Kissimmee watershed and diversion of Okeechobee's waters to the coasts, often in ecologically damaging heavy flows to the St. Lucie and Caloosahatchee estuaries. Containment of the lake by the Hoover Dike and water control has constrained the important littoral zone to the smaller lake, unable to move outward in response to changing water levels. More recently, nutrient enrichment, mostly from agriculture, has changed the lake's chemistry and threatened its biological functions. Nutrient enrichment has probably also been exacerbated by canalization of, and levees along, the Kissimmee River and by the Hoover Dike. These modifications have served to retain entrained sediments in the river and lake rather than allowing them to exit and settle in the larger, natural wetlands and littoral zone, including the northern Everglades. The phosphorus load now embodied in lake sediments may require centuries to assimilate, and unless loading of phosphorus from the watershed is substantially reduced, the internal recycling will never decline. Clearly, ongoing and planned phosphorus reductions and improved watershed storage are critical to salvaging the lake's health.

ELEVATION SURVEYS, NGVD, AND NAVD

Water-level determinations on Lake Okeechobee relative to sea level were first attempted in the 1880's in a survey from the gulf coast following the Caloosahatchee River to the lake. To provide for navigation, including boat locks and channel depths, it was important to know the lake "stage" (water elevation) as well as the difference in water elevation between the lake and tidal waters. That initial survey established a benchmark (elevation reference marker) in a notch cut into a buttressed root in the famous "Sentinel Cypress" near the lake's shore (see footnote, The Modern Lake section). Later surveys found errors in that benchmark elevation. Part of the problem involved running elevations on peat soils, which are unstable and move even in response to surveyors' footsteps. In addition, the Sentinel Cypress itself apparently settled by at least a foot after establishing the benchmark due to subsidence of the peat soil around the tree. Needless to say, the lake stages determined by early surveys were inaccurately referenced.[300]

The first accurate survey for water elevations on Lake Okeechobee was conducted in the mid-1920s, following completion of Conner's Highway (now U.S. 98), the first paved road from the east coast to the lake. Using the highway as a stable platform, the survey determined the lake's elevation relative to sea level on the Atlantic Coast. Shortly thereafter, a national standard (datum) was developed, initially called "Sea Level Datum of 1929." Later changed to "NGVD29"(National Geodetic Vertical Datum), it was an elevation scale based on mean sea level in 1929 as zero elevation. Twenty-six locations (stations) were used around the coastline of the U.S. and Canada in that effort. Elevations in this book are referenced to NGVD29. After the Conner's Highway survey was corrected to NGVD29, it was shown that the earlier "Okeechobee datum" was about a foot and a half higher than the NGVD29-based elevations. For a discussion of this history, see pages 195-198 of Lamar Johnson's Beyond the Fourth Generation.[300]

NGVD29 was a great improvement, but as more data became available, as well as computers to handle huge data sets, there was a movement to modernize the system, and a new datum was born. Called the North American Vertical Datum or "NAVD88," it incorporates Mexico and Canada and was actually completed in 1991. NAVD88 ties all North American elevations to a single tidal record at Quebec, Canada, which is then used to establish elevations throughout the geographic area using theoretical ellipsoid geometry for the earth's shape — not the network of 26 tidal records used for NGVD29. It lowered most Florida elevations by 8 to 18 inches, depending on location. For example, a tree island elevation of 8.00 feet NGVD in the Everglades changed to 6.50 feet NAVD88. Most of the change is due to the different geometry used. Only a portion is from the rise in sea level. One should be aware that NGVD and NAVD are merely standards and do not necessarily reflect actual nearby mean sea level, which may be affected by: 1) local and regional influences on tides, such as winds, currents, and landform effects; and 2) long-term changing sea level (see Chapter 10). In evaluating changes of land or water levels over a period of time, it is not important what standard (datum) is used, so long as all of the data are corrected to the same standard. Bottom line — the reader should be aware that the various elevations, especially those in older literature, are not necessarily referenced to the same datum, and should be researched carefully to establish their relationship to NGVD29 (still widely used for the sake of continuity in continuous records such as water levels of Lake Okeechobee) or the newer NAVD88. For more information, see Reference 658.

Part III

The Flora and Fauna
of Southern Florida

12 Origins of the Flora and Fauna*

The preceding chapters have detailed many aspects of the flora in the context of plant communities and environments of the region. Little mention has been made of the region's animal life — the fauna. This chapter explores the biogeographic origins of the flora and fauna. Subsequent chapters deal with specific animal groups.

Before mankind emerged as a travel agent for a myriad of species ranging from the Norway rat and Asian carp to Brazilian pepper and Australian pine, flora and fauna depended on natural means of dispersal for colonizing new regions. Biologically, South Florida is a "new land" because of its history of alternating submergence and emergence, and because of its major changes in climate, both of which have eradicated many species of plants and animals through the glacial cycles. The important factors that determined the region's "new" flora and fauna were its geographic setting at the edge of the tropics, from which it is isolated by water and its land connection by a long peninsula to temperate North America. The flora and fauna present when Columbus arrived in the "New World" (the Western Hemisphere) had arrived in the region by their own modes of dispersal** and had succeeded based on their tolerance of the climate, the habitats, competition, and numerous other factors.

TROPICAL VERSUS SUBTROPICAL[128, 599]

The climate of southern Florida is presented in Chapter 2. In spite of any climatic labels, the proof of tropical climate in southern Florida is demonstrated by the vegetation. The mangrove swamps and upland hammocks of the extreme southern Florida mainland support a prevalence of plants derived from the Caribbean tropics. The climate for these tropical representatives is significantly more favorable near the coast, where the Florida Current and coastal marine waters contribute protective warmth. It is this maritime influence that has extended the climate for tropical plants north of the Tropic of Cancer, which is normally the defined northward limit of the tropics. Inland, or coastally only a few miles north of the latitude of Miami, tropical plants are commonly killed by cold weather, or pruned to a low stature, so it is useful to distinguish the climate of those areas as subtropical based on vegetation. The short difference between Miami and Fort Lauderdale, only 25 miles, shows a marked difference in survival of many tropical plants.

The marine environment of the Atlantic side of the Florida Keys and northward beyond Fort Lauderdale is obviously tropical, evidenced by the prevalence of reef-building corals. Inshore waters of southern Florida, however, are not tropical. Cold fronts can blast through southern Florida and plunge temperatures from the mid-70's F. one day to freezing the next. Notable freezes that killed many tropical plants as well as tropical marine fishes trapped in shallow inshore waters occurred in 1894, 1895, 1899, 1917, 1962, 1977, and 1989, and damaging frosts recur about every other year.[623, 657]

---

* The main sources for this chapter are references 333, 365, 436, and 642.
** Native Americans inhabited central Florida at least 14,000 years ago and southern Florida at least 10,000 years ago, and more recently began travel to and from Cuba.[98, 109, 414] While it is possible that American Indians aided the dispersal of tropical plants and animals to southern Florida, it is doubtful that they had a significant impact because the vast majority of the tropical species are not kinds they would have used. There would have been no reason to carry these species about in their travels.

The warm maritime effect is demonstrated in the distribution of native tropical trees. As an example of extreme cold sensitivity, the manchineel (*Hippomane mancinella*), renowned for its poisonous fruit, thrives in the Florida Keys but only in highly protected locations on the southernmost mainland. Two abundant tropical hammock species in Everglades National Park, the pigeonplum and West Indian mahogany, become rare only as far north as Fort Lauderdale. The least sensitive tropical trees, such as gumbo limbo and strangler fig, are abundant at the latitude of Miami but become dwarfed northward and restricted to the immediate coast. Some individuals of these species are found as far as Cape Canaveral on Florida's east coast and Tampa Bay on the west coast (see Figure 2.2). The varying sensitivity of tropical plants to cold is the factor that determines their northern limits, and a northward restriction to the coast is the standard pattern of their distribution.

ELEMENTS OF THE FLORA [365, 464]

During the relatively recent glaciation, when the Florida peninsula was cooler and drier than today, a temperate flora thrived. The tropical flora has only invaded the region in the last 5000 years. The warming trend following the last glacial period had apparently progressed enough by then that the climate resembled that of today and a tropical flora began to flourish.

Speaking only of "vascular plants" (those with roots and stems, as opposed to algae and fungi), about 1650 species now exist south of the latitude of Lake Okeechobee. Just over 60 percent of these are allied to the tropics, with about 56 percent occurring elsewhere in the Caribbean region. The remaining 40 percent of southern Florida's vascular flora is composed of temperate species derived mostly from the coastal plane of the southeastern states. Within this overall flora, there is a significant number of endemic plants, that is, species found in Florida, which have evolved there due to isolation (see Chapter 6). They compose about 9 percent of the flora of the region and their distant origin is a mix of temperate and tropical sources.*

Other interesting statistics are that 65 percent of the flora is herbaceous and 35 percent is woody types, namely trees and shrubs. Among the herbaceous species, about half are tropical; among the woody species, about 77 percent are tropical.[365] Thus, the strongest tropical heritage resides among the woody species, and among them this heritage is most pronounced in the larger stature species: the trees. There are 130 species of trees in southern Florida, and 112 of them (87.5 percent) are related to the Caribbean tropics. The remaining 12.5 percent are from temperate North America.[599]

ORIGIN OF THE TROPICAL FLORA [365, 366, 599]

How did tropical plants get to southern Florida? Three modes of dispersal account for the distribution of most plants; their seeds or spores are transported by wind, water, or animals. Dispersal by animals across water is nearly always by birds.

TREES [452, 505, 599]

Almost all of the tropical trees that occur naturally in southern Florida are also found in the West Indies.** Most species occur in hardwood hammocks and only a few in mangrove swamps (four species including buttonwood) and freshwater wetlands (two species). Seeds of the hammock species are commonly carried by birds; thus migrating birds, eating seeds while in the tropics and

---

* The percentages presented here are from reference 365. They include a significant number of introduced "naturalized" plants (species introduced by man and reproducing in the wild, outside of cultivation), not just the native flora.

** It is interesting to note that dispersal of trees has essentially been one way. Very few temperate trees are found in the tropical islands of the West Indies, although fossil pollen studies indicate that a larger representation occurred in the distant past. The limited occurrence of the Virginia live oak in western Cuba is one of the few examples existing today, and probably represents a population remaining from times when climate was more favorable for temperate plants there.[599]

then releasing them in droppings in Florida, have probably accounted for the arrival of many tropical trees. The white-crowned pigeon, for example, is a prolific disperser of tropical seeds throughout the West Indies and southern Florida, documented to carry the seeds of over 20 tropical hammock tree species.[55, 505] West Indian mahogany, however, is an example of a wind-dispersed species, having winged seeds.

The mangroves, as well as many other coastal wetland plants, have floating seeds and obviously were dispersed to Florida via ocean currents. The two tropical trees of freshwater wetlands — pond apple and cocoplum — are an interesting biogeographic problem. Both occur in tropical West Africa and tropical America, including the West Indies. Both have fruits that are eaten by animals, and birds probably accounted for their dispersal to Florida, perhaps from Cuba. Nevertheless, their occurrence on both sides of the Atlantic is puzzling: far outside the range of reasonable dispersal by flotation or by migrating birds.

PALMS [599, 661]

Palms are universally recognized as being tropically related, yet many species are adapted to warm temperate climates. Worldwide, there are about 2600 species. Many do not reach tree stature, and all have internal structure (like grasses and other monocotyledonous plants) that does not produce growth rings as do typical trees so that their age cannot be easily determined. Probably the most widely recognized species is the coconut palm, which is not native to southern Florida although it has been here for centuries. Most palms produce small fruits that are eaten and dispersed by birds and mammals.

Eleven palm species* are native to Florida, three of which are mostly restricted to central and northern areas. Eight occur in southern Florida, with four of these limited to rocky pinelands and the Florida keys — two known to be dispersed by the white-crowned pigeon.[55] Thus, only four species occur in Everglades habitats, namely the royal palm, cabbage palm, saw palmetto, and paurotis or Everglades palm (*Acoelorraphe wrightii*). The three former species are described in chapters 5 and 6 (hammocks and pinelands). The Everglades (or paurotis) palm reaches heights of 30 feet and grows with numerous, clustered stems. This aspect and its fan-shaped leaves make it popular as an ornamental landscape plant, but its cold sensitivity restricts it to the southern peninsula. In the wild, it naturally adorns many southern Everglades tree islands where it is typically conspicuous, growing at the edge next to the marsh or open water. The cabbage (or sabal) palm (see Figure 4.1), the state tree of both Florida and South Carolina, is unique it its occurrence in many habitats and also common as a landscape plant throughout the state and into the southeast. Its seeds are dispersed by raccoons and other animals. In very moist locations, it is often a host for the whisk-fern (*Psilotum nudum*), an ancient vascular plant, as well as other epiphytes.

EPIPHYTES

Another prominent feature of the tropical vegetation of the region is the abundance of plants that grow upon other plants: the epiphytes. Although some parasitic plants live in the region, epiphytes are not parasitic by definition, and are wind-dispersed. The region's major epiphytes are ferns, bromeliads, and orchids.

Ferns [125, 326, 438]

Numerous types of ferns grow as epiphytes, particularly in very damp locations. Of about 100 species of ferns found in Florida, almost 60 are tropical species and are limited to southern Florida. In contrast to bromeliads and orchids, which generally occupy fairly well-lit sites, ferns abound even in the dense forest shade. They grow on trees with rough bark, dead logs, or rocky walls of

---

* A twelfth species, *Sabal miamiensis*, has been recently described but is probably extinct in the wild.

solution holes (see Chapter 5). The golden polypody or serpent fern (*Phlebodium aureum*), a common tropical species in the Everglades region and central Florida, normally grows on the trunks of cabbage palms. Ferns reproduce by spores, which are technically not seeds but serve the same purpose and are wind-blown.

Bromeliads[125]

South Florida's most obvious epiphytes, the bromeliads (often called "airplants"), are relatives of the pineapple hence the alternate name "air-pines." All native bromeliads are normally epiphytic, but they may grow on the ground if they happen to fall. Due to its widespread occurrence in the South, the best-known species is Spanish moss (*Tillandsia usneoides*), which has an unusual growth form. Most other species resemble the top, leafy part of a pineapple; however, close inspection of Spanish moss reveals that it is a "string" of small plants attached by a stem. Spanish moss is not tropical and is more common farther north in Florida with its range extending as far north as Virginia. Most of Florida's 15 other bromeliads are tropical, are restricted to southern Florida, and their seeds are distributed by the wind. A spectacular and common species in the Everglades region is the cardinal airplant (*Tillandsia fasciculata*) (see cover). Its leaves may exceed three feet long (although usually half that), and its small flowers are enclosed by conspicuous, attractive bracts that are usually bright red but variable in color. It blooms from January through the summer.

Orchids [86, 125, 371]

Less obvious epiphytes unless they are in bloom, are the orchids. There are about 20,000 species of orchids worldwide, and they occur in both tropical and temperate climates. By far the greatest diversity of species occurs in the tropics where most are epiphytic; temperate orchids are mostly terrestrial. Florida harbors about 100 types of orchids. Temperate, mostly terrestrial species, are predominate in northern Florida, and tropical species (many of which are epiphytes) are predominate in southern Florida. The most common epiphytic orchid in the region is the beautiful butterfly orchid (*Encyclia tampensis*) (see Figure 12.1), a tropical species that blooms primarily in the late spring and early summer.

Orchids are best known for their spectacular flowers. The vegetative parts of many species are rather small, and often go unnoticed. Intricate flowers usually indicate specialized means of pollination, and the degree to which many orchids have evolved in their means of pollination is amazing. This specialization, which usually employs specific insect pollinators, is most developed in tropical species and has undoubtedly prevented many from becoming established in southern Florida. Because distribution of the tiny orchid seeds is by wind, they are able to reach South Florida from the nearby islands of the West Indies, germinate, and grow into mature orchids. The problem is pollination: such specialized species cannot reproduce without the necessary types of pollinators. As a result, most of the tropical orchids of the region are types that are normally capable of self-pollination. While southern Florida does have numerous tropical orchids (about 60 species are shared with Cuba), many more could have established if the proper pollinators were present.

MARINE FLORA [366, 589, 661]

All of southern Florida's marine flora reached the region from the West Indies on ocean currents. Although many species of tropical algae occur, only seven species of marine "grasses" are found, the most prominent being turtlegrass (*Thalassia testudinum*), followed by shoalweed (*Halodule wrightii*) and manateegrass (*Syringodium filiforme*). The brief pulses of cold air that invade the Florida peninsula in winter have little effect on the marine flora, except for low-tide exposures. In fact, tropical marine species extend much farther north than the tropical land vegetation, some even reaching the coast of North Carolina.

FIGURE 12.1 Butterfly orchid in bloom in the mangroves along West Lake boardwalk, Everglades National Park. (Photo by T. Lodge.)

HURRICANES AND DISPERSAL

The violent, destructive impacts of hurricanes were discussed in Chapter 9, but hurricanes (and tropical storms) also assist in the inland transport of floating seeds on storm surges, and probably serve to transport airborne seeds. The cyclonic motion makes any hurricane that enters the Straits of Florida a potential carrier of airborne seeds from Cuba and the Bahamas to southern Florida. Hurricane Inez (October 1966) was a good example. The eye crossed Cuba from the Caribbean into the straits and moved northeastward to a stopping point in the Bahamas just east of Miami. Of minimal hurricane strength at that point, it reversed direction and followed the Florida Keys into the Gulf of Mexico. Its particular track was nearly ideal for carrying airborne seeds from Cuba and the Bahamas to Florida (see Figure 9.1).

PROXIMITY AND DISPERSAL [365, 599]

The success of wind-blown dispersal in particular is dependent upon the proximity of land masses. It is evident that South Florida and the West Indies have been close enough during the past 5,000 years to facilitate the influx of tropical plants. Before that time, southern Florida was too cold to have supported sensitive tropical plants. During earlier, warm interglacial periods, the southern tip of the Florida peninsula may have been so far north that dispersal of tropical plants to Florida from the West Indies may have been more restricted. At those times, the continued existence of many tropical plants that had become established in Florida would have been more dependent on their local reproduction because re-seeding would have been limited. Modern sea level and climate may be special, transient conditions for Florida. If sea level continues rising (see Chapter 10), the shoreline of Florida will continue to retreat northward, perhaps again out of the reach of natural colonization by many tropical plants.

ORIGIN OF THE TEMPERATE FLORA [365, 366, 599]

While much interest in the Everglades region is focused on tropical vegetation, the plants that originated in temperate North America also lend character to the ecology of the area. They are particularly important in freshwater wetland communities of the region, including the lower elevation pinelands, wetland tree islands, and marshes.

TREES

The temperate trees of southern Florida dispersed southward into the region by a variety of means. Bald and pond cypress have floating seeds, which enable them to spread through freshwater wetlands and streams. Red maple and coastalplain willow have wind-blown seeds. Other species, such as Virginia live oak, sweetbay, and swamp bay have the kinds of fruits and seeds that are dispersed by mammals and birds.

MARSH VEGETATION

The seeds of most marsh plants are dispersed both by wind and by floating on water. The plants of the Everglades itself — the freshwater marshes — are mostly temperate. However, many of the marsh plants do not have a pronounced tropical or temperate affinity. Sawgrass, the most abundant plant of the Everglades, occurs from Virginia to Florida as well as in tropical areas of Central America and on the islands of the West Indies. Marsh plants, mostly grasses and similar herbaceous (non-woody) species, are adapted to regular adverse conditions that may kill their top portions. Cold weather, drought, and fire are periodic rigors of the interior freshwater marshes from which the vegetation normally recovers by regrowth from the roots, which are protected in the soil. This ability to "escape" adverse conditions allows a great many herbaceous plants to be distributed across the tropical/temperate boundaries so important to woody and epiphytic plants. Thus, southern Florida has an intriguing mix of native temperate flora that has been naturally "invaded" in geologically recent time by a great many tropical species due to unique and transitory circumstances of geography and climate.

ORIGINS OF THE FAUNA [183, 373, 436, 642]

The mobility of animal life makes it impractical to define species as "tropical" or "temperate" based on their tolerance of cold weather. The actual freezing point is less important for animals than it is for most plants because many can avoid exposure to adverse temperatures by seeking some form of shelter. In this respect, animals are like the marsh plants discussed previously. Nevertheless, patterns of distribution of individual species, or of groups of related species such as genera or families, make it possible to determine most origins.

The distributions and relationships of the native animals in southern Florida show that the terrestrial and freshwater species are almost completely of temperate, North American origin. However, fossil evidence shows that Florida has been invaded by South American land animals several times over the past 3 million years since the land bridge through Central America came into existence, only to perish during times of cooler, drier climate. In today's native fauna, forms that can disperse through the air (flying or drifting) show a mixture of origins. Some are from the tropics, but most are of temperate origin, except for the isolated islands of the Florida Keys where, among spiders for example, tropical species predominate. Only in the marine animal life of the region is there a strong overall, natural tropical heritage.

Beginning with invertebrates and continuing with fishes, amphibians, reptiles, mammals, and birds, origins of South Florida's fauna, as well as the roles played by these animals in the ecosystem, are revealed in the following chapters. While the biogeographic approach is useful for organizing knowledge, it is the ecological functions performed by animal life that are critical to the performance of the Everglades ecosystem.

13 Invertebrates

Invertebrates do not have a backbone and include all animals except fishes, amphibians, reptiles, mammals, and birds. While non-biologists may think that these exceptions rule out most forms of animal life, such is not the case. Some familiar invertebrate groups are: sponges, jellyfish, corals, worms, clams, snails, octopuses and squids, spiders, scorpions, insects, crabs and lobsters, shrimp, starfish, and sea urchins. This list only scratches the surface by naming well-known groups, leaving out such creatures as rotifers, bryozoans, copepods, ostracods, milli-pedes, and arrow worms, among innumerable others. Invertebrates include such a wide variety of animal life that all conceivable modes of dispersal apply to invertebrates of one kind or another.

MARINE INVERTEBRATES [112, 302, 340]

Most marine invertebrates have planktonic larval forms that are dispersed by ocean currents. A tropical starfish need not crawl from Cuba to Florida's tropical waters through the depths of the Straits of Florida. Such migration would be impossible for an adult, but the larval form can make the trip passively. In fact, this type of dispersal accounts for the prevalence of tropical marine invertebrates in the waters surrounding southern Florida, including the famous reef-building corals. At the same time, larval dispersal greatly complicates studies of the effects of overfishing on invertebrates, such as the highly valued Caribbean spiny lobster (*Panulirus argus*), which is the tropical epicurean equivalent of the American lobster of the northeast's cold marine waters. Effective regulation of fishing requires an understanding of larval move-ments. Studies have shown that the spiny lobster population of Florida's southeast coast is established by larvae that arrive from various local, as well as distant Caribbean, sources via the Florida Current. It is also interesting that in 1993, when huge floods occurred in the Mississippi River system, larval transport was greatly affected in the Gulf of Mexico and was apparently the cause of abundant slipper lobster and sand crab larvae from the northern Gulf of Mexico arriving in the Florida Keys. The important generalization is that varying conditions can make large differences in the marine invertebrate larval recruitment that establishes adult populations. [143, 665, 666, 667]

Not all southern Florida marine invertebrates are strictly tropical. Two commercially important examples are the blue crab (*Callinectes sapidus*) and pink shrimp (*Farfantepenaeus duorarum*) (often mixed with the similar pinkspotted shrimp, *F. braziliensis*). Blue crabs are found from Cape Cod all the way southward around Florida and the Gulf of Mexico. Pink shrimp are found as far north as Chesapeake Bay but are much more common in the warmer waters from the Carolinas southward. Florida Bay and the mangrove swamps in Everglades National Park are extremely important nursery grounds for the huge population of pink shrimp harvested near the Dry Tortugas, a group of islands about 70 miles west of Key West. Furthermore, freshwater discharges from the Everglades through the Shark River Slough have been shown to be important to the success of such shrimp harvests.[81, 83]

FRESHWATER* INVERTEBRATES

A list of some freshwater Everglades invertebrates is presented in Table 13.1. Most freshwater invertebrates of southern Florida are not from the tropics, but rather from the freshwaters of North America. The Everglades does not have a great diversity of freshwater invertebrates due to its limited habitats and to its stressful conditions of seasonal drying and its high summertime temperatures, which many temperate species cannot tolerate. A total of 174 different kinds of invertebrates have been found in the Everglades, and their importance in food chains of the region has recently gained attention.[88, 362, 405, 477, 478] They include: oligochaete worms (small, aquatic relatives of earthworms), which live in sediments and feed on sediment debris; leaches, which are mostly predatory on other invertebrates and occasionally parasitic on vertebrates; and many crawling and free-swimming clams, snails, crustaceans, and insects. Most aquatic insects are larval forms that emerge to become adults living in the air. Some important invertebrate examples follow.

FLORIDA APPLESNAIL

The Florida applesnail** (Figure 13.1) is an important freshwater mollusk in the Everglades. This large, globose, dark brown snail, whose adults measure over two inches in diameter, is found in Florida's wetlands, lakes, and rivers. It is restricted to the warm water of spring runs in the northern part of the state. Despite its apparent tropical affinity, fossils show that the applesnail has survived continuously in peninsular Florida since Pliocene time.[593]

The applesnail is peculiar in its dual ability to extract oxygen from water using gills, and from air using the equivalent of a lung. To breathe air, it crawls up a plant stem to a point near the surface, extends a tube called a siphon, pumps air in and out of a chamber with a noticeable rocking motion, and then returns to its underwater browsing on algae. Because of their air-breathing habit, they avoid open water (lacking plant stems needed to reach the surface) and depths greater than about 20 inches. The applesnail's eggs are deposited in masses on plant stems at night, typically six to eight inches above the water line. Initially pink, they turn white as the shells harden, and are most often seen in summer. The eggs hatch in about two weeks and are intolerant of submergence except within a few days of hatching. This sensitivity provides water managers with an important criterion in raising water levels where important snail populations occur — a maximum rise of about a half-inch per day can be tolerated, but significantly faster rates will submerge unhatched eggs, killing most of them. On the other end, baby applesnails must enter water immediately upon hatching. If the water level has receded below the soil, they die quickly. Although not well documented, adults can temporarily survive drydowns by aestivating within damp algal mat or mud. As a result of the various water-level tolerances, Florida applesnails are most abundant in or very near shallow water of long hydroperiod and with abundant emergent vegetation.[137, 138, 139, 478, 608]

The Florida applesnail is an important topic in the Everglades, due to its predation by a variety of wildlife including young alligators and numerous birds. The most publicized example is the snail kite, a hawk-like bird that feeds exclusively on the applesnail and is thus completely dependent upon water levels that maintain the snail's habitat. The limpkin (Figure 13.2), a wading bird related to cranes, is also heavily dependent on the applesnail.[307]

SEMINOLE AND MESA RAMS-HORNS

Much smaller than the Florida applesnail, the Seminole rams-horn (*Planorbella duryi*) is common throughout The Everglades. Its shell is a flattened coil but variable in shape, reaching a maximum

---

* Fresh water is defined as having salinity (saltiness) low enough that there is no appreciable sea water mixed into it. Sea water contains about 3.5 percent salt by weight. Ecologists usually express salinity in parts per thousand (ppt), so sea water is about 35 ppt. The upper limit of salinity normally used to define fresh water is 0.3 ppt.[323] Between these values, water is termed *brackish*, typical of estuaries and coastal wetlands.

** This "standardized" name follows reference 606, but "apple snail" (two words) is the normal name in Everglades literature.

TABLE 13.1
Representative freshwater invertebrates of the Everglades.

Common name[a]	Scientific name[a]
segmented worms – oligochaetes	
unnamed aquatic-sediment worm	*Bratislavia unidentata*
unnamed aquatic-sediment worms	*Dero* spp.
segmented worms - leaches	
unnamed leach	*Mooreobdella* sp.
snails (family)	
cockscomb hydrobe (Hydrobiidae)	*Littoridinops monroensis*
carib fossaria (Lymnaeidae)	*Fossaria cubensis*
mimic pondsnail (Lymnaeidae)	*Pseudosuccinea columella*
physa (Physidae)	*Physella* spp.
Florida applesnail (Pilidae)	*Pomacea paludosa*
ghost rams-horn(Planorbidae)	*Biomphalaria havanensis*
Seminole rams-horn (Planorbidae)	*Planorbella duryi*
mesa rams-horn (Planorbidae)	*Planorbella scalaris*
crustaceans (group)	
Everglades crayfish (decapods)	*Procambarus alleni*
slough crayfish (decapods)	*Procambarus fallax*
riverine grass shrimp (decapods)	*Palaemonetes paludosus*
scud or side-swimmer (amphipods)	*Hyalella azteca*
copepod (harpacticoids)	*Cletocamptus deitersi*
copepod (cyclopoids)	*Microcyclops rubellus*
insects – odonates (group)	
American bluet (damselflies)	*Enallagma* sp.
Rampur's forktail (damselflies)	*Ischnura ramburii*
common green darner (dragonflies)	*Anax junius*
eastern pondhawk (dragonflies)	*Erythemis simplicicolis*
blue dasher (dragonflies)	*Pachydiplax longipennis*
Halloween pennant (dragonflies)	*Celithemis eponina*
four-spotted pennant (dragonflies)	*Brachymesia gravida*
insects – mayflies (family)	
mayfly (Baetidae)	*Callibaetis floridamus*
mayflies (Caenidae)	*Caenis diminuta*
insects – hemipterans or true bugs (family)	
water bug (Belostomatidae)	*Belostoma* spp.
giant water bug or electric-light bug (Belostomatidae)	*Lithocerus americamus*
creeping water bug or alligator flea (Naucoridae)	*Pelocoris femoratus*
insects – beetles (family)	
predaceous diving beetles (Dytiscidae)	*Celina* spp.
whirligig beetle (Gyrinidae)	*Gyrinus elevatus*
water scavenger beetles (Hydrophilidae)	*Berosus* spp.
insects – dipterans or flies/mosquitoes/midges (family)	
punkies or biting midges (Ceratopogonidae)	*Dasyhelea* spp.
midge (Chironomidae)	*Goeldichironomus holoprasinus*
midge (Chironomidae)	*Larsia decolorata*
midge (Chironomidae)	*Polypedilum trigonus*
midge (Chironomidae)	*Pseudochironomus* sp.
green midges (Chironomidae)	*Tanytarsus* spp.

[a] Listings, and common and scientific names follow references 179, 270, 362, 459, 477, 606, and 653.

FIGURE 13.1 Florida applesnail and its eggs on a cattail leaf. (Photo by T. Lodge.)

FIGURE 13.2 A limpkin dines on a Florida applesnail. (Photo by R. Hamer.)

of about three quarters of an inch in diameter, larger than the also common and similar mesa rams-horn (*P. scalaris*), which has a slightly raised, flat-topped spire. These two species are endemic to the Florida peninsula and confined to freshwater. Their shells are good markers for soils developed in freshwater.[*, 593]

The Seminole rams-horn, like the Florida applesnail, grazes on algae. In turn (and together with young applesnails), it is an important prey of the redear sunfish, a snail "specialist" that earned the vernacular "shellcracker." This snail can survive for an undetermined time by burrowing in moist soil. Because it can survive in very limited water in underground limestone cavities, it often is found in short-hydroperiod marshes (especially where solution holes retain water) as well as more permanently flooded habitats.[252, 358] Furthermore, it can disperse rapidly during high water by "riding" on the underside of the surface. This ability, best exhibited by smaller individuals, involves attachment of the snail's foot to the water surface. Wind-blown currents can then propel them hundreds of times farther than their own slow crawling would allow. Other snails, including juvenile Florida applesnails, also travel by this means, and large numbers of snails are sometimes observed along the windward sides of tree islands or sawgrass patches in the Everglades, long distances from their apparent sources. It is important, however, that the water be deeper than just a few inches for this mode of transport to be effective, and that the surface be free of restrictions. Dense sawgrass, for example, greatly limits this kind of passive dispersal.

CRAYFISH

About 300 species of crayfish, which look like miniature facsimiles of the marine American lobster, occur in the freshwaters of North America. Florida hosts about 50 species, but only five occur in the southern half of the peninsula.[**] Crayfishes are highly important in food chains and documented to be preyed upon by at least 40 species of vertebrates in the Everglades, including largemouth bass, pig frogs, young alligators, and all wading birds, particularly the white ibis. Interest in their importance has spawned a great amount of recent research.[2, 3, 4, 207, 270, 277, 494, 620]

Two nearly identical species occur in the Everglades — the Everglades crayfish (Figure 13.3) and the slough crayfish. The Everglades crayfish is adapted to the region's alternating wet and dry seasons. It lives in underground burrows during the dry season and browses on algae and small invertebrates over the marsh bottom during the wet season. Females carry eggs while still underground at the end of the dry season, and young crayfish populate the newly flooded marshes at the inception of the rainy season. This timing makes them an important prey for wading birds in the early wet season in short hydroperiod areas where there is a considerable time lag in the growth of other prey species such as mosquitofish. The slough crayfish inhabits deeper, more permanent water, often sharing this habitat with the Everglades crayfish. It is probable that the slough crayfish has only recently invaded the Everglades, taking advantage of canals for dispersal and more permanent deeper habitat in the impounded water conservation areas.[620]

RIVERINE GRASS SHRIMP AND SIDE-SWIMMER AMPHIPOD

Two other abundant crustaceans in the Everglades are the riverine grass shrimp[***] and the side-swimmer amphipod, both considered "keystone species" because of their importance in food chains.[405] The grass shrimp belongs to the genus *Palaemonetes*, which has many species. Several

---

[*] The term "Helisoma marl"[128] refers to these species. The genus *Helisoma* is closely related and very similar to *Planorbella*, and older literature includes our species of *Planorbella* in the genus *Helisoma*.

[**] Early literature recognizes only four species from the southern peninsula.[276] In the late 1960's, a fifth species was discovered from fragments that appeared in well water drawn from the Biscayne Aquifer, the specimens having been dismembered by the pump impellers. In the 1990's, living specimens were finally collected. Known as *Procambarus milleri*, it is a colorful, peaceful, subterranean species and has become popular in the aquarium trade.[481] There are many cave-dwelling and subterranean crayfish species known from other regions of the country, including northern Florida.

[***] This "standardized" name follows reference 653, but "freshwater prawn" is its usual name in Everglades literature.

FIGURE 13.3 A large adult Everglades crayfish. (Photo by T. Lodge.)

occur in South Florida's estuarine and marine habitats, but the riverine grass shrimp is the only abundant species in the freshwater habitats of the Everglades region. It resembles the edible pink shrimp, except that it grows only about an inch long and is so transparent that it is almost invisible. Unlike the Everglades crayfish, it cannot survive a complete drydown, and therefore its populations are low in short hydroperiod wetlands, except near refugia such as alligator holes. In more permanently flooded habitats, such as sloughs, however, its abundance can exceed 100 individuals per square yard. Riverine grass shrimp feed primarily by shredding dead plant material,[405] and, in turn, nearly all predatory fishes eat them. However, except for the white ibis and little blue heron, which catch significant numbers, most wading birds do not feed on them because of their small size and cryptic guise, yet through the food chain, the riverine grass shrimp is a highly important link to all fish-eating birds in freshwater habitats.[357, 478]

The common Everglades species of side-swimmer or scud (technically an amphipod) is a quarter-inch crustacean that superficially resembles a miniature shrimp but lacks the extended tail. There are about 90 species of freshwater amphipods in the United States, but only *Hyalella azteca* occurs in the Everglades. Scud populations can be as abundant as 9000 per square yard, but are usually only a few percent of that density. They feed chiefly by scraping periphyton from surfaces, and in turn are eaten by other aquatic invertebrates and small fish.

Aquatic Insects

Aquatic insects are also important in the food chains of the Everglades, as they are in all freshwater habitats. Numerous insects (e.g., water scavenger beetles and giant water bugs) are primarily aquatic but have retained the ability to fly and will relocate if their environment dries up. Most species that inhabit aquatic habitats, however, live in the water only as larvae and emerge to live in the air as adults. Prominent examples are mosquitoes, mayflies, damselflies, dragonflies, and a large number of midges, tiny insects related to flies. Their dual lifestyles represent a significant transfer of energy from aquatic to terrestrial environments, where flying insects are important in the diets of many birds.[459, 478]

Dragonflies (also called darning needles, mosquito hawks, and snake doctors) represent highly visible and interesting examples of the insects associated with aquatic habitats. Their larvae are predatory, feeding on anything they can catch and overpower, including other aquatic insects (and sometimes their own kind), worms, snails, and occasionally even small fish and polliwogs. In turn, they are eaten by larger fishes and by certain wading birds, notably the little blue heron and the white ibis. Adult dragonflies feed on insects. They splendidly consume large numbers of mosquitoes but also eat larger insects including beetles, wasps, and other dragonflies. The eastern pondhawk,* which is two inches in size, is a voracious example that is sometimes cannibalistic.

Of the 86 species of dragonfly that reside in the Florida peninsula, only about a dozen reproduce in the Everglades, primarily because the limited kind of habitat (marsh) is not suitable for the species that require the moving water of streams in their larval stages, or forested habitat as adults. The Everglades species are mainly temperate, and common examples are: the Halloween pennant (Figure 13.4); the powerful green darner, which is unusual in that it is a migratory species; the four-spotted pennant; and the eastern pondhawk. Within all the habitats of southern Florida, about half of the dragonflies are tropical and half are temperate. Because they are strong fliers, long distances do not limit most adult dragonflies in colonizing new areas, at least for those species that do not require shade. The important factor that does limit their distribution is suitability of the aquatic habitat for their larvae.[179, 180, 458]

Among the many insects listed in Table 13.1, some require additional comments. Hemipterans (true bugs) are predatory, attacking prey with forelegs that hold the prey and a piercing mouth that injects digestive enzymes and toxins before sucking out the prey's body fluids. This feeding style inflicts occasional consequences on humans: the sting-like bite of several species is painful. Careless handling of specimens in a dip net, for example, can be most unpleasant. The quarter-to-half-inch-long "alligator flea" has a bite that feels like a hot needle. Its vernacular comes from human perspective (including the author's): it does not bite alligators. The nearly three-inch "electric-light bug" — so named for the occasional collection in huge numbers beneath bright lights — purportedly has a very painful bite that is normally inflicted on its prey, typically small fish and pollywogs (tadpoles).

Among the species of Diptera (flies, mosquitoes, and their relatives) are many species that have small, worm-like aquatic larvae, which are important in aquatic food chains as prey for fish and other aquatic invertebrates. As adults, the chironomids or midges are small flying insects that do not bite. The ceratopogonids, however, include the tiny no-see-ums or punkies, which, as adults, feed on the blood of people and other vertebrates. No-see-ums are not prevalent in the freshwater Everglades, but mangrove swamps and salt marshes may support intolerable numbers, most obnoxious in the early morning and evening, or during cloudy weather.[290]

TERRESTRIAL INVERTEBRATES

SPIDERS

Although they are terrestrial invertebrates, many spiders utilize a means of dispersal that is more normally associated with plants, namely wind. Young spiders of many species are known to climb to a high point on a plant or obstacle and release a length of silk into the air. When the trailing silk is long enough, a breeze can pull the spider into the air. The process is called ballooning, and it accounts for the occurrence of many tropical spiders in southern Florida. Most of species in the Florida Keys are of tropical, West Indian origin and probably arrived there by ballooning. The southern mainland hosts a mixture of tropical and temperate North American spiders.[338, 372]

---

* Florida's dragonflies have been assigned "English names," in distinction to "common names." Few people distinguish between kinds of dragonflies (or most invertebrates for that matter), so specific common names have not evolved as they have, for example, for birds and fishes.[179]

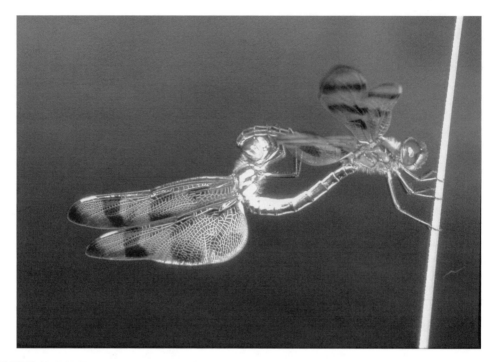

FIGURE 13.4 Halloween pennants, one of the common Everglades dragonflies, perched on sawgrass. This pair is mating — the female at left, her neck held by "tweezers" on the tip of the male's abdomen is taking sperm from the male's second abdominal segment. Seemingly awkward, the procedure obviously works: dragonfly fossils appear in coal deposits laid down ancient swamps. (Photo by T. Lodge.)

Scorpions

Worldwide, most scorpions are found in warm, dry climates, but the hundreds of species range from the tropics to cool temperate regions. Two scorpions occur in southern Florida. Both can deliver painful but not dangerous stings (unlike an Arizona species that is potentially fatal) from their large, tail-tip stinger, which is used for stinging prey as well as defense. Their relatively delicate crayfish-like pinchers are harmless and used for holding prey, such as insects and spiders. Named for their genus, the larger is the brown centruroides (*Centruroides gracilis*), which lives on the ground under leaf litter, stones, and logs. It is dark brown and grows to about five inches long. The smaller *Centruroides hentzi* (less than three inches) is light tan and commonly lives under the loose bark of dead trees. Ecologists and arborists working with dead trees sometimes experience a Hentz's centruroides falling into their shirt — it is not fun.[338, 348]

Insects

Dispersal of tropical insects to southern Florida has accounted for the presence of many of the species. Flying insects are frequently carried by storms, and a few actually migrate from location to location. However, even though the wind may assist tropical species in their dispersal to southern Florida, other factors determine whether or not they can become established. Several butterflies demonstrate cases in point.

Butterflies

The zebra butterfly (*Heliconius charitonius*) is of tropical origin, but it occurs widely throughout southern Florida and reaches northern Florida during summer, where it may overwinter in mild years. This handsome, slow-flying species has black wings with bold yellow stripes. Its distinctive elongated wings give it a much greater wingspan than is typical for its size among North American butterflies, although this feature is common among related tropical butterflies of Central and South America. Its occurrence in Florida is based on the presence of several species of tropical passion-flower vines. Butterflies as a group depend upon specific kinds of food plants for their larval development, and the caterpillars of the zebra butterfly feed on passionflower leaves. The close relationship of the butterfly to this food plant goes beyond nourishment. Passionflowers contain a bitter substance that the caterpillars accumulate without harm. The caterpillars, and the butterflies in turn, are then unpalatable to potential predators such as birds and lizards, which reject them after their first taste. The relatively tough butterflies often show the wear of an attempted taste or two, with older specimens often having frayed wings.[289, 572] The biogeographic principle is that some species cannot relocate without the presence of another species. In this case, the zebra butterfly could not have successfully invaded Florida without the previous establishment of its food plant.

Several other tropical butterflies are present in southern Florida, due to the presence of certain tropical plants. In a few cases, the rarity of the food plant accounts for very limited distribution of the respective butterfly. Examples are the endangered Schaus' swallowtail (*Heraclides aristodemus*), which feeds only on sea torchwood (*Amyris elemifera*), tropical hammock tree, and the rare Florida atala (*Eumaeus atala*), which feeds on coontie, a type of cycad that superficially resembles a fern. Although previously abundant, the coontie is now highly restricted in distribution due to land development, and thus the butterfly has become rare.[161, 206]

FLORIDA TREE SNAIL

The Florida tree snail (*Liguus fasciatus*) (see Color Figure 10), a tropical species, is a terrestrial mollusk of great interest in Florida. It is restricted to the extreme southern portion of Florida, but it also occurs in Cuba. It feeds on small epiphytes, such as lichens, and on tropical hammock trees, usually those with smooth bark such as wild tamarind. These large snails grow to a length of about two and a half inches and are highly variable in coloration, which is the basis for the keen interest among shell collectors and other observers. As many as 50 distinct varieties have been found, ranging in color from pure white to various banding and streaking patterns of green, yellow, pink, and brown, to individuals almost completely colored in brown and yellow tones.[161, 206, 275]

The Florida tree snail undoubtedly originated in the tropics, but how it invaded Florida is a mystery. Perhaps the explanation lies in its ability to aestivate through several months of the dry season by cementing its shell closed against the bark of a tree. It is possible, for example, that it could travel on a large tree limb that broke off in Cuba and floated to southern Florida with its branches above the water. The exact circumstance that allows a tree snail to land successfully by this rafting mode requires considerable imagination.

IMPORTANCE OF INVERTEBRATES

Because of their lower position in the food chain, invertebrates are of great importance in the diets of many Everglades predators, including fishes, amphibians, young alligators and crocodiles, wading birds, otters, and many others. Important studies have only recently begun regarding the roles of invertebrates in the Everglades, as cited earlier in this chapter. The effects of nutrient enrichment on their roles in the Everglades has also received attention.[479, 480] There are so many kinds of invertebrates that ample opportunity awaits future students of the Everglades ecosystem.

14 Freshwater Fishes*

I first visited Everglades National Park in September 1966 during the height of the annual rainy season. Out of curiosity, I stopped at one point to observe water flowing through a culvert under the road to Pa-hay-okee Overlook. The area was teeming with fishes, all easily visible in the clear, shallow water. I was fascinated watching gar, bass, several kinds of sunfish, and innumerable smaller species, all jockeying about in the sunlight. The sight was a delight to a budding ichthyologist from Ohio who was beginning graduate school in Florida.

Freshwater fishes are a mainstay of the Everglades food chains. They provide the diet for alligators, otters, wading birds, and other predators. The historic Everglades succeeded as a wildlife exhibition because its annual cycles promoted the growth and availability of a large freshwater fish biomass (the total weight of living matter in a given volume or area of environment, e.g., pounds per acre of fish). The representative species that occur in freshwaters of the Everglades are listed in Table 14.1.

PRIMARY FRESHWATER FISHES

Among freshwater fishes, there is a wide range in tolerance to salinity. Species that are intolerant of salty water are unable to stray into estuaries and are termed *primary freshwater fishes* by ecologists. Their dependence on freshwater restricts their ability to invade and colonize new areas, limiting their travels to freshwater routes. Where two river systems have no freshwater connection, primary freshwater fishes cannot move between the systems without help, such as a flood, a geologic event causing a stream to change course, or the hand of man.

All of the native primary freshwater fishes of southern Florida arrived from temperate North America after the emergence of the region from the sea. None is from the American tropics. Largemouth bass, bluegills, redear sunfish, golden shiners, and yellow bullheads are common in the Everglades. These species occur widely in the United States, even as far north as the Great Lakes. On the other hand, relatively few of the many primary freshwater fishes in the Southeast have been able to colonize the Florida peninsula, despite convenient river and interconnecting wetland corridors. Only 20 species, which are tolerant of warm, slowly moving waters, have reached southern Florida naturally. Among them, several are distinctive Florida varieties, often recognizable by color patterns differing from their northern counterparts.[79]

The Florida largemouth bass, found only in peninsular Florida, is the most famous of the primary freshwater fishes of the region. It is a subspecies of the largemouth bass (Figure 14.1), which occurs naturally in North America from the Great Lakes southward to the Gulf and lower Atlantic coasts and is now widely stocked around the world. The largemouth bass is a spectacular game fish, and the Florida subspecies is especially renowned. It frequently grows to 10 pounds and occasionally exceeds 20, averaging substantially larger than the northern subspecies. Because it is more wary, the Florida subspecies is also more difficult to catch. Research has indicated that this behavior may have a genetic basis. The "catch per unit effort" is lower for Florida bass than for the northern subspecies, even when the two are raised in identical, side-by-side ponds.[668] Perhaps this innate wariness is an adaptation to the prevalence of large predators — alligators, herons, and egrets — in Florida's extensive wetland habitats.

---

* The main sources for this chapter are references 165, 354, 356, 360, and 604.

TABLE 14.1
Representative fishes found in freshwaters of the Everglades

Common name[a]	Scientific name[a]	Ecological group[b]
Florida gar	*Lepisosteus platyrhincus*	2
bowfin (mudfish)	*Amia calva*	1
tarpon	*Megalops atlanticus*	3
American eel	*Anguilla rostrata*	3
golden shiner	*Notemigonus crysoleucas*	1
coastal shiner	*Notropis petersoni*	1
lake chubsucker	*Erimyzon sucetta*	1
yellow bullhead (butter cat)	*Ameiurus natalis*	1
tadpole madtom	*Noturus gyrinus*	1
walking catfish	*Clarias batrachus*	1-i
sheepshead minnow	*Cyprinodon variegates*	3
golden topminnow	*Fundulus chrysotus*	2
marsh killifish	*Fundulus confluentus*	3
Seminole killifish	*Fundulus seminolis*	2
flagfish	*Jordanella floridae*	2
bluefin killifish	*Lucania goodie*	2
pike killifish	*Belonesox belizanus*	2-i
eastern mosquitofish	*Gambusia holbrooki*	2
least killifish	*Heterandria formosa*	2
sailfin molly	*Poecilia latipinna*	3
brook silverside	*Labidesthes sicculus*	3
common snook	*Centropomus undecimalis*	3
Everglades pygmy sunfish	*Elassoma evergladei*	1
bluespotted sunfish	*Enneacanthus gloriosus*	1
warmouth	*Lepomis gulosus*	1
bluegill (bream)	*Lepomis macrochirus*	1
dollar sunfish	*Lepomis marginatus*	1
redear sunfish (shellcracker)	*Lepomis microlophus*	1
spotted sunfish (stump-knocker)	*Lepomis punctatus*	1
largemouth bass	*Micropterus salmoides*	1
swamp darter	*Etheostoma fusiforme*	1
oscar	*Astronotus ocellatus*	2-i
Mayan cichlid	*Cichlasoma urophthalmus*	3-i
blue tilapia	*Oreochromis (Tilapia) aurea*	2-i
spotted tilapia	*Tilapia mariae*	2-i
striped mullet	*Mugil cephalus*	3

[a] Common and scientific names follow reference 508, except those in parentheses, which are common Florida vernaculars.[96] "Bream," here used for the bluegill, is often applied to any similarly shaped sunfish, such as the redear sunfish, and is therefore equivalent to the lay term, "panfish."

[b] Ecological group refers to salinity tolerance described in the text: 1 = primary freshwater fish; 2 = secondary freshwater fish; 3 = peripheral freshwater fish. Those numbers followed by "i" signify introduced, exotic species

SECONDARY FRESHWATER FISHES

Freshwater fishes with more tolerance to salt water and the ability to live there temporarily, are termed *secondary freshwater fishes*. They can colonize new areas by swimming from the mouth

FIGURE 14.1 Predator takes predator: a great white heron eats a young largemouth bass at the Anhinga Trail, Everglades National Park. (Photo by T. Lodge.)

of one river system into the next through interconnecting estuaries, salty coastal wetlands, or even seawater. This ability allows them to colonize faster than primary species.

Most of the eight species of secondary freshwater fishes in southern Florida are the killifishes and their relatives. Generally small and colorful, many of these fishes make good aquarium specimens. They originated in the American tropics and long ago spread into North America where they evolved separately and produced many North American species. Although their distant ancestors lived in the tropics, the southern Florida species came from temperate North America. Examples include the inch-long eastern mosquitofish (the most abundant and ubiquitous fish in Florida, sometimes reaching nearly 100 juveniles and adults per square yard in the Everglades), the least killifish (North America's smallest fish and also abundant), the golden topminnow, and the flagfish (named for its beautiful pattern of red stripes resembling the United States flag).

PERIPHERAL FRESHWATER FISHES

The last group, the *peripheral freshwater fishes*, may live part or most of their lives in marine waters but regularly move or stray into freshwater. While most species in this group remain near saltwater, some can be found anywhere in southern Florida's freshwater habitats. The sailfin molly, a popular aquarium fish, is such an example; however, it is most abundant in brackish, coastal waters. Other examples of the more than 80 South Florida species in this group include the American eel, tarpon, snook, several killifishes (other than those categorized as secondary), needlefishes, mullet, several gobies and sleepers, and occasionally even bull sharks (mostly juveniles). In the historic Everglades, some peripheral species may have been able to reach far into the Everglades through the Shark River Slough, even as far as Mullet Slough of the Big Cypress Swamp, which was possibly named for the occurrence of striped mullet. Today, some of these species, notably tarpon and mullet, reach far inland through canals.

Many peripheral freshwater fishes are tropical species, being intolerant of cold weather. Tarpon and snook are often killed when confined to shallow waters that cool quickly during winter cold fronts, and many a sad fisherman has seen a trophy-sized snook floating dead following a severe cold spell.

THE FLORIDA GAR

The strangest fish in the Everglades is the Florida gar, whose bizarre appearance includes sharp, needle-like teeth that fill a long snout. Young gar have an intriguing, highly variable pattern of numerous dark spots and patches on an olive to yellow, long slender body. They darken with age so that adults appear mostly dark brown, especially when seen from above. Several kinds of gars exist in eastern and central North America, some of which are extremely large. The aptly named "alligator gar" (*Lepisosteus spatula*) is occasionally mistaken for an alligator and occurs from the lower Mississippi drainage basin to the rivers of the western panhandle of Florida.[229] Only the relatively small Florida gar, seldom as long as two feet, lives in the Everglades. The much larger longnose gar (*Lepisosteus osseus*) historically only occurred in Lake Okeechobee and northward but has occasionally been found in canals in the Everglades water conservation areas. As with all gars, the Florida gar is predatory and is adept at catching smaller fish from schools by using a fast sideways snap. It is also capable of catching individual prey, pursuing them along the bottom or in dense tangles of vegetation. Using a slow, stealthy approach, this technique is effective on fish and grass shrimp.[292]

Florida gar are sometimes seen in huge numbers, which is the result of low water confining individuals from the expanses of Everglades marshes into limited aquatic habitat during the dry season. At these times, gar become prey for the alligator. The sight of a gar held in an alligator's jaws looks prehistoric (Figure 14.2). In fact, gar have changed little from ancestors that dominated the earth's waters when the dinosaurs flourished, and they have primitive, interlocking scales that differ greatly from most fish. They also have the dual ability to breathe air and water, and can be observed regularly rising to the water's surface to renew the air in their swim bladders. The Florida gar should not be confused with the similarly shaped, but unrelated needlefishes, which are marine (though three species commonly enter freshwater in upper estuarine locations). Needlefishes are greenish, bluish, or silvery and have a translucent appearance, in marked contrast to the darker and opaque Florida gar.

INTRODUCED FISHES

The Everglades now hosts several firmly established exotic species, most of which have "escaped" from the tropical fish trade. It was in the mid-1970's that I first saw one within Everglades National Park: a walking catfish in the jaws of an alligator at the Anhinga Trail. Today, introduced kinds are prevalent there, with the blue tilapia, oscar, and Mayan cichlid being among the most visible. Five introduced species are regularly seen at the Anhinga Trail, not including the walking catfish, which usually goes unseen — except by anhingas, cormorants, herons, and egrets. Since the late 1970's countless walking catfish have been observed in the beaks of these birds.

FRESHWATER FISHES AND THE FOOD CHAIN

Food-chain relationships of the Everglades fishes (also see Chapter 20) are important in under-standing the freshwater ecology of the region. Herbivorous species include the golden shiner, sailfin molly, least killifish, and flagfish (Figure 14.3). Other fishes, such as golden topminnows, marsh killifish, mosquitofish, bullheads, redear sunfish, and bluegills feed on invertebrates, many of which

FIGURE 14.2 An American alligator eats a large Florida gar at the Anhinga Trail, Everglades National Park. (Photo by T. Lodge.)

graze on plants. Adult Florida gar, warmouth, and largemouth bass feed principally on other fish. In turn, all of these fishes are the vital food supply for other predators (Figure 14.4).[292]

The important role that fishes play in the food chain can only occur where fishes are able to maintain populations, which is a problem in a region where aquatic habitats dry seasonally. The single species known to survive a complete drydown is the marsh killifish.* Actually, adults perish, but the buried eggs survive even during a lengthy drydown and hatch rapidly when flooding returns. Other species are only able to repopulate the Everglades marshes from an aquatic refuge, such as an alligator hole or a solution cavity in the rocky glades (also a canal in the modern Everglades). Mosquitofish are particularly adept at survival in very limited space, possibly even using crayfish burrows, so that they seem to come from nowhere upon the return of summer rains. For small species such as the mosquitofish, only a few inches of water may suffice for their dispersal through the marshes from an aquatic refuge. Larger species such as the golden shiner, sunfishes, and Florida gar probably require closer to a foot of water depth, especially for spawning. However, little is known about depth requirements for marsh utilization by freshwater fishes, although recent research has begun to explore the mobility of fishes in marsh habitats.[604] For reestablishing populations, it is important for numerous species to have unrestricted access to the open expanses of marsh with the return of each wet season, because freshwater fish are the main prey of most wading birds and numerous other wildlife species in the Everglades. Freshwater fish populations are the key ingredient in the overall success of Everglades wildlife.[153]

---

* Observations indicate that the bowfin may aestivate in algal mat or mud upon drying in the Everglades, possibly explaining why individuals sometimes appear far from permanent water soon after reflooding. However, the species probably cannot withstand much desiccation, so it may only survive a short drydown.[165]

FIGURE 14.3 Common forage fishes of Everglades marsh habitats: sailfin molly (top), an adult male least killifish (below molly), golden topminnow (left center), flagfish (lower left), and eastern mosquitofish at right. The larger specimens are about an inch-and-a-half long. (Photo by T. Lodge.)

FIGURE 14.4 Predator takes predator: an anhinga swallows warmouth, a species that commonly invades shallow marsh habitat. (Photo by R.Hamer.)

THE FISHERMAN'S PERSPECTIVE

For a freshwater fisherman experienced in the eastern or central United States, the general impression in southern Florida is familiarity. Most of the freshwater species found in the interior are familiar, except perhaps the gar and now several introduced exotic species. The surprise is in the peripheral species: it is not unusual to be fishing for bluegills and see a three-foot tarpon swim past.

15 Marine and Estuarine Fishes*

While the Everglades is a freshwater habitat, its waters drain into and nourish the prolific estuarine and marine portions of Everglades National Park; the boundaries of which extend far out into Florida Bay and the Gulf of Mexico. Marine and estuarine fishes in these waters command abundant interest from aquarists, connoisseurs, and fishermen, but they are also an important food supply for many fish-eating birds. This supply does not form an annually concentrated biomass like the freshwater Everglades, but it is more consistently available on daily tidal cycles.[472]

DIVERSITY OF MARINE AND ESTUARINE FISHES

A vast number of marine and estuarine fishes occur in southern Florida. If the Florida Current and its deeper "counter-current" are included together with the nearby portions of the Gulf of Mexico and habitats of the various bays and estuaries around the mainland, the species number about 1000. If only near-shore waters are included, the number is well over 500, which is far greater than any other vertebrate group of southern Florida.

The richness of the marine fauna results from geographic setting and habitat diversity. The geographic setting in the edge of the tropics brings a commingling of tropical and temperate species to the waters of southern Florida. Tropical marine fishes have no problem in reaching the area. In contrast to freshwater fishes, most marine species have minute planktonic larvae that are dispersed by ocean currents, and the Florida Current has been the highway for the arrival of tropical marine fishes in southern Florida. If freshwater species began life as drifting plankton, many would be carried out to sea to perish. Adaptation has enabled most freshwater fishes to swim soon after hatching from eggs that stick to the bottom or to vegetation.

The marine and estuarine habitats of South Florida are highly diverse, including:

• Deep, open ocean waters of the Florida Current in the Straits of Florida
• Coral reefs along the Florida Keys
• Continental shelf waters, which are narrow on the east coast but very broad in the Gulf of Mexico
• Estuaries and mangrove swamps
• Shallow-water grass flats of bays and coastal waters
• Protected, shallow, coastal embayments prone to dry season hypersalinity

The number of species of fish that occur in each of these habitats differs considerably. The richest habitat is the coral reefs where a prominent tropical ichthyofauna is present. A study of Alligator Reef, located east of Everglades National Park, revealed 517 species.[587] The marine waters within the park include the last four habitats listed and harbor 247 species, 164 of which are strictly marine and 83 of which are the peripheral species that may enter freshwater. Diversity is lower in these inshore waters, mainly because relatively few tropical reef species find acceptable habitat there, and are unable to tolerate the rigors of changing temperature, salinity, and water clarity that characterize these waters.

---

* The primary sources for this chapter are references 341, 354, 509, 566, and 594.

GAME FISHES [587, 594]

Species of interest to Everglades National Park's sport fishermen are listed in Table 15.1 Biogeographically, these game species include both temperate and tropical types. However, several tropical species, such as tarpon and great barracuda, travel far north into temperate waters, especially in the summer. Tarpon and snook regularly enter freshwater. The gray snapper, known locally as the mangrove snapper, represents the mostly tropical snapper family, but this particular species is extremely widespread in inshore waters from southern Brazil to New York, including the entire Gulf of Mexico and the Caribbean. The goliath grouper (formerly called the jewfish) is a giant tropical grouper found in the tidal channels of Florida Bay as well as offshore reefs. This species, which was formerly depleted by overfishing but now protected and rebounding, can reach a length of eight feet and may weigh nearly 1000 pounds. The bonefish is a tropical species that is restricted to the warmer portions of Florida Bay near the tropical waters of the Keys.

Three members of the drum family are important temperate species in Everglades National Park. The black drum is the largest, but the red drum and the spotted seatrout (unrelated to freshwater trouts) are more numerous and among the most popular of all species with fishermen. All three occur around the entire Gulf coast and commonly up the Atlantic coast to Chesapeake Bay, with smaller numbers found farther north. The red drum (locally called redfish and known as channel bass along the Atlantic coast) grows considerably larger, to 90 pounds and nearly five feet long in more northern parts of its range. The redfish population in the shallow waters of the park is made up mostly of individuals two years old and younger and seldom as long as 30 inches. Like many other species, redfish use the marine and estuarine habitats of the area as a nursery, with larger individuals moving into deeper waters of the Gulf of Mexico.[596] Numerous studies of marine fishes have shown the importance of the nursery function of the area. The spotted seatrout spends its entire life cycle in the park's waters.[517]

Several migratory species occur in Everglades National Park. The Spanish mackerel is a temperate species that visits the park's waters in significant numbers in winter, it migrates from the northern Gulf of Mexico and from north of Cape Canaveral, where it resides during the summer.[302] It follows a pattern characteristic of a few other temperate species, which is the opposite of that exhibited by the tropical tripletail (*Lobotes surinamensis*). The tripletail drifts on warm currents near the surface and visits the park only in the summer. The tarpon also migrates, and large numbers of this spectacular game fish — often exceeding five feet long — move north for summer. Prime fishing for tarpon in the park occurs in April, at the beginning of the migration.

MULLET [525]

Mullet are abundant in Florida Bay as well as other marine, estuarine, and coastal freshwaters. Their importance is indicated by the fact that they are prey for every gamefish listed in Table 15.1, except bonefish. Mullet are often the first fish handled on a fishing trip: they are commonly purchased for bait.

Four species of mullet occur in South Florida, but the striped mullet is the most abundant. Mullets are essentially vegetarians, feeding by scraping algae and microorganisms from surfaces, by grubbing on the bottom, and by consuming surface film. They are an essential link in the food chain for game species and occur in enormous numbers in Florida Bay. The striped mullet is the species that commonly enters freshwater, and it occurs almost worldwide in warm temperate and tropical inshore waters, especially in and near estuaries.

TABLE 15.1
Fishes of importance in Florida Bay and adjacent marine and estuarine waters

Common name[a]	Scientific name[a]
game fishes:[b]	
ladyfish	*Elops saurus*
tarpon	*Megalops atlanticus*
bonefish	*Albula vulpes*
common snook	*Centropomus undecimalis*
goliath grouper (jewfish)	*Epinephelus itajara*
crevalle jack	*Caranx hippos*
gray (mangrove) snapper	*Lutjanus griseus*
sheepshead	*Archosargus probatocephalus*
spotted seatrout	*Cynoscion nebulosus*
black drum	*Pogonias cromis*
red drum (redfish)	*Sciaenops ocellatus*
great barracuda	*Sphyraena barracuda*
Spanish mackerel	*Scomberomorus maculates*
abundant nongame species:[c]	
scaled sardine	*Harengula jaguana*
Atlantic thread herring	*Opisthonema oglinum*
hardhead catfish	*Arius felis*
inshore lizardfish	*Synodus foetens*
gulf toadfish	*Opsanus beta*
hardhead halfbeak	*Chriodorus atherinoides*
silverstripe halfbeak	*Hyporhamphus unifasciatus*
redfin needlefish	*Strongylura notata*
goldspotted killifish	*Floridichthys carpio*
rainwater killifish	*Lucania parva*
mangrove mosquitofish	*Gambusia rhizophorae*
fringed pipefish	*Anarchopterus criniger*
dwarf seahorse	*Hippocampus zosterae*
gulf pipefish	*Syngnathus scovelli*
silver jenny	*Eucinostomus gula*
tidewater mojarra	*Eucinostomus harengulus*
pinfish	*Lagodon rhomboids*
striped mullet	*Mugil cephalus*
white mullet	*Mugil curema*
fantail mullet	*Mugil gyran*
code goby	*Gobiosoma robustum*

[a] Common and scientific names follow reference 508, except that vernaculars commonly used in Florida are included (in parentheses) for some game fishes.
[b] Selected from reference 594.
[c] Selected from reference 566.

IMPORTANCE OF THE REGION'S MARINE
AND ESTUARINE FISHES

To the sportsman, the marine and estuarine waters of South Florida support an interesting mix of tropical and temperate fishes, composed of prominent migratory populations as well as permanent residents. The mix is attractive, demonstrated by the large number of visitors to Everglades National Park whose only intention is a good day's fishing. However, the important ecological functions of the area are its support of fish-eating birds and its habitat as a nursery for a great many marine fishes, all based on the high productivity that arises from freshwater inputs from the Everglades as well as the neighboring Big Cypress Swamp ecosystems.[112] Wading birds utilize the mud flats of Florida Bay and the shallows of the mangrove swamps and other coastal habitats, and ospreys fish in the more open waters. All of these species prey heavily on the young of larger game fishes and on the high diversity of nongame species, the more common ones listed in Table 15.1. Wading birds normally feed during low tide, which renders prey trapped and accessible and affords a daily food source for the species, such as the roseate spoonbill,[368] that typically live in marine and estuarine habitats. The marine and estuarine areas also are a critical alternative in years when little is available in freshwater for those birds that typically utilize the Everglades.

16 Amphibians*

The biogeographic dispersal patterns of amphibians and primary freshwater fishes are very similar. The fact that all amphibians depend on fresh water for at least the first stages of their life cycles accounts for this similarity. Amphibian eggs are laid in freshwater, and larval stages, such as the frog's tadpoles, live only in fresh water. Some amphibians, such as the large eel-like sirens, are completely aquatic for their entire life histories. Others, such as toads and treefrogs, live on land or on vegetation as adults and are able to move from one stream system to another more easily than freshwater fishes. Nevertheless, tropical amphibians have not dispersed to Florida via the overland route, northward through Central America and around the Gulf of Mexico, or across salt water by rafting on floating objects through the West Indies. As a result, no tropical American amphibians have become naturally established in Florida.

Nearly all of the native amphibians of South Florida and three widely established exotic species are listed in Table 16.1. Of the native species, most occur throughout the southeastern states, although many are secretive and seldom seen. Only the pig frog is restricted to the extreme southeast, although its range includes all of Florida. Several native species occur in southern Florida as distinctive varieties of species that are common widely in the eastern United States. Examples include the peninsula newt, a subspecies of the common newt of the eastern United States, and the Everglades dwarf siren, an eel-like salamander that lives in dense tangles of floating aquatic vegetation.

The few types of salamanders that have been able to reach southern Florida are naturally adapted to marsh and swamp habitats. They include the peninsula newt and three eel-like species — the two-toed amphiuma, the greater siren, and the Everglades dwarf siren. Because of adverse environmental circumstances, none of the four-dozen woodland and brook salamanders of the eastern states has been able to reach southern Florida. This restriction also applies to most of the frogs and toads, with only a limited number of species able to find habitats conducive to their survival. Thus, southern Florida has few native amphibians, although many tropical species could live in the region if they could get there. With human assistance, several exotic species have become successful residents.

The strange ecology of the eel-like two-toed amphiuma (or congo eel) and greater siren require some comment. Both are wetland inhabitants and reach a very large size (slightly over three feet) and are predators primarily on clams, snails, and crayfish as well as other amphibians such as tadpoles. Sirens have red external gills on each side of the head and small forelegs with fingers, and they often swim in open water where they are occasionally seen. The amphiuma lacks such gills and has four thin vestigial, filament-like legs, each with a tiny pair of toes. It is secretive, moving slowly through dense aquatic vegetation and through soft sediments and mud. When their wetland habitats dry, both species will aestivate in the mud. Burrowing through mud helps these amphibian predators find crayfish and clams. Both species are sometimes caught by fisherman using worms, and they are difficult to unhook, partially because they writhe intensely, but also because they have a strong bite — no significant teeth but jaw strength that is not dangerous but unpleasant when clamped on a finger. Both species are hunted by certain water snakes, with the mud snake being the native Everglades specialist. The mud snake, named for its habit of hunting through the mud of bottom sediments for its prey, may grow over six feet long but still faces a fierce battle in subduing amphiumas and sirens, which are both capable of eating young mud snakes. Thus, there comes a point in the life of a mud snake where the battle could go either way — a tough way to make a living. The much smaller (eight-inch) Everglades siren is not nearly so formidable.

---

* The primary citations for his chapter are references 36, 59, and 409.

TABLE 16.1
Amphibians of the Everglades region[a]

Common name[b]	Scientific name[b]	General habitats[c]
two-toed amphiuma or congo eel	*Amphiuma means*	A, Sm
greater siren	*Siren lacertina*	A, Sm
Everglades dwarf siren	*Pseudobranchus axanthus belli*	Sm
peninsula newt	*Notophthalmus viridescens piaropicola*	Sh, Sm
pig frog	*Rana grylio*	A, Sl, Sm
southern leopard frog	*Rana sphenocephala*	A, H, M, Sl, Sm
Florida cricket frog	*Acris gryllus dorsalis*	A, Sh, Sm
green treefrog	*Hyla cinerea*	A, H, M, P, M, R, Sm
squirrel treefrog	*Hyla squirella*	A, H, M, P, R, Sm, Wp
Cuban treefrog — i	*Osteopilus septentrionalis*	A, H, M, P, Sl
little grass frog	*Limnaoedus (Pseudacris) ocularis*	Sm
Florida chorus frog	*Pseudacris nigrita verrucosa*	H, P
greenhouse frog – i	*Eleutherocactylus p. planirostris*	R
eastern narrowmouth toad	*Gastrophryne c. carolinensis*	H, M, P, Sm, Wp
southern toad	*Bufo terrestris*	H, P, Sm
oak toad	*Bufo quercicus*	H, P, Sm
marine (giant) toad – i	*Bufo marinus*	R

[a] Listings are primarily based on reference 409, with additions from reference 59.
[b] Common and scientific names follow reference 36 and 59. "i" signifies introduced species.
[c] General habitats key: A= artificial waters (canals/lakes) including shorelines; H= hammock; M=mangroves (low salinity); P=pine rockland; R=residential/urban; Sm=sawgrass marsh; Sh=solution hole; Sl=slough; Wp= wet prairie

The squirrel treefrog may be the most commonly encountered amphibian for visitors to the Everglades. Like the somewhat larger green treefrog (Figure 16.1), it can change color from green to brown, with several intermediate variations, including spots and blotches. During daytime, individuals are apt to crowd together in protected locations, often using man-made shelters. Their occasional loud, scratchy outbursts may mystify listeners. The location of the sound is elusive, and to the unfamiliar, indiscernible. The squirrel treefrog and green treefrog, both highly photogenic as amphibians go, were once common in the Miami area, but have almost completely disappeared. Their elimination was probably due to predation by the much larger, introduced Cuban treefrog, which is well established in urban areas and has recently invaded a variety of natural habitats. A similar situation has apparently resulted in the elimination of the southern toad from urban areas. The much larger marine toad or giant toad, which was introduced into the Miami area from Mexico, preyed on the southern toad. The giant toad is highly successful in residential neighborhoods where some unfortunate dogs meet their deaths every year after biting this species. Large poison glands located on the sides of the neck area exude a white, milky toxin when the toad is disturbed.

Because of its edible qualities, the pig frog (Figure 16.2) is of economic interest. Its name derives from the close resemblance of its call to the casual, contented grunt made by farmyard pigs. Pig frogs are common, but not easily seen because of their cryptic coloration and their tendency to stay away from shore; they remain dispersed throughout the deeper freshwater marshes. The pig frog grows very large, its body length sometimes reaching just over six inches, nearly the size of its larger relative, the bullfrog (*Rana catesbeiana*), which is found farther north, not reaching southern Florida. Like the bullfrog, the pig frog is a voracious predator. It typically sits motionlessly awaiting an unsuspecting target to come within lunging distance and will eat almost any moving prey that can be swallowed whole. The abundance of pig frogs in the Everglades would indicate an important ecological role, only recently receiving attention.[609] Frog legs served in southern

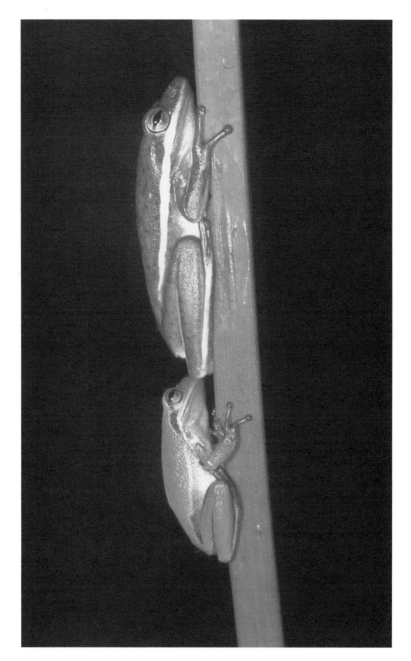

FIGURE 16.1 A green treefrog (above) and a squirrel treefrog, humorously positioned on a sedge stem. Where treefrogs are abundant, sometimes several line up in a location protected from the sun during the heat of the day. (Photo by T. Lodge.)

Florida are usually those of pig frogs, and they have been hunted extensively in the Everglades (see Recollections). However, pig frogs are known in the trade as bullfrogs for the obvious reason: legs from a "pig frog" might not market well, even though they probably taste and look the same as those of the widely consumed bullfrog.

FIGURE 16.2 A pig frog, the largest of South Florida's frogs sits on a spatterdock leaf, surrounded by maidencane, at the Anhinga Trail, Everglades National Park. (Photo by R.Hamer.)

THE IMPORTANCE OF AMPHIBIANS

The importance of amphibians in the Everglades ecosystem is inadequately documented and in need of research. However, while the list of species is short, amphibian populations are often so large that they must play a major role in the food chains. Evidenced by the din of their calls, most species start breeding upon the return of summer rains, and tadpoles immediately populate the newly flooded wetlands. Feeding principally on algae, they rapidly form a biomass of potential prey for fishes and wading birds. They are probably important foods for wading birds in short hydroperiod wetlands and in isolated wetlands. In these areas, neither time nor access is sufficient for fish populations to invade and expand into a major biomass. In contiguous marsh areas of long hydroperiod, fishes form a much greater biomass, but, even there, the tadpoles of some amphibians such as pig frogs, leopard frogs, and green treefrogs are common. In areas of very short hydroperiod (down to a minimum of about one month needed for the most rapidly developing species to mature from egg to young adult), larval amphibians may be more important than fish. Because many adult amphibians disperse from wetland habitats into uplands, they are important to the food chains of uplands as well as wetlands.

17 Reptiles*

In contrast to amphibians, many reptiles can tolerate salt water and dry conditions. Sea turtles, for example, spend their entire lives in marine habitats, while alligators and some freshwater turtles can live in estuarine or even marine waters, at least temporarily. At the other extreme, many reptiles are completely terrestrial, spending their entire lives out of the water, and even inhabiting deserts. Their ability to get along without a regular supply of drinking water far exceeds that of amphibians. A major factor in this independence from water is the reptilian egg, its hard shell and internal organization allow for incubation away from water. The live birth capability of a few (such as rattlesnakes) is similarly advantageous.

Reptilian tolerances for salt water and drought confer a superior potential over amphibians for crossing geographic barriers, but terrestrial and freshwater reptiles are still constrained by relatively small expanses of ocean. Only a few non-marine reptiles (namely some lizards) have invaded southern Florida from the tropics, probably by "rafting" on floating objects. Nearly all native, non-marine reptiles of the region came from the temperate land mass of North America, just as freshwater fishes and amphibians. Like those groups, numerous tropical exotic reptiles have been introduced by man (see Chapter 21). Miami and other urban areas now harbor over a dozen exotic species of lizards from tropical parts of the world, and six exotic reptiles are established in Everglades National Park. The most notable among them is the huge Burmese python, now established and becoming an ecological concern in and near water in mangrove and freshwater habitats.[408]

REPTILES OF THE EVERGLADES REGION

Representative reptiles of the Everglades region are listed in Table 17.1. All native terrestrial and freshwater species are allied to the temperate, North American continent except for the green anole and the reef gecko. Marine/estuarine species such as the American crocodile and marine turtles are tropical. Because of the secretive habits of many reptiles, relatively few are commonly seen by visitors to the Everglades. The American alligator — a grand exception — is easily seen because of its lifestyle and large size, and certain large turtles, notably the Florida redbelly turtle (Figure 17.1) and the Florida softshell (Figure 17.2) are also commonly seen basking or swimming in the clear waters of the Everglades. Softshell turtles are easily eaten by alligators but are incredibly fast swimmers, thus not easily caught. Much less visible are the highly venomous eastern coral snake and its diminutive and harmless near-mimics, the scarlet kingsnake and the Florida scarlet snake. These particular snakes are uncommon and tend to be nocturnal, living underground (*fossorial* habit) by day. Other reptiles, such as rattlesnakes, are difficult to see because of their cryptic color patterns. Rattlers have become much less common, but are still a consideration when hiking. The huge eastern diamondback (in 1966 the author observed a seven-foot specimen near Flamingo in Everglades National Park) is an obvious concern but even the pygmy rattler, whose adults are less than two feet long, can inflict a dangerous bite. The list contains three often-seen water snakes (freshwater members of the genus *Nerodia*) that are nonpoisonous but resemble the poisonous Florida cottonmouth, which is now much less common.

Of the few tropical reptiles that have reached southern Florida by natural means, most are marine, including the American crocodile (discussed in a following section) and five species of sea

---

* The main sources for this chapter are references 35, 37, 333, and 409.

TABLE 17.1
Representative reptiles of the Everglades region[a]

Common name[b]	Scientific name[b]	General habitat(s)[c]
crocodilians		
American alligator	*Alligator mississippiensis*	f,b
American crocodile	*Crocodylus acutus*	m,b
turtles		
loggerhead	*Caretta caretta*	m
Florida snapping turtle	*Chelydra serpentina osceola*	f
common musk turtle (stinkpot)	*Sternotherus odoratus*	f
Florida mud turtle	*Kinosternon subrubrum steindachneri*	f
striped mud turtle	*Kinosternon baurii*	f
diamondback terrapin	*Malaclemys terrapin*	b
Florida box turtle	*Terrapene carolina bauri*	t
Florida chicken turtle	*Deirochelys reticularia chrysea*	f
peninsula cooter	*Pseudemys floridana peninsularis*	f
Florida redbelly turtle	*Pseudemys nelsoni*	f
gopher tortoise	*Gopherus polyphemus*	t
Florida softshell	*Apalone ferox*	f
lizards		
green anole (chameleon)	*Anolis carolinensis*	t
brown anole	*Anolis sagrei sagrei*	t, i
southeastern five-lined skink	*Eumeces inexpectatus*	t
reef gecko	*Sphaerodactylus notatus*	t
snakes		
Burmese python	*Python molurus bivittatus*	b,f,t, i
mangrove salt marsh snake	*Nerodia clarkii compressicauda*	b,m
Florida (banded) water snake	*Nerodia fasciata pictiventris*	f
Florida green water snake	*Nerodia floridana*	f
brown water snake	*Nerodia taxispilota*	f
striped crayfish snake	*Regina alleni*	f
peninsula ribbon snake	*Thamnophis sauritus sackeni*	t,f
southern ringneck snake	*Diadophis punctatus punctatus*	t
rough green snake	*Opheodrys aestivus*	t
mud snake	*Farancia abacura*	f
Everglades racer	*Coluber constrictor paludicola*	t,f
eastern indigo snake	*Drymarchon corais couperi*	t
corn snake (red rat snake)	*Elaphe guttata guttata*	t
Everglades rat snake	*Elaphe obsoleta rossalleni*	t
Florida kingsnake	*Lampropeltis getulus floridana*	t
scarlet kingsnake	*Lampropeltis triangulum elapsoides*	t
Florida scarlet snake	*Cemophora coccinea coccinea*	t
eastern coral snake	*Micrurus fulvius fulvius*	t
Florida cottonmouth (water moccasin)	*Agkistrodon piscivorus conanti*	f
dusky pygmy rattlesnake	*Sistrurus miliarius barbour*	t
eastern diamondback rattlesnake	*Crotalus adamanteus*	t,f,b

[a] Listing is compiled from references 59 and 409.

[b] Common and scientific names follow references 35 and 37.

[c] The kind of habitat(s) of normal occurrence: b=brackish water; f=fresh water; m=marine; t=terrestrial. (i=introduced)

FIGURE 17.1 A large group of Florida redbelly turtles bask with an alligator at the Anhinga Trail, Everglades National Park. (Photo by T. Lodge.)

FIGURE 17.2 A Florida softshell takes air at the surface. Its large webbed feet and flattened body make it a fast swimmer. (Photo by T. Lodge.)

turtles. The green turtle (*Chelonia mydas*) was formerly abundant but is now an endangered species. The loggerhead is most often seen today, and regularly nests on Cape Sable, one of the Everglades National Park's few beaches. Of terrestrial reptiles, only two are natural tropical invaders of the region.* One is the reef gecko, a very small and secretive lizard. It occurs in debris along beaches as well as inland in pinelands, hammocks, and in residential neighborhoods but is seldom seen due to its nocturnal life style. The other is the green anole (Figure 17.3), a common lizard also known as the *chameleon* because its color changes between green and brown. Its widespread occurrence throughout the South indicates it has been there for some time and probably persisted in the area through the last glacial cycle.

Like many freshwater fishes and amphibians, southern Florida populations of several species of reptiles are distinct and recognizable as a variety or subspecies of their relatives living farther north. The subspecies of the common kingsnake has a very pale, speckled color pattern in contrast to those that live farther north in Florida or elsewhere in the range of the species. The Everglades subspecies of the rat snake is orange, in contrast to the dull yellow of most that live outside of the Everglades region. In addition, a number of reptiles, such as the Florida redbelly turtle, are confined to Florida even though they are closely related to species found farther north. These examples indicate that the environmental factors found in the Florida peninsula differ from those of the southeastern states and are favorable for special Florida or southern Florida varieties, or even separate (but closely related) species.

It is no accident that the Florida redbelly turtle is the most commonly seen turtle in the Everglades. This large, especially handsome water turtle has a dark shell up to 13 inches long, colored with large red markings. The lower shell (plastron) is red-orange in juveniles, fading to orange or yellow in adults. This species is highly adapted to coexistence with the alligator. Its shell is unusually hard and highly arched, making the Florida redbelly relatively resistant to alligator predation (Figure 17.4), and it preferentially lays its eggs in alligator nests. Upland (non-flooding) nesting locations are limited in the Everglades, but alligator nests provide a common opportunity. Redbelly turtles lay their eggs in alligator nests by day when the alligator is not apt to be present, and most female gators are too small to break the turtle's shell. By their location in gator nests, the eggs get the same evening and nighttime predator protection as the alligator eggs, provided by the female alligator. Also, the hatching baby turtles are temporarily protected by the resident alligator's instinct as if they were her own. The Florida redbelly turtle obviously benefits from this relationship, with no apparent ill affect on the alligator.

The gopher tortoise is unusual in the Everglades region, but is fairly common in other areas of Florida where sandy soils are dry enough for it to construct its deep burrow. A population of gopher tortoises occurs on the higher ground on Cape Sable[419] in Everglades National Park. This area resembles an island, isolated from the nearest alternative habitat by many miles of mangrove swamps and shallow tidal waters. The existence of the gopher tortoise on Cape Sable is probably the result of human introduction because this coastal feature is of relatively recent origin (see Chapter 10).

THE AMERICAN ALLIGATOR [127, 200, 387, 396, 600]

Most people would define an alligator as a large dangerous reptile found in the swamps of Florida. While that definition has some merit, it has some inaccuracies, and ignores the fact that the alligator is the most important species in the Everglades. Those who love the sight of wading birds, or enjoy fishing for bass in the Everglades owe a debt of gratitude to the alligator. In reality, driving a car to the Everglades is far more dangerous than encountering an alligator there.

---

* A third possible natural invader, disputed by herpetologists, is the bark anole (*Anolis distichus*), a small lizard that occurs in the Miami area where it typically lives on tree trunks, cleverly keeping out of view on the opposite side from the observer.[565]

FIGURE 17.3 A green anole on the seed head of a southern cattail, Loxahatchee National Wildlife Refuge. (Photo by T. Lodge.)

ALLIGATORS AND OTHER CROCODILIANS COMPARED

Alligators are part of the crocodilian group within the reptilian class of vertebrates. There are at least 22 species* of crocodilians worldwide; 14 are true crocodiles, only 2 of which are responsible for the infamous reputation of all crocodilians. The culprits are the enormous Indopacific (or

FIGURE 17.4 A young adult Florida redbelly turtle in the jaws of a seven-foot alligator. This turtle species is adept in coexisting with alligators, which includes protection by a particularly thick shell (see text for details). This individual slipped away like a bar of soap and escaped when the alligator tried to break the shell. Anhinga Trail, Everglades National Park. (Photo by T. Lodge.)

---

* The number of species recognized here is conservative; older literature often recognizing more. Differences lie in interpreting what constitutes a species, a problem agonized by taxonomists working with nearly all groups of plants and animals. Most often, it is geographic variation that may be interpreted in two ways: separate species or merely variation within a species. For example, the spectacled caiman, counted here as one species (*Caiman crocodylus*), varies widely over its range and has been considered as several species by some authorities.

saltwater) crocodile of India, southeast Asia, and Australia and the Nile crocodile of Africa. Both of these species are huge, often exceeding 15 feet.* Their sheer size and technique of ambushing large terrestrial animals along the water's edge pose a considerable threat to man. Other types of crocodilians include the slender-snouted gharial (or gavial) of India (one species), the alligator-like caimans of South and Central America (five species), and the alligators. The American alligator of the southeastern United States and the smaller Chinese alligator are the only two types of alligators in the world.[460, 513]

In addition to the American alligator, the similarly sized American crocodile (see the next section) and occasionally an escaped caiman inhabit South Florida. Most species of caimans are substantially smaller than the American alligator and crocodile. The spectacled caiman (*Caiman crocodylus*) is the species usually sold in pet shops. Because of releases by unhappy owners who thought it was a novelty when it was only a foot long, the spectacled caiman is now locally established in canals east of Homestead, Florida. It is unusual to see a caiman or the secretive American crocodile.

SIZE AND DANGER TO MAN

The largest American alligators ever known were about 17 feet long, and there is a dubious record of 19 feet 2 inches. In 1997, an individual 14 feet and five-eights inch long, weighing 800 pounds, was removed from Lake Monroe in the St. Johns River system, north of Orlando. This specimen is the largest well documented record for Florida.[32] However, if 12-foot or larger alligators were common, the species might deserve its dangerous reputation, but any alligator over about 11 feet is unusual. Moreover, alligators in the Everglades are not noted for large size compared with records from the river systems farther north in Florida.

In general, an adult human is too large to be of interest to an alligator, but precautions should be taken nonetheless. It is obviously dangerous to swim where a large alligator may be present, and children and even large dogs should be kept a safe distance from *any* alligator. Individual alligators accustomed to being fed are a particular danger. They become aggressive, associating people with food. For this reason, feeding alligators is illegal in Florida. Imagine an innocent visitor to a park, with toy poodle on leash, walking up to the pond for a closer look at the "tame" alligator. Through the gator's eyes, it looks like dinner is being served.

ALLIGATORS AS PREDATORS AND AS PREY

Raccoons are important predators on eggs and young gators. Largemouth bass, the larger water snakes, otters, wood storks, large herons and egrets, and bull alligators add to the peril. However, gators longer than six feet appear to be safe from predation except by man and larger bull gators. In turn, gators attempt to catch and eat virtually anything that moves unless it appears too large. Juveniles feed mostly on invertebrates. In freshwater, snails and crayfish are important items on the menu, and in brackish water, blue crabs are commonly taken. Gators over six feet long begin to take birds and mammals, including the species that would have preyed on them when they were smaller. Turtles and snakes — even poisonous species — are taken without hesitation and apparently without harm to the gator. Larger hard-shelled turtles, especially the Florida redbelly turtle, normally escape until an alligator is large enough to have jaw strength capable of crushing the shell. Such strength is within the capability of a ten-foot or larger gator.

---

* Size records for crocodilians vary widely, in part because reptiles do not reach a terminal size as do mammals, but also because of exaggeration, making it difficult to verify the authenticity of records. However, to be conservative, the Nile crocodile reaches a maximum length of about 20 feet, and the Indopacific crocodile has exceeded 25 feet.[460, 513] To dramatize, a live Indopacific crocodile recently delivered for exhibition at Parrot Jungle in Miami was 20 feet long and weighed a ton[104] — about 1200 pounds greater than Florida's recent record alligator.

ALLIGATOR DISTRIBUTION

The American alligator (known in the vernacular as "gator") is found throughout all of Florida and Louisiana, as well as southeastern Texas, southern Arkansas, and coastal areas of Mississippi, Alabama, Georgia, and North and South Carolina. Although primarily an inhabitant of freshwater, it can tolerate brackish water and live for months in seawater. Its numbers have been reduced by land development, especially the draining of wetlands, and, in the past, by hunting for gator hides and even meat.

ALLIGATOR PROTECTION

Most people are familiar with the use of alligator hides for wallets and shoes, but few are aware that the meat, notably the tail, is considered a delicacy in some circles. In the 1950's and 1960's the rate at which alligators were disappearing was alarming. The market for hides supported innumerable gator "poachers." The rate of decline pointed to near certain extinction, at least for wild populations. Several states began prohibiting the taking of alligators, and the species was listed as endangered in 1966 under the federal Endangered Species Preservation Act. Initial provisions for enforcement against poaching proved ineffective however, and gators continued to disappear almost as rapidly as ever. Policing the wilderness was fruitless. Subsequently, numerous state laws, a 1969 amendment of the federal Lacey Act, and finally the federal Endangered Species Act (1973) made it illegal to deal in any identifiable alligator "parts" (skulls, teeth, claws, meat, etc., as well as hides) in interstate commerce, providing the needed authority to enforce alligator protection. Products could no longer be legally sold, and the value of alligator hide plummeted. The population rebound of the alligator was rapid, first in Louisiana, then in Florida, and later throughout its range. As a result, the alligator was removed from the listing as endangered (except for purposes of its similarity of appearance to the American crocodile). However, alligator hunting is now strictly regulated through licensing, designated hunting periods and areas, and serial-number tracking of hides.

THE ALLIGATOR'S LIFE CYCLE

A walk through an alligator's life cycle will provide for a better understanding of the importance of this complicated reptile in the Everglades. Males, or bulls, are much larger than females and are highly territorial during the mating season. A successful bull thwarts competition, sometimes in blood-chilling battles, but usually by intimidation. Less aggressive bulls flee, an interaction that is commonly seen during the spring. Also, smaller alligators of both sexes are part of the diet of large bulls. While cannibalism may seem counter-productive to alligator success, it is really the opposite. It did not prevent the rapid rebound of alligator numbers when poaching was stopped, because large bulls were so rare that juveniles were nearly free of a major predator. However, as the new bulls matured and became cannibalistic, they helped prevent overpopulation. This type of population control is an example of self-adjusting feedback.

Large bulls typically patrol territories of about five square miles. They wander over regularly used pathways, and studies have shown that they know where they are going, even as they travel long distances over land during dry periods. Authorities are often called upon to relocate potentially dangerous bulls, but some have been known to find their way back to their former territory, in one case a one-eyed individual returned 35 miles. Females, on the contrary, prefer home territory and are not prone to wandering great distances.

Female gators become receptive to males in April and May, near the beginning of the summer rainy season. Bulls can be heard "bellowing" at that time, although bellowing is not strictly confined to the mating season. The sound somewhat resembles the low rumble of the engine of a bulldozer. Bulls answer each other's bellowing as well as loud noises from trucks and aircraft, especially sonic booms. The female occasionally bellows and will answer a bull's bellow with her higher pitched, softer voice.

Alligators mate in the water, following a period of amorous nudging and cavortive swimming. After mating, bulls merely go on to the next gator hole in a polygamous lifestyle. A fertilized female begins building a nest by heaping soil and vegetation. Nests are usually built in marshes or on low tree islands near (but not next to) the gator pond (see next section). A nest may be as high as two feet above the water level. She lays an average of 30 eggs after the summer rains have reflooded the marshes in late June or early July. During the two-month incubation period, she normally tends the nest at night and leaves it alone during daylight. Young, "chirping," 9-inch-long gators are normally first seen and heard in late August in the Everglades (but timing is highly variable, with occasional hatching as late as December). They are self-sufficient upon hatching although the mother normally helps them out of the nest by gently uncovering the hatching eggs. An occasional female gator vigorously defends her nest against approaches by man, and there have been instances where intruders have been chased.

Female gators are tolerant of juveniles, even those of other families. It is not uncommon to see babies and even year-old gators sunning on the backs of a female (Figure 17.5) but *never* on a bull. Gators grow at a maximum rate of about a foot a year, but growth depends on food supply and stress, and larger gators grow more slowly.

ALLIGATOR HOLES AND THEIR IMPORTANCE [93]

Alligators are pond-builders. Given a depressed wetland area during low water, alligators dig up vegetation and soil, piling the material around the edge, expanding the depression, and creating a pond called an *alligator hole* (see Figures 3.10 and 17.6, and Color Figures 8 and 9). Much of the behavior is obviously foraging activity.[348] Alligators not only scoop sideways with their jaws open, but use their body and tail to "herd" fish into an edge or blind end of the depression (Figure 17.7). Alligators also forage while submerged (Figure 17.8) and uproot vegetation. Thus, it is not convincing that alligators intentionally build ponds, but rather that their normal activity results in ponds. However, in addition to alligator holes, alligators do intentionally construct dens by tunneling 10 or 15 feet into the sides of the ponds, with the entrances below normal water level and an enlarged den at the end. Often, the alligator dens have air holes that reach up to the ground surface. The dens are used by gators during the dry season for aestivation, and were used by poachers to catch alligators. Vivid descriptions of this risky work have been provided by Glen Simmons.[537]

An active alligator hole can often be recognized by the low ring of trees, shrubs (often willows), and/or cattails that grow on the material pushed up around the edge of the pond. Many are at the edges of tree islands, natural or artificially created. Some holes, however, have no ring of vegetation and are surrounded by marsh. The holes are typically about three feet deeper than the surrounding marsh elevation, but range from eight inches to five feet, usually limited by excavation to the bedrock. The diameter of open water ranges from 15 to 50 feet. Gator holes have two or three trails connecting to adjacent marshes. Many alligator holes are probably centuries old and have been maintained by successive generations of alligators.

It is well-established that alligator holes are important to the success of the Everglades as a haven for water birds.[*] The relationship between alligator holes and wading birds is tied to the seasonal cycle of regional rainfall. Late in the dry season, from November into May in normal years, when vast areas of the Everglades become dry, gator holes act as refugia for aquatic life.[360] Fish, frogs, turtles, water snakes, and other wildlife inhabit the ponds and use the alligator trails for access. Many become food for the resident gator, but they are also food for herons, egrets, ibis, storks, and anhingas. Furthermore, only in very severe droughts do the gator ponds become completely dry. Thus, some aquatic life normally remains to "seed" new cycles of life when water

---

[*] Also, at the southern end of the Everglades where freshwater flows meet tidal water, numerous creeks provide habitat very much like the alligator holes and, in parallel, are maintained by alligators. These creeks are the headwaters of the tidal rivers, such as the Shark River. Interconnecting alligator trails between these creeks and the adjacent marshes are thought to be similarly important to many species of wildlife (see Chapter 8).

FIGURE 17.5 A female alligator with young probably about nine months old basking on her back. Shark Valley, Everglades National Park. (Photo by T. Lodge.)

FIGURE 17.6 Alligator hole nearly concealed in a small bayhead, probably induced by generations of alligator activity. Note the cattail patch at left center, probably induced by nutrients from animal activity during the dry season.

FIGURE 17.7 An alligator foraging in a drying gator hole. The great egret and tricolor heron take advantage of the disturbance, which exposes fish that evade the alligator. Corkscrew Swamp Sanctuary. (Photo by T. Lodge.)

FIGURE 17.8 An alligator foraging on the bottom in three feet of water, while largemouth bass watch for evading prey (and caught several crayfish while the author watched). Anhinga Trail, Everglades National Park. (Photo by T. Lodge.)

levels again flood the marshes and swamps during the wet season. During extremely dry or cold weather, gators seem to disappear; actually they retreat into their dens to aestivate leaving the gator hole to the other wildlife.

The preceding discussion has provided a broad description of the alligator in order to enhance understanding and appreciation of its importance. Consideration of this most interesting member of the Everglades wildlife should be at the heart of any discussion of the ecology of the Everglades. The alligator is an exemplary "keystone species."

THE AMERICAN CROCODILE [324, 388, 419, 614]

Unlike the alligator, the American crocodile (Figure 17.9) in Florida is almost completely confined to the brackish and salt waters of the southern tip of the state, including the Keys. Vero Beach on the east coast and Tampa Bay are historically its northernmost limits (see Figure 2.2). The species is also found in coastal habitats of the larger Caribbean islands as well as parts of Central America and northern South America. It is not adapted to cold weather like the alligator and, therefore, could not survive in northern Florida. Thus, the American crocodile is a tropical reptile with southern Florida as the northern edge of its range.

American crocodiles attain roughly the same size as American alligators but, in general, crocodiles are shyer, this is probably because they live in habitats where they do not often encounter people and have less opportunity to become accustomed to the presence of man. Due to their preference for saline tidal habitats and their relative rarity, crocodiles are seldom seen. Their coloration is brown to olive green, in contrast to adult alligators, which are mostly black, with yellow banding particularly prominent in juveniles. The crocodile's snout, seen from above, is much more tapered and comes to a narrower tip. Also, the large fourth tooth, counting back from the front end along either side of the lower jaw, is conspicuously exposed on the outside when the mouth is closed. All the teeth of the lower jaw of the alligator fit into pockets in the upper jaw, leaving only the teeth of the upper jaw visible when the mouth is closed. Juvenile American crocodiles feed on fish and invertebrates such as crabs, while adults feed mainly on fish — mullet being an important species here. The narrower jaw appears to be an adaptation for catching fish. Their technique of using a rapid sideways snap at potential prey is similar to that used by the slender-snouted gars and needlefishes.

Like alligators, crocodiles also build nests, but their nests are made of sand, marl, and peat piled up on dry land near water. Typically, the female lays 35 to 40 eggs during late April or early May, after which she checks the nest on regular nightly visits. About three months later, she assists the 10-inch hatchlings, digging them out of the nest, and carrying them in her jaws to the water's edge; she remains with them for days or even weeks. Crocodiles do not build ponds as alligator do, at least in Florida; instead, they construct dens in the banks of tidal creeks or canals and wear trails through mangroves. Preferred habitat is moderate salinity, about half seawater strength, and protection from wind and waves. Baby crocodiles are susceptible to predation from many of the same predators faced by baby alligators such as the larger wading birds, otters, and raccoons. However, the estuarine habitat also brings predation from larger fishes such as sharks and jacks and from the amazingly aggressive blue crab, which is a common prey of larger crocodiles. A baby crocodile was once rescued from the grips of a blue crab by a researcher, who found the crocodile little damaged but unconscious. He picked up the baby, squeezed water out of its lungs, and blew air in orally. The crocodile rapidly recovered, thanks to mouth-to-mouth resuscitation.*

The scarcity of the American crocodile in southern Florida prompted its listing as endangered in 1975 under the Endangered Species Act. However, peculiarities of crocodile behavior make the species difficult to manage. For example, crocodiles often cross roads in their wanderings through coastal waters, resulting in road-kills. The section of U.S. 1 between the mainland and Key Largo

---

* The effort of Dr. Frank Mazzotti, University of Florida.

FIGURE 17.9 An American crocodile using an artificial embankment created for crocodile use in the Harrison Tract, Crocodile Lakes National Wildlife Refuge, North Key Largo. This project was part of the wetland mitigation for planned modifications to the "18-mile stretch" of U.S. 1, which may never be completed due to environmental concerns for human impacts to the Florida Keys. (Photo by T. Lodge.)

has been a particular problem and is one of the few places on earth where signs warn motorists, "crocodile crossing." Properly engineered culverts, with a substantial air space that crocodiles can see through, facilitate their passage. However, submerged culverts beneath roads have proved to be killers. Crocodiles traveling through coastal habitats are attracted to the current in seeking new habitat but are reluctant to enter the submerged passageway; instead they crawl out of the water and over the road — unsatisfactory for crocodiles and motorists alike. Numerous projects have been successful using a combination of modified culverts and fencing.

Most crocodiles remain within several miles of a "home" location, but some individuals under wildlife management surveillance have moved more than 60 miles, adding to management difficulties. One specimen, originally found off the beach at Fort Lauderdale, was relocated to suitable habitat near Florida Bay. His subsequent travels included a visit to Big Pine Key, some 70 miles from the point of release. Such wanderings through marine waters undoubtedly explain how the American crocodile managed to reach southern Florida from the tropics.

Protection and management efforts have been successful for this species. Maximum numbers of nesting crocodiles have increased from about 20 in 1975 to more than 50 today. Part of this success is due to the artificial habitat provided by the extensive cooling-canal system of Florida Power and Light's Turkey Point power plant located on southern Biscayne Bay. Its warm waters, artificial nesting sites, and protection have fostered a considerable portion of the crocodile's population increase. Protection and habitat modifications in the Crocodile Lakes National Wildlife Refuge on north Key Largo and protected habitats in Everglades National Park have also been successful.

18 Mammals*

Despite the great adaptability exhibited in mammalian evolution, temperate North America is the origin for all of South Florida's native terrestrial mammals. The only native species having an origin distantly traceable to the tropics, yet also having entered Florida from the north, is the Virginia opossum, North America's only marsupial or pouched mammal. However, fossils prove that land mammals from the South American tropics had invaded Florida by about 2.5 million years ago,** after the emergence of the land bridge between North and South America (see Chapter 1). Despite this connection, most land mammals of tropical origin perished in Florida during glacial periods when climatic conditions were cool and, more importantly, dry, thus eliminating their wet-forest habitat.[183, 373, 642]

Although not naturally reaching southern Florida, a distinctive tropical mammal that has reached the Florida panhandle naturally is the nine-banded armadillo. This species is now common nearly throughout Florida wherever sandy soils occur including the northern Big Cypress Swamp, and there are sporadic records for Everglades National Park. The armadillo populations of the peninsula have been shown to be from human introduction.[84, 332]

Marine and flying mammals have tropical representation in Florida. The marine access route is obvious just as with tropical marine reptiles. Marine mammals are covered in a separate section at the end of this chapter. Bats, the only flying mammals, could have entered Florida from the tropics as well as from temperate North America. Six species have been recorded in South Florida, two of which apparently came directly from the tropics, and only one, the mastiff bat (*Eumops glaucinus*), is represented by a viable but very small population, which is primarily confined to developed areas around Miami where its roosts are in houses and buildings. Overall, however, bat populations in the region are so small that their ecological role is negligible.[84, 275, 291, 331]

In the faunal history of Florida there is another ecological factor of more than academic interest. Numerous prehistoric animals, including what are dubbed "megafauna" (horses, mammoths, mastodons, giant ground sloths, giant tortoises, and others), were present when Florida began warming after the last ice age. Most were mammals of temperate, Asian origin that had crossed into North America during past glacial periods of low sea level. The last of the megafauna perished by 8500 years ago, apparently abetted by the most influential pre-Columbian mammalian invader of all — *Homo sapiens*. Native Americans first moved into the Florida peninsula at least 13,000 years ago and were present in South Florida soon thereafter. Spear points imbedded in mastodon and other bones indicate that they hunted large animals and contributed to the massive extinctions of Florida's post ice-age megafauna, perhaps more effectively than the changes in climate (see Chapter 21).[94, 109, 248, 373, 390, 413, 414]

LAND MAMMALS OF THE EVERGLADES

Representative terrestrial mammals of the Everglades region are listed in Table 18.1. Most native mammals are of widespread distribution in North America. Common examples include the Virginia

---

* The main sources for this chapter are references 291, 331, 410, and 647.
** A few mammals of tropical origin first appear in Florida's fossil record nearly 9 million years ago, probably having reached North America by "island-hopping." While the land bridge was forming, closely spaced islands would have emerged as forerunners of a complete land connection.[373] The effect of the land bridge in facilitating the movement of land animals was gradual, not abrupt.

TABLE 18.1
Representative mammals of the Everglades region.[a]

Common name[b]	Scientific name[b]
terrestrial and semi-aquatic species:	
Virginia opossum	*Didelphis virginiana*
short-tailed shrew	*Blarina brevicauda*
marsh rabbit	*Sylvilagus palustris*
eastern cottontail	*Sylvilagus floridanus*
gray squirrel	*Sciurus carolinensis*
fox squirrel	*Sciurus niger*
cotton mouse	*Peromyscus gossypinus*
marsh rice rat	*Oryzomys palustris*
hispid cotton rat	*Sigmodon hispidus*
round-tailed muskrat	*Neofiber alleni*
black rat or ship rat (introduced)	*Rattus rattus*
gray fox	*Urocyon cinereoargenteus*
black bear	*Ursus americanus*
common raccoon	*Procyon lotor*
Everglades mink	*Mustela vison mink*
striped skunk	*Mephitis mephitis*
eastern spotted skunk or civit cat	*Spilogale putorius*
northern river otter	*Lutra canadensis*
Florida panther	*Felis concolor coryi*
bobcat	*Lynx rufus*
feral or wild pig (introduced)	*Sus scrofa*
white-tailed deer	*Odocoileus virginianus*
marine/estuarine species:	
Atlantic bottlenose dolphin	*Tursiops truncatus*
Florida manatee	*Trichechus manatus latirostris*

[a] Listing is primarily compiled from references 331 and 410.
[b] Common and scientific names follow references 291, 614 and 647.

opossum, the gray squirrel, common raccoon, northern river otter, common gray fox, and the white-tailed deer. Other examples of widespread species that are present but now uncommon in South Florida are the short-tailed shrew, black bear, eastern fox squirrel, mink, and panther. The latter three are represented in our region by subspecies discussed below. Other common mammals are species widespread in the southeast coastal plain, such as the marsh rabbit, cotton mouse, hispid cotton rat, and marsh rice rat. A regional pattern observed in several mammals is the occurrence of subspecies in the Florida Keys that have probably evolved very recently, after isolation by rising sea level (see white-tailed deer section). Examples are the Key Largo cotton mouse, Key Largo woodrat, lower keys rabbit, and key deer, all of which are listed as endangered and are discussed individually below.

The round-tailed muskrat (or "Florida water rat") is the only Everglades mammal species that is almost wholly confined to the Florida peninsula. It is smaller and not closely related to the common muskrat that occurs north of Florida. It is an adept swimmer, well adapted to marsh habitat. Round-tailed muskrats eat roots and stems of marsh plants such as pickerelweed and arrowhead and, in flooded habitat, build a mounded nest of cut grasses and sedges up to about two feet in diameter. In dry weather, they burrow. The species avoids sawgrass marsh habitat, thus its populations decline where management promotes the expansion of sawgrass into slough habitat (see Chapter 3). The round-tailed muskrat's populations are highly variable, sometimes growing to the point that their nests are commonly seen, and then dwindling to rarity. It is of interest that they often become abundant in sugarcane fields, where they damage crop roots. Upon sugarcane harvest, they are exposed to predation by hawks, owls, and foxes, and the population locally plummets.

The most abundant native mammal in southern Florida is the cotton mouse, a small rodent. It is represented by three subspecies in the region, which probably evolved from the parent species rather quickly after flooding of the Everglades and isolation of the Keys by rising sea level, separating populations from one another. One subspecies occurs on the west side of the Everglades in the Big Cypress Swamp, another on the Atlantic Coastal Ridge, and the endangered third subspecies (*Peromyscus gossypinus allipaticola*), restricted to the Florida Keys, now only on northern Key Largo.

Two medium-sized rodents are the hispid cotton rat and the marsh rice rat. The name "rat" conjures up the unpleasant image of the introduced rats that invade human habitation, but both of these native species are much more endearing in appearance — most would actually say "cute." The cotton rat is the most ubiquitous mammal in the region, occupying a wide variety of lowland habitats including saline coastal flats. The marsh rice rat is nearly as pervasive. Both species occur on Everglades tree islands, where their population dynamics have been studied.[220] Both species are also represented in the region by various subspecies. The lower keys population of the marsh rice rat, often called the "silver" rice rat, was listed as endangered in 1991. Its status as a possible subspecies or even separate species is debated.[275, 291, 614]

Although squirrels, also rodents, are probably the most commonly seen mammals in eastern North America, they are uncommon in the Everglades. Two species, the gray squirrel and the fox squirrel, live in southern Florida. Gray squirrels are locally common among Virginia live oaks on the Atlantic Coastal Ridge and regularly seen at Royal Palm in Everglades National Park. The threatened Big Cypress (or mangrove) fox squirrel (*Sciurus niger avicennia*), a handsomely colored subspecies of fox squirrel, primarily inhabits pinelands and cypress but also enters hammocks and mangrove swamps of the Big Cypress region, and is occasionally seen at Corkscrew Swamp. On the Atlantic Coastal ridge there is another rare subspecies, Sherman's fox squirrel (*S. n. shermani*). The few present on the southern end of the ridge in Miami-Dade and Broward counties are thought to be intergrades between the Big Cypress and Sherman's subspecies.

Four members of the weasel family occur in South Florida: two skunks, the northern river otter, and the Everglades mink. Both skunks are seldom seen (or smelled). The spotted skunk is the rarer of the two, and neither inhabits wetlands. Northern river otters, which are more at home in water than on land, are reasonably common but usually seen only in the dry season when they use roadside canals, particularly along the Shark Valley Road (Figure 18.1). The Everglades subspecies of mink is rare and most of what we know about it — distribution, body measurements, diet, etc. — is from examination of occasional road-kills along the Tamiami Trail in the Everglades and Big Cypress Swamp. It is separated by more than 100 miles from other Florida populations of mink. Like the otter, the mink is semi-aquatic and occurs widely in North America all the way to northern Canada and Alaska. Both species feed on insects, crayfish, fish, frogs, small reptiles, and some birds, with the much larger otter catching larger prey.

Raccoons and marsh rabbits (Figure 18.2) are probably the most commonly seen mammals in the Everglades. The relatively small, dark-colored marsh rabbit often forages along roadsides and, unlike most other rabbits, is a willing swimmer. An isolated lower keys subspecies of the marsh rabbit (*Sylvilagus palustris hefneri*)* is listed as endangered. The eastern cottontail is an upland species and is not common in South Florida, occurring on the Atlantic Coastal Ridge but not the western side of the Everglades. Raccoons are omnivores, eating various fruits and berries, thus dispersing the seeds of many plants, but also eating animals, mostly aquatic life. They are so common that they are problematic for certain species because of raiding nests such as those of

---

* Funding of the research that led to the designation of the lower keys marsh rabbit population as a separate subspecies was provided through Hugh M. Hefner, founder and publisher of the magazine, *Playboy*.

FIGURE 18.1 Northern river otters keep track of the author/photographer, Shark Valley, Everglades National Park. (Photo by T. Lodge.)

FIGURE 18.2 A marsh rabbit at the Anhinga Trail, Everglades National Park. (Photo by T. Lodge.)

alligators, sea turtles, and occasionally even crocodiles. With the populations of panthers and large alligators low, raccoons have few predators. Like the raccoon in its omnivorous habits, the black bear is one of South Florida's rarest inhabitants. Small numbers occur in the Big Cypress Swamp, where their huge tracks are not infrequent.

There are several introduced mammals in South Florida. Two are common enough to be of ecological significance: the black rat, and the domestic pig. The black rat (far more common than the also introduced Norway rat, *Rattus norvegicus*) is a significant competitor of native rodents and is thought to be one of the reasons that the Key Largo woodrat (*Neotoma floridana smalli*) is endangered. Feral pigs, originally from domestic stock, have reverted to the "wild boar" appearance with large tusks. They may weight over 400 pounds. They are abundant many places in Florida, including the Big Cypress Swamp where they are hunted by man and Florida panthers alike. Small numbers occur in the Everglades. Where their populations are high, they can have significant impact on vegetation by their rooting habit. Many people have witnessed their extensive earthworks, appearing as if gardening machinery had been used to till the soil.

THE WHITE-TAILED DEER [370, 562]

White-tailed deer (Figure 18.3) are fairly common in southern Florida. In the northern United States, males may reach as large as 300 pounds and females 200 pounds, but South Florida mainland specimens are far smaller, and those from the lower Florida Keys, considered to be a separate subspecies called the Key deer (*Odocoileus virginianus clavium*), normally weigh less than 80 pounds and stand only 25 to 30 inches high at the shoulder. The key deer's largest numbers are on Big Pine Key, but some occur on neighboring keys. Key deer are not only noted for their small size but also for their lower reproductive rate and their low water requirement. Both conditions may be adaptations to the lack of predators and to their salty environment where freshwater is often scarce. These adaptations have likely evolved very recently — the result of sea level rise that finally isolated their habitat from connection with the mainland only about 4000 years ago. [342]

The population of white-tailed deer in the Everglades is thought to be much larger today than historically because drainage has provided more dry land.[166] Willoughby[656] noted deer in his 1895 canoe voyage through the Everglades, which followed a series of dry years. Population estimates for the post-drainage Everglades have ranged from about one to four deer per square mile, and for the Big Cypress Swamp over seven per square mile. In general, the Everglades supports few deer compared to good upland habitats. Especially in dry years, it is not unusual to see deer feeding on the flooded marsh in the Everglades during early morning hours. However, in excessively flooded areas of the Everglades, such as the recent condition of Water Conservation Area 3A, their numbers are depressed, possibly more closely resembling the historical condition. Promoting their special interest, hunters regularly request lower water levels to improve conditions for deer — an awkward issue in "single-species management." Key visual indicators of the common presence of deer are "munched" leaves of swamplily, and the absence of swamplily may indicate an over abundance of deer.

THE FLORIDA PANTHER [376, 562, 614, 615]

By far the rarest mammal in Table 18.1 is the panther. The species is known as the puma, cougar, or mountain lion in western parts of North America where it still thrives. The endangered Florida population is considered a separate subspecies (*Felis concolor coryi*), originally occurring northward into South Carolina and westward into Arkansas and to the eastern edge of Texas. In 2002, the entire population was composed of 80 known individuals, primarily confined to the Big Cypress Swamp and northward just across the Caloosahatchee River. The Florida panther has undoubtedly been the focus of more research, and perhaps also program controversy, than any other animal in Florida.

FIGURE 18.3 White-tailed deer, doe and fawn wade through a shallow patch of wet prairie walled by dense sawgrass in the background. The large leaves in the foreground are alligatorflag. Shark Valley, Everglades National Park. (Photo by T. Lodge.)

Adult Florida panthers grow to seven feet, including the tail, and weigh from 60 lbs (small female) to 160 pounds (large male). Their footprints, like those of other cats in lacking claw marks, are over three inches in diameter, which is far larger than footprints of the moderately common bobcat. The home range of a panther is typically about 250 square miles for males and less than 100 square miles for females. Travel is normally during the night, but in cooler winter weather they may move also during daylight. Panthers are mostly solitary, except for the 12 to 18 months when a female is raising her one to four kittens. An adult male or female is generally intolerant of the presence of another adult panther of the same sex in its territory, and intrusion occasionally results in a fight to the death.

Deer and feral hogs are the Florida panther's preferred prey, but any small to medium-sized animal may be taken, including armadillos, rodents, rabbits, raccoons, and even small alligators. This panther is not considered to be dangerous to man and is currently the focus of an intensive program to understand, protect, and reestablish populations in Florida. Many individuals have been fitted with a collar containing a radio transmitter. This program has received much criticism by adversaries who feel that the animals should not be disturbed. The objective of the program is to monitor panthers' travels in order to better understand and protect this nearly extinct subspecies. The information has helped, for example, in improving highway design in critical areas, most notably Interstate 75 through the Big Cypress Swamp, so that road-kills are less likely. Wildlife underpasses and roadway fencing have proved effective for panthers as well as other wildlife.[202] In 1995, eight female Texas cougars — not the Florida subspecies — were introduced into the Big Cypress Swamp to correct observed genetic problems in the South Florida population, another controversial but probably helpful program. Present efforts to save the subspecies is a landscape-level strategy focused on identifying habitat and movement corridors that can be protected and enhanced. Good habitat along Fisheating Creek, for example, could be made available to cats in the Big Cypress Swamp by improving travel corridors. The bottom line: it is estimated that a population of at least 240 individuals (far larger than the present population status) would be required to be demographically stable and large enough to effectively maintain genetic integrity.

MARINE MAMMALS

Two marine mammals are of note in the nearshore waters of southern Florida, including the shallow bays and estuaries of Everglades National Park. These are the Atlantic bottlenose dolphin and the endangered Florida manatee. Numerous other marine mammals, including whales and other dolphin species, would be listed if offshore waters were included, but most of these are rarely encountered. It is also of interest that the West Indian monk seal (*Monachus tropicalis*) once occurred in marine waters around peninsular Florida and the Keys, but it was hunted to extinction early in the 20th century. Bottlenose dolphins are common in the waters of the park, where they feed on fish. The Florida manatee, by contrast, is a rather sluggish vegetarian and is one of the few tropical marine mammals. It is a subspecies of the West Indian manatee, which occurs coastally through much of the Caribbean region. The Florida subspecies, which is mostly confined to the Florida coast, is separated from the one other Caribbean subspecies by the Florida Current. It is sometimes killed by cold water farther north in Florida and, in some places, now takes advantage of heated water from power plants during the winter months. Manatees commonly enter freshwater, and some reach Lake Okeechobee through the canal connections. The endangered status of the Florida manatee is primarily due to boat collisions, and many individuals show the telltale series of scars from encounters with boat props.

19 Birds *, **

With close to 400 species of birds having been documented as occurring naturally in southern Florida, the overall list of birds, or "avifauna," is rich. Within the total avifauna, just under 300 species are considered to be of regular occurrence, other species being known from only a few records.

Before purchasing binoculars and a field guide to birds, exploring some details of this rich avifauna are in order. About 60 percent of these species are winter residents, migrating into South Florida from the north, or are hurried visitors, stopping only briefly in the spring or fall in their migration between tropical America and points farther north in the United States or Canada. The remaining 40 percent of the species of birds that regularly occur in southern Florida are the true natives — the birds that breed there. Breeding range can be thought of as the true home of a bird species from the standpoint of biogeography.

The list of species of birds breeding in southern Florida, which includes about 116 species, is substantially smaller than lists for areas of similar size in the northeastern United States. Furthermore, not all of these south-Florida breeders are year-round residents. For example, wood storks breed in southern Florida in the winter and spring, and then they move farther north in Florida and adjacent areas for the summer. In an alternate life style, the exquisitely graceful American swallow-tailed kite breeds in the region in late spring, then leaves for tropical America in August and does not return until February. For most people, however, the dry season is an excellent time to see birds in southern Florida. The optimum is from mid-December into April.

An analysis of the geographic origins of the breeding birds of southern Florida reveals some interesting trends. Given the ability of birds to fly long distances over water, one would think that the region, with its predominance of tropical vegetation and proximity to tropical areas such as Cuba, would also be a haven for tropical birds. It may come as a disappointment that this is not generally the case, but the tropics have indeed enriched bird life in southern Florida. To explain the situation, it is helpful to discuss South Florida's breeding birds by dividing them into two categories: land birds and waterbirds.

The distinction between land birds and waterbirds is based on habitat requirements, rather than their relationships by evolution. Natural groupings of birds tend to include species that are mostly all either land birds or waterbirds, and therefore the land and water categories have few misfits. Land birds are those type of birds that normally live in terrestrial habitats, without particular dependence on aquatic or marine habitats. Examples of groups of birds that can be categorized as land birds are hawks, hawk-like birds including vultures and kites, doves, owls, woodpeckers, and songbirds (sometimes called perching birds, but technically known as the passerine birds). The songbirds or passerine birds include many species such as swallows, jays, the mockingbird, thrushes, flycatchers, warblers, and blackbirds. It is convenient to distinguish between passerine and non-passerine land birds in some comparisons.

---

* The main sources for this chapter are refrences 507 and 578.

** Common names of birds have been meticulously standardized, in periodic checklists by the American Ornithologists' Union (which most guidebooks follow, e.g. references 427, 463, 532), that the use of scientific names is not necessary in this text. As a result of standardization, authorities have decided that the English common names are proper nouns and should be capitalized, but there is wide disagreement on this point. For the sake of uniformity in this text, they will not be capitalized (it is standard practice not to capitalize common names of plants and animals).

There are some misfits within this land bird category. Obvious examples are the osprey (or fish hawk) and the bald eagle, both of which are seen regularly in the Everglades. Both are associated with water but are included as land birds. However, it is difficult to find more than a handful of exceptions, and they are unimportant for the purpose of this discussion.

Waterbirds, then, are those species that are strongly associated with freshwater, estuarine, or marine habitats. They include grebes, pelicans, cormorants, the anhinga, ducks, herons and egrets, the limpkin, ibises and the related spoonbill, the wood stork, cranes, gallinules, rails, shorebirds, gulls, and terns. Again, there are only a few exceptions. The cattle egret, an insect eater, prefers dry grasslands and an association with cattle or even noisy tractors to help stir up a meal. The killdeer (in the shorebird group by natural relationships) prefers upland fields, rather than wetlands or shores like its relatives. Nevertheless, these few exceptions are minor.

BREEDING LAND BIRDS

Just over 70 species of land birds breed in southern Florida, half of which are songbirds or passerine types. Some interesting geographic trends in these land birds are shown in Table 19.1. It is obvious that southern Florida has relatively few breeding land birds, and the increase in numbers when comparing southern Florida to all of Florida and then to comparably sized states farther north (Kentucky and Michigan) is in passerine birds. The number of species of remaining non-passerine land birds in each region is about equal.

Within the Florida peninsula, there is a pronounced decrease in the number of breeding passerine birds from north to south. Part of this trend is understandable because of habitat limitations. In southern Florida, wetlands (especially the Everglades) account for most of the area, which leaves little habitat for land birds. However, the breeding ranges of many land birds end in areas where habitat does not appear to be limiting. Furthermore, research has shown that the breeding ranges of numerous species have been moving north, farther and farther from southern Florida. It is thought that much of this northward movement is not due to human influences.

A hypothetical explanation is that this northward shift is a continuing biological response to the overall climatic changes that ended the last glaciation. These particular passerine birds of temperate origin were probably much more numerous in Florida during the height of the glacial age when Florida was much larger, cooler, and drier than today. Presumably, their populations have been shifting slowly northward over the past 15,000 years, in response to climatic change. Further, there is abundant evidence in Europe and North America that the northern breeding limits of many species of birds are moving north. However, the hypothesis is complicated by several species, including the indigo bunting, gray catbird, and American robin, whose breeding ranges have been moving southward in Florida in recent years.

With relatively few breeding land birds present in southern Florida, why have tropical species not moved in from Cuba, Hispaniola, and the Bahamas? Very few species of tropical land birds are naturally present in southern Florida even though many species occur in these adjacent West Indian islands. Before the turn of this century, Florida had a large population of the Carolina parakeet, a beautiful, small, yellow and green parrot species. Its range actually extended throughout the eastern half of the United States. This species, now extinct due to hunting, represented the single success of the parrot group in naturally reaching Florida, where it had been resident long enough to evolve into an endemic species. On the other hand, a few tropical land birds are naturally expanding populations and breeding ranges in Florida. Notable examples are two non-passerines (the smooth-billed ani and the white-crowned pigeon) and two songbirds (the gray kingbird and black-whiskered vireo). Other species also appear to be ready to make the move, with occasional visits. The process has been slow, and has resulted in an apparent deficit of breeding land birds in southern Florida.

One last word on this topic: people have successfully introduced many tropical land birds into southern Florida, including the spot-breasted oriole, the red-whiskered bulbul, and several parrots

TABLE 19.1
Geographic trends in numbers of breeding land birds in the eastern United States.[a]

Region	Breeding passerines (songbirds)	Numbers of species Breeding non-passerines	Total breeding land birds
South Florida	36	37	73
Florida	61	43	104
Kentucky	85	34	119
Michigan	109	43	152

[a] Source: modified from reference 507.

(the red-crested amazon, monk parakeet, and white-winged parakeet lead the list). The success of these species is likely helped by the weak competition resulting from the relatively few native species of breeding land birds in the region, and their adaptability to man's altered and disturbed habitats.

BREEDING WATERBIRDS

A selection of Everglades waterbirds is provided in Table 19.2. In contrast to the geographic trend in land birds, more waterbird species occur in Florida than in states farther north. Southern Florida alone boasts almost 120 species, with 43 breeding in the region. The breeding waterbirds of southern Florida differ markedly from the land birds in both geographic distribution and general origin. First, the number of breeding species is the same for southern Florida as for the remainder of the state, although the list of species differs somewhat. A trend of decreasing numbers begins north of the state. Georgia, for example, has 29 breeding species.

Second, a large percentage of the waterbirds that breed in South Florida are also common to the West Indies, where substantially more waterbirds occur. These species are generally considered to be tropical, although many of them have ranges that extend far north into the continent. Species that prefer shallow tidal habitats appear to treat South Florida just like another tropical island. Groups of the greater flamingo of the Caribbean region occasionally visit the shallows of Florida Bay (but most of the few flamingos seen in the past few decades are thought to have escaped from captivity). The reddish egret, found in shallow estuarine and marine waters, and especially Florida Bay, also uses shallow tidal waters around islands of the West Indies and Bahamas. The roseate spoonbill, which frequents similar habitats, moves about the region in an opposite pattern, with many individuals migrating northward from Cuba to southern Florida to breed during the winter months.

Pelagic, or open-ocean seabirds, also treat southern Florida like any other Caribbean island. Examples that breed in the region include the magnificent frigatebird and two kinds of terns. However, several other pelagic seabirds, regularly seen at sea off the coast, could potentially nest in southern Florida except for lack of proper habitat. Florida has none of the rocky cliffs preferred by many pelagic seabirds. The use of such inhospitable nesting sites helps to prevent predation of eggs and nestlings by land animals, in particular the raccoon. Because the magnificent frigatebird is accustomed to using mangrove trees on isolated islands as nesting sites, it finds South Florida acceptable, although actual nesting is restricted to the westernmost keys.

Of the species with ranges that extend far north into the continent, most swell the southern Florida population with wintering individuals. The great blue heron ranges as far north as central Canada during summer months. While some remain relatively far north in winter, most migrate to tropical America. A substantial number winter in southern Florida together with resident Florida individuals and other similar migrants.

TABLE 19.2
Representative waterbirds of the Everglades ecosystem.

Common name[a]	General habitat(s)[b]	Scientific name[a]
diving birds and ducks		
pied-billed grebe	f	*Podilymbus podiceps*
Anhinga	f	*Anhinga anhinga*
double-crested cormorant	f,m*	*Phalacrocorax auritus*
fulvous whistling-duck	f	*Dendrocygna bicolor*
mottled duck	f	*Anas fulvigula*
wood duck	f	*Aix sponsa*
red-breasted merganser	f,m*	*Mergus serrator*
wading birds		
least bittern	f	*Ixobrychus exilis*
American bittern	f*,m	*Botaurus lentiginosus*
black-crowned night-heron	f*,m	*Nycticorax nycticorax*
yellow-crowned night-heron	f,m*	*Nyctanassa violacea*
green heron	f,m	*Butorides striatus*
tricolored heron	f,m	*Egretta tricolor*
little blue heron	f*,m	*Egretta caerulea*
reddish egret	m	*Egretta rufescens*
snowy egret	f,m	*Egretta thula*
great egret	f,m	*Casmerodius albus*
great blue heron	f,m	*Ardea herodias*
great white heron	f,m*	*Ardea herodias*
wood stork	f*,m	*Mycteria americana*
glossy ibis	f*,m	*Plegadis falcinellus*
white ibis	f,m	*Eudocimus albus*
roseate spoonbill	f,m*	*Ajaia ajaja*
Limpkin	f	*Aramus guarauna*
birds of prey[c]		
bald eagle	f,m	*Haliaeetus leucocephalus*
snail kite	f	*Rostrhamus sociabilis*
Osprey	f,m	*Pandion haliaetus*
other birds of interest		
purple gallinule	f	*Porphyrula martinica*
common moorhen	f	*Gallinula chloropus*
American coot	f	*Fulica americana*
Cape Sable seaside sparrow[c]	f	*Ammodramus maritimus mirabilis*

[a] Common and scientific names follow reference 427.
[b] The kind of habitat(s) of normal occurrence: f=freshwater; m=marine and estuarine; asterisks (*) indicate the more common habitat.
[c] Birds of prey, as a group, and the Cape Sable seaside sparrow are not categorized as water birds, but these particular examples are associated with Everglades wetland and aquatic habitats.

Two species of pelicans can be seen in southern Florida. The more common brown pelican breeds in Florida and is generally tropical. It seldom ventures inland and feeds on fish by diving into the water head first. Its lower beak supports a huge throat pouch that expands like a parachute on hitting the water. How the bird manages this technique without breaking its thin neck is mystifying and is in marked contrast to the technique of the American white pelican, which feeds while swimming and frequently cooperates in large groups. American white pelicans do not breed in southern Florida but winter there, commonly roosting on islands in Florida Bay. Banding studies have shown that Florida migrants are often found in North Dakota and Saskatchewan for the summer breeding period. They are an exciting addition to the Florida's wintering wildlife.

The superficially similar double-crested cormorant and anhinga add only two more breeding waterbirds to the southern Florida list, but they are particularly interesting fish-eating birds. The cormorant is primarily a coastal species and catches fish in its beak during underwater dives. It is easily distinguished from the anhinga by the prominent hook on its beak. The anhinga (sometimes called the snakebird, water turkey, or darter) normally uses interior fresh waters where it is frequently seen with head and snake-like neck alone protruding from the water. This species catches its prey by spearing it with its long, pointed beak. When successful, it surfaces holding the impaled fish in the air. Then with a quick motion, the anhinga flips the fish up and catches it in the beak. A few adjustments are made to maneuver the fish, and it is swallowed head first. The use of the beak as a dagger is a chilling reminder of how dangerous some wildlife can be if cornered or approached when injured. This bird, as well as the herons and egrets, can put out an eye with one sudden jab. Left alone, these birds are harmless and highly entertaining in their peculiar feeding styles.

The anhinga is probably the most memorable waterbird seen in Everglades National Park, and the most popular boardwalk viewing area in the park, the Anhinga Trail, is its namesake. The anhinga's body feathers are structurally incapable of trapping air, unlike most diving birds. Consequently, a diving anhinga's buoyancy is close to neutral. These birds are able to swim about slowly under water — through vegetation and around obstacles — with little effort. However, the wet feathers are very poor insulators, and a great deal of body heat is lost during their underwater work as well as through evaporation afterwards. This leads to a commonly stated misconception. After foraging underwater, anhingas conspicuously perch with wings outstretched (Figure 19.1). People commonly remark that they are "drying their wings so they can fly." While this pose aids in drying, its importance is in raising the bird's body temperature or *thermoregulation*. Their back is always to the sun, thus capturing its warmth (a behavior also commonly exhibited by vultures prior to flight in the morning). Furthermore, anhingas can fly even when soaking wet. Given the right stimulus, such as an approaching alligator, an anhinga can burst from the water into flight.

Ducks and shorebirds are very poorly represented in southern Florida as breeding populations. By stretching the numbers to include infrequent occurrences such as the breeding records of the American oystercatcher, there are only about six shorebird species and three ducks, including the resident but secretive wood duck. However, wintering species in each of these groups number about 20.

FEEDING BEHAVIOR OF WADING BIRDS [255, 307, 510]

The large wading birds of the Everglades region attract much attention. Fourteen species breed in southern Florida, and their feeding activities are of particular interest, each with its own style. A short diversion into some theoretical ecology will add insight to understanding why each species is unique.

Ecologists have rather abstractly defined the entire scope of activities of a given species as its ecological niche. This niche is not just a place (the term *habitat* is used for the place or places used by a species). Instead, an ecological niche is more comparable to a profession, although the habitat where the profession occurs is part of the picture. It is theorized that direct, total competition between two species (the condition in which they share an identical niche) would result in the disappearance of one species. The concept is called *competitive exclusion*, and it probably explains why some introduced species are unable to become established in a new region even though the available habitat appears appropriate for their survival. Severe competition from one or more established species is usually thought to be the negative factor in these situations. In ecological terms, the required niche is already filled by one or a combination of species established in the region.

FIGURE 19.1 An anhinga thermoregulates (warming up, not drying out) after a dive at the Anhinga Trail, Everglades National Park. (Photo by T. Lodge.)

While competitive exclusion probably does occur, species can and do change through the evolutionary process of adaptation by natural selection. The world is full of examples where natural selection appears to have changed the physical shape, size, or behavior of species. A famous showcase of such adaptations involves the finches of the Galapagos Islands, located in the Pacific Ocean off the coast of Ecuador.[245] Inheritable variations that enable the individuals to use a slightly different niche reduce competition. Such advantageous variations improve the *fitness* of an individual, which is defined by its chances of survival and reproduction; thus, the changes become more common through succeeding generations (i.e., the species evolves). For example, a change may allow a species to obtain slightly different foods not readily available to competitors. This specialization would be an adaptive change in the niche of the species. On careful observation, it is usually easy to identify physical and behavioral specializations that reduce competition among even very similar species.[246, 385]

Few areas on earth allow a casual observer to see the subtle differences between competitors better than in the feeding behavior of wading birds in South Florida parks and refuges. These locations offer this unusual opportunity because of the large number of species that can be seen and because they can be observed closely — the protection makes many of the birds unafraid of people. The best opportunities for observing wading birds arise when drying conditions force large concentrations of food (mostly fishes and aquatic invertebrates) into shallow pools, and huge numbers of waterbirds descend on the concentrated prey. At these times, the birds are forced into their highest level of competition, and specializations are easily seen. The Anhinga Trail and the Shark Valley in Everglades National Park, a series of ponds at Loxahatchee National Wildlife Refuge, and a boardwalk at Corkscrew Swamp Sanctuary are examples known for their particularly long periods of wading bird activity. Numerous other locations also offer very good opportunities, with optimum times differing from place to place and year to year. A famous, but short annual feast occurs at Mrazek Pond, not far from Everglades National Park's facilities at Flamingo. The pond lies at the edge of the road, making it easily accessible (see Figure 8.1). Activity generally lasts only a few weeks each year, with the heaviest use lasting only about a week. It may occur anytime between late December and April, depending on water levels. South Florida wading bird nesting and other activity is now reported in detailed annual reports.[224]

The most obvious difference in feeding techniques that reduce competition is the touch-oriented feeding of the wood stork, white ibis, and spoonbill, versus the visually oriented feeding of the herons, egrets, and bitterns. Some behavioral traits of several prominent Everglades species will further explain.

WOOD STORK [307, 448, 510, 614, 616]

Of the three species that locate prey by touch, the wood stork commands respect as the true fisherman; it takes mostly fish and few invertebrates. Often feeding in cooperative groups, its technique, called groping, is to probe its long, heavy, partially open bill into the water while walking slowly about (Figure 19.2). If a living organism is detected by the highly sensitive receptors on the inside of the bill, the bill snaps shut by reflex action in less than 0.03 of a second. If something is caught, it is picked up and, with a quick snap, tossed to the back of the mouth and swallowed. While most of the bird's motions are relatively slow, the catch-and-swallow sequence is often so fast that one must watch very closely to see what is being eaten. Only when the prey is large, or spiny like a catfish, does the stork take some cautious preparations. Normal feeding behavior also involves assistance with one of its flesh-colored feet (pink in breeding condition), using a slow stirring motion that disturbs any motionless fish. In murky or weed-choked water between six and 20 inches deep, wood storks have a considerable advantage over sight-oriented wading birds. On the other hand, the technique is not effective in clear water where potential prey can see and evade these large, awkward birds. The stork's alternative sight-oriented feeding behavior is inefficient compared to herons and egrets.

FIGURE 19.2 An immature wood stork forages in Mrazek Pond, Everglades National Park. (Photo by T. Lodge.)

Most of the wood stork's diet consists of fish measuring between one and ten inches long, but it also eats a variety of other creatures, including larger invertebrates and sometimes even water snakes and baby alligators. Wood storks may also feed effectively even when all surface water is gone and prey is concealed in soft mud.

The specialized grope feeding behavior of the wood stork has rendered it endangered in the United States (although it occurs commonly in much of South America). It is highly dependent upon a concentrated food source. In years that lack dry weather, or when water is kept too deep by artificial management, nesting failures result. Nesting also fails in years when the drydown conditions proceed too rapidly, so that the required food source is largely depleted before the wood stork young are fledged. The entire reproductive period, from prenesting courtship until chicks are self-dependent, is 110 to 150 days (the longest of all Florida wading birds). An orderly sequence of drydown in freshwater wetlands is the key to success, and a probable cause of the wood stork's demise in the modified Everglades (see Chapter 21). However, breeding populations of the wood stork have recently shown significant recovery in the Everglades region,[224] and improvement is expected with further restoration.

WHITE IBIS

The narrow, downward-curving beak of the white ibis (Figure 19.3) is an indicator that something is unique about its mode of feeding. In contrast to the wood stork's slow, cautious approach, the white ibis rapidly probes the water — sometimes with its head completely submerged — and explores in, around, and under obstacles. The results of these efforts net a much higher percentage of invertebrates, typically crayfish and insect larvae in fresh water and small crabs in saline coastal areas. Much of this prey is extracted directly from burrows or other hiding places. Most of the relatively few small fish caught by this technique are probably those that have taken shelter in some nook or cranny where they are safe from other predators — but not from the white ibis.

FIGURE 19.3 A breeding condition white ibis at Eco Pond, Everglades National Park. (Photo by T. Lodge.)

The white ibis was historically the most abundant wading bird in southern Florida, but its populations declined precipitously from the 1980's into the mid-1990's. Their nesting success has recently increased.[224, 516]

GLOSSY IBIS

A puzzling question in comparing ecological niches is the similarity between the white ibis and the dark-colored glossy ibis, which is more common farther north in Florida and along the Atlantic coast. Its populations fluctuate in southern Florida, but it now more frequently visits Everglades National Park. While the slight difference in the ranges of these two species may constitute a difference in their respective niches, skillful observations could probably also reveal some subtle differences in feeding techniques. Both of these species also feed on land, including fields and golf courses, especially following heavy rains, when they forage for earthworms. The glossy ibis is locally abundant, and its populations have been expanding since it first started breeding in Florida in the 1880s. It now breeds in the Loxahatchee area in small numbers.[224, 516]

ROSEATE SPOONBILL [30, 178, 368]

The bizarre spoon-shaped beak of this species obviously indicates that its feeding technique differs radically from other wading birds (Figure 19.4). The spoonbill's normal feeding behavior is to swing the mostly submerged opened bill in a wide arc from side to side as the bird walks about in shallow water. When several birds are present, they often team up, forming a line. The mature spoonbill's brilliant pink wings, white back and neck, and pale greenish head, make these cooperative feeding efforts a sight to behold. The technique is successful in catching small fish and invertebrates at mid-depth in the shallow water column, or even strained from loose sediments at the bottom. Spoonbills nest primarily on mangrove islands in Florida Bay, but normally fly to brackish wetlands of the mainland to feed. The juvenile birds are very pale (almost white) and obtain their adult pink color by extraction of a red carotenoid pigment from crustaceans in their diet, the same pigment group also deposited in the greater flamingo's feathers.

GREAT BLUE HERON

The author watched a great blue heron impale a largemouth bass, perhaps 16 inches long and weighing three pounds, which illustrates the deadly force with which this sight-oriented heron can attack its prey. That event, at the Anhinga Trail in Everglades National Park, was unusual, and an opportunistic alligator ended up with the prize, which the heron was barely able to carry, much less eat. The technique of spearing fish from 3 to 12 inches long is common for "great blues," which are most abundant in southern Florida in winter (when migratory individuals join the resident population), in both interior freshwater wetlands and in saline coastal habitats. This species objects to the close presence of other members of its species and tends to feed alone. Aided by its large size, the great blue heron often fishes in water up to about two feet, which is much deeper than other herons except the closely related great white heron. If attracted by abundant prey, great blues will even swim — like ducks — in water too deep to touch bottom.

GREAT WHITE HERON

There is an ongoing disagreement among ornithologists as to the status of this bird, whether it is merely a color form of the great blue (its current status) or a separate species. Whatever the case, the great white heron is almost completely confined to saline habitats of southern Florida and the West Indies. Close to extirpation (biologists use this term for local extinction) in Florida after the 1935 Labor Day hurricane, it has rebounded well, but Hurricane Donna (1960) set it back. It is now seen often, usually standing alone in the shallows of Florida Bay. Its habitat

FIGURE 19.4 A roseate spoonbill poses at Mrazek Pond, Everglades National Park. (Photo by T. Lodge.)

specialization makes its niche quite different from the niche of the more ubiquitous great blue. These two large herons occasionally interbreed and produce offspring that were once thought to be yet another kind of heron, called Wurdemann's heron, recognized as being like a great blue but with a nearly pure white head and neck. Although rare, Wurdemann's herons are most often seen in the Florida Keys.

During the spring, juvenile and some adult great white herons may move northward from Florida Bay into the freshwater Everglades. At the Anhinga Trail, located in a freshwater area of Everglades National Park, a voracious individual was observed while spearing and swallowing a young largemouth bass (see Figure 14.1) and about 20 six-to-eight-inch golden shiners, which were abundant that year (see front cover). On other days, it caught and consumed a dragonfly and a small turtle and swallowed the remains of a dead anhinga, complete with beak intact. In dry coastal habitats, great whites have been seen catching and eating cotton rats. Thus, while most great whites live and feed in Florida Bay and eat fish similar in size to those selected by the great blue, they will also eat nearly anything else they can swallow. This ability to improvise in unusual situations is part of the niche of this species. Many species lack such versatility and such voracity.

TRICOLORED HERON

Formerly called the Louisiana heron, this species ranges widely over the Caribbean region, throughout Florida, and into closely neighboring areas of the southeast. Farther north it is most common in saline tidal areas, but in Florida it also frequents freshwater habitats. Compared to the great blue heron, this smaller, beautifully colored species is a restless and agile athlete, actively running after the inch-long fish that form the bulk of its diet. Its preferred habitat is a wide expanse of very shallow water; probably two to five inches deep being optimal for its highly active style. Tricolored herons normally feed alone but will take advantage of disturbances caused by other animals. Along a tidal creek near Mud Lake in Everglades National Park, a tricolored was seen walking quickly along the shore while a red-breasted merganser (a fish-eating duck) swam a parallel course just offshore. The small fish between the two were in a state of panic. Both birds benefited from the confusion. Sailfin mollies and killifishes are perhaps the most common prey of the tricolored heron (Figure 19.5). The exquisitely colored breeding adults, with blue beaks and dark, ruby-red eyes, are the most striking in appearance among southern Florida's herons and egrets.

FIGURE 19.5 An immature tricolored heron holds a sailfin molly, Nine Mile Pond, Everglades National Park. (Photo by T. Lodge.)

Reddish Egret

Slightly larger than the tricolored heron, the reddish egret is a specialist of shallow marine flats at low tide and is seldom found inland. It comes in two color phases: a rusty-brown head and neck with a slate-gray body that is similar to the smaller little blue heron; and a white form, which can be confused with the smaller snowy egret and immature little blue heron. The antics of this highly active species include the athletic ability of the tricolored heron in addition to the regular use of wings as a sun visor for shading and for maneuvering. Its athletic performance, however, requires fairly regular rests, during which the bird stands motionless for a short time before resuming its agile endeavors. Experienced observers can recognize the reddish egret at a distance purely by its behavior. When this species "fluffs" its feathers, the display of shaggy head, neck, and back feathers is quite a surprising treat. During feeding, these feathers are normally held flat against the body. The species' decorative feathers brought it closer to extinction by plume hunters than any other species. Its recovery since the early part of this century has been very slow.

Great Egret

Through its series of official name-changes from the misnomer, "American egret" (the species is found not only throughout most of North, Central, and South America but also in Europe and much of Africa, Asia, and Australia) to "common egret" (but it was almost hunted to extirpation in North America because of its beautiful plumes) to the current "great egret," the bird has remained the same. "Great patient egret" would be a more descriptive name for this species. Periods of nearly motionless stance followed by a slow, careful walk describe its normal style. It prefers to feed alone in open marsh habitat. However, in times when food is concentrated into pools by low water, it will feed near others of its own species, as well as other kinds of wading birds. Its posture of leaning far forward, with neck outstretched and head held horizontally rather than pointed downward like most herons, is typical. In final efforts of locating prey, it is apt to move its head from side to side, as if to enhance its depth perception. It mainly eats fish, but also takes amphibians, reptiles, and small mammals when the opportunity arises. Many people in residential areas have witnessed great egrets foraging for lizards (anoles) along shrubbery edges, and even taking handouts such as bologna. Based on its wide diet and non-specific habitat preferences, the great egret is probably the best "generalist" among the region's wading birds. Its prey is substantially smaller than prey selected by the somewhat larger great blue and great white herons (Figure 19.6).

Snowy Egret

What the snowy lacks in size, this smaller heron offsets by aggressive feeding. Yellow slippers and fast footwork are the trademarks of this gregarious species, which is commonly seen feeding in flocks with various other species. Its behavior varies more than any of its competitors. When wading in shallow water, this swiftly reacting bird often uses its feet in creative ways: stamping, stirring, and probing to flush out potential prey or even chattering its bill in the water. Some of these actions attract its principal food, the mosquitofish, an inch-long species that is apt to investigate disturbances in seeking its own food. At times, a snowy will search slowly and carefully, but more often it acts nervously, turning quickly and jabbing at prey, especially when competing with other waterbirds. It may even feed on the wing, like a sea gull.

Little Blue Heron

The most characteristic feeding behavior of the little blue can be described as meticulous investigation. This species is most at home along the edges of bodies of water where it walks very cautiously and carefully examining any obstacles such as stones, plants, and logs. While the little blue will feed competitively with other herons in shallow, open water, it looks awkwardly out of

FIGURE 19.6 A great egret with a very small meal, a young golden topminnow. (Photo by T. Lodge.)

place, like it has been pushed into an activity for which it has limited enthusiasm or training. The little blue's preferred techniques give it a strikingly different diet than the previously described species. Small frogs and polliwogs, insects, grass shrimp, crayfish, spiders, and other invertebrates form a much larger proportion of the little blue's diet than is normal for the other herons and egrets (Figure 19.7). Much of its prey is caught on vegetation above water level, but like all the other wading birds, it will take available fish. Identification of juveniles is tricky because they are white and resemble snowy egrets.

FIGURE 19.7 A little blue heron with a crayfish, Corkscrew Swamp Sanctuary. (Photo by T. Lodge.)

GREEN HERON

With limited observation, the green heron (formerly the green-backed heron) looks like a bird with no neck. Even experienced observers are surprised when this squat little heron suddenly thrusts forth a neck far longer than conceivable from the well concealed S-shaped coil beneath a cover of feathers. Its legs are relatively short for herons but deceptively longer than they appear during the bird's normal crouched stance. Found in numerous varieties throughout the world's temperate and tropical climates, green-backed herons are most at home perched on rocks, roots, or tree branches, waiting for prey to come within reach. They seldom wade but will stand quietly in very shallow water in protected areas. When potential prey comes within range, this little predator may begin to stretch out, with neck and legs extended, in anticipation of striking and may even leap from its perch into the water.

This species is one of the few in the heron/egret group that prospers in years when no pronounced drydown occurs. It is easily understood why the green heron is a poor competitor in a shallow, dry season pond full of other, much larger wading birds. However, what this little heron lacks in physical ability is compensated by innovation: some individuals learn to attract prey with bits of food. This phenomenon has been observed where fish food is available such as in zoos and outdoor public aquariums. The birds will find a piece of food, drop it in the water, and catch the fish that arrive to eat it. In the wild, individuals may use berries or other objects to attract curious small fish. This learned, specialized skill helps the green heron feed in locations where fishing might otherwise be poor and where there is little competition.

BLACK-CROWNED AND YELLOW-CROWNED NIGHT-HERONS

While most other herons will feed at night, the night-herons, which are common in southern Florida, are specialists at nocturnal feeding and are seldom seen in daytime feeding groups of the other wading birds. Both species have large eyes, which are an incredible deep red in adults. Both feed heavily on crabs, crayfish, and other large aquatic invertebrates but will take fish without hesitation.

FIGURE 19.8 A juvenile black-crowned night-heron with a damselfly, Anhinga Trail, Everglades National Park. See text for explanation. (Photo by T. Lodge.)

In a display of creativity similar to the green heron, a juvenile black-crowned night-heron was observed catching a damselfly, which it did not eat (Figure 19.8). After killing the insect with a pinch, it placed the damselfly on the water surface and then slowly backed away six or eight inches to watch motionlessly. A minute passed as the damselfly drifted about a foot, when the heron carefully retrieved the floating insect, minced it slightly, and repeated the process — several times. No fish approached, but the bird obviously appeared to by fishing, a behavior not normally associated with this species. Better known are the black-crowned night-heron's occasional infamous raids of nesting colonies to feed on the chicks of other wading birds.

WADING BIRD ROOKERIES [224, 449]

Most of the wading birds nest in large colonies, typically in trees on small islands such as mangrove islands in Florida Bay or tree islands in the Everglades (Figure 19.9), but white ibis have been known to nest in sawgrass. The social nesting aggregations are called *rookeries* and usually contain a mixture of species, with great egrets perhaps nesting close to tricolored herons and white ibis. Some birds fly great distances from their rookery to foraging locations. Wood storks, unusually adept at soaring on rising air currents (thermals), often commute as far as 35 miles, and a maximum of 80 miles has been observed.[448, 510] Great egrets normally fly relatively shorter distances, and the majority travel less than five miles; however, small numbers do fly more than 15 miles and occasionally as far as 25 miles. Most white ibis commute less than six miles, and while much longer flights have been recorded, they are usually by birds from failing (starved) colonies.[57]

Rookeries commonly contain anhingas, cormorants, and brown pelicans in addition to wading birds. A few wading birds, such as green herons and reddish egrets, do not join rookeries but instead nest in dispersed locations. Outside of the breeding season, most of these species, and other wading birds disperse, using shifting sites for roosting.

FIGURE 19.9 A snowy egret nest braced on a cardinal airplant growing on a small cypress, western Water Conservation Area 3A. (Photo by T. Lodge.)

Most rookeries of the Everglades were formerly located in the southern part of the ecosystem near the mangrove-marsh ecotone, now in Everglades National Park. Rookery Branch, located at the head of the Shark River and once home to the largest of the region's rookeries, is now abandoned. In the 1930's, it harbored as many as 200,000 birds, mostly white ibis. Such huge rookeries have been called "super colonies." Many of the birds that formerly nested in the park have moved north into the water conservation areas where new rookeries were initially unstable. The numbers of nesting birds there has increased greatly beginning in the late 1990's and continued into this century. Corkscrew Swamp, the most famous existing rookery in southern Florida, is located northwest of the Big Cypress Swamp. It is almost entirely dominated by wood storks. Chapter 21 includes a discussion of the major changes in rookeries. [208, 210, 224, 325, 516]

THREATENED AND ENDANGERED SPECIES [307, 364, 510, 614]

Several Everglades birds are listed as *threatened* or *endangered* under Florida and/or federal regulations. In addition to these categories, Florida regulations also contain a status of lesser jeopardy called *species of special concern*, which has been given to several wading birds such as the white ibis, little blue heron, and tricolored heron. Prominent issues in management of the Everglades have focused on three endangered birds: the wood stork (discussed elsewhere in this chapter), the snail kite, and the Cape Sable seaside sparrow.

SNAIL KITE [63, 64, 157, 312, 421, 576]

The snail kite (formerly called the Florida Everglade kite) is related to hawks and eagles and is about the same size and shape as the commonly seen red-shouldered hawk. It differs from the hawk in that it is generally dark brown (female) or slate black (male), with white at the base of the tail. The beak of the snail kite is narrow with a pronounced hook. This adaptation allows it to eat its primary food, the applesnail, which it snatches in its talons from the water while hovering. Snail kites seldom take other prey, but they have been known to catch small turtles and crayfish when snails are scarce.[614]

The specialized diet renders the snail kite extremely habitat-specific to shallow, long-hydro-period marshes that contain sufficient applesnail populations, where the snails are visible and the water surface is not obstructed by dense vegetation. Sloughs with scattered emergent vegetation and long-hydroperiod wet prairies are optimal habitats, provided that trees are locally available for perching. Snails are most vulnerable while breathing; they are usually caught while on vegetation within about four inches of the surface and are probably safe from predation much deeper than that.

It was formerly thought that snail kites were not apt to relocate in search of suitable habitat when local conditions deteriorated, but tracking studies have shown that individuals often move great distances throughout their range in Florida, which extends from the southern Everglades northward to the St. Johns marshes east of Orlando. However, the Florida population does not migrate and is therefore separate from other populations of the species living in Cuba and Central and South America.

CAPE SABLE SEASIDE SPARROW [136, 157, 347, 383, 384, 444, 467, 471, 617, 629]

The seaside sparrow is a small, cryptically colored habitat specialist, seldom found outside of coastal salt marshes. It builds nests woven of grasses and suspended in dense grass tussocks and feeds mostly on insects and spiders in a secretive life style. Four subspecies have been recognized. The two most abundant occur along the northern gulf coast from just north of Tampa, Florida, westward through Texas, and along the Atlantic coast from northeast Florida to New England. A third subspecies is the Cape Sable seaside sparrow ("CSSS" is its widely used acronym), named for its former occurrence in salt marshes on Cape Sable. It is non-migratory and endemic to the extreme southern Florida mainland where its current restriction to freshwater wetlands is unusual for the species. The entire range of the CSSS is within Everglades National Park and the southern Big Cypress National Preserve — the most restricted range of any North American bird. Its small populations are confined to areas of marl prairie habitat dominated by muhly grass on both sides of the Shark River Slough where its highly sedentary life style is sensitive to fire and water levels. Also, it avoids sawgrass habitat.

This Cape Sable subspecies attracted attention when a population monitored between the 1950's and the 1970's decreased by an estimated 95 percent. Further declines occurred in the mid-1990s because of high water. The total population of this endangered subspecies is now on the order of a thousand individuals, with some estimates as low as several hundred. Nests are placed in the sturdy bases of muhly grass tussocks, close to the ground, in the dry season. Accordingly, flooding deeper than about four inches during the breeding period has been shown to inhibit nesting or cause nest abandonment. Fire is required for maintaining optimum conditions by preventing succession to shrub habitat, but fire also renders habitat unsuitable for about a year due to the short stature of recovering grasses — not suitable for cover or nesting. Given the small areas of remaining populations and the constraints of water level and fire, the plight of the CSSS is understandable, and a closely related fourth subspecies, the dusky seaside sparrow of the Cape Canaveral area, was officially recognized as extinct in January 1991. The CSSS issue may be the most difficult legal and technical constraint in Everglades restoration, discussed in "Dangers of the Endangered Species Act."

A CONTEST OF BEAUTY

Details of southern Florida's avifauna can go on endlessly; however, a contest of beauty presents a fitting finale. The most striking of the waterbirds (and not formerly mentioned) is the purple gallinule. This bird has a general shape similar to a small chicken, but it exhibits truly unbelievable colors, which are most enhanced in breeding plumage. The bright red beak is tipped in yellow. The forehead is light blue, and the remainder of the head and the entire breast are a deep purple-blue. The back is an iridescent dark green. The sight of one of these long-toed birds walking across lily pads in the sunlight is breathtaking. The most striking land bird is a winter resident, the male painted bunting. Sparrow-sized with somewhat secretive habits, it is not easily found but is surprisingly worth the search. With fire-red breast and rump, bright green back, and rich, dark-blue head, it seems almost mythical. How such color can survive in the wild stretches the imagination. However, if this contest were based on form and grace alone, the species that stands out is the elegantly graceful American swallow-tailed kite (Figure 19.10), a relative of hawks, which is a spring and summer breeding resident of Florida.[660]

DANGERS OF THE ENDANGERED SPECIES ACT

Most conservationists view the federal Endangered Species Act (ESA) of 1973 — and its numerous state equivalents — *as powerful tools that help in the protection of disappearing wildlife species. Indeed, the track record of protection has been impressive, with many species saved from nearly certain extinction and many others recovering to the extent that they have been "down-listed" or even "de-listed" (removed from listed status such as threatened or endangered). Although the goal of the act is to preserve ecosystems, its framework is focused on preventing the extinction of individual species. These more narrowly defined legal requirements can force collision courses between species having conflicting ecological requirements,*

---

* Under Florida law, selected animals are listed by the Florida Fish and Wildlife Conservation Commission under Rules 39-27.003 through .005, Florida Administrative Code (FAC) under the categories of "endangered", "threatened", or "species of special concern." State-listed plants are categorized as "endangered", "threatened", or "commercially exploited" under authority of the Florida Department of Agriculture and Consumer Services in Chapter 5B-40, FAC. The state's listing process is independent of the federal process, adding complexity to decisions where species on both federal and state lists are treated under two sets of regulations. The intent is admirable, but it is easy to understand that the complexity can be counterproductive by encumbering decisions by well meaning responsible parties.

FIGURE 19.10 A swallow-tailed kite carries nesting material, Spanish moss, over Coral Gables. (Photo by T. Lodge.)

or can become obstacles to restoring ecosystems because of conflicts between the specialized requirements of individual species and the broader goals of ecosystem restoration. Ecosystem manipulation to protect one listed species may further jeopardize other listed species, or may even imperil relatively common species to the extent that their listing must be considered. Accordingly, the ESA and its state equivalents are "two-edged swords."

The ESA's statutory requirement to avoid "jeopardy" is at the heart of the matter. No federal action can be taken that would likely jeopardize the continued existence of any federally threatened or endangered species, or result in the destruction or adverse modification of its critical habitat.[60] Thus, if the U.S. Fish and Wildlife Service (or the National Marine Fisheries Service, depending on the particular species) deems an action to potentially jeopardize the existence of a listed species, the action cannot be taken even if the intent is to provide benefits to other listed species. Relocations and/or captive breeding programs provide possible but rarely authorized exceptions.

Use of the ESA as a tool by an outside interest is also a danger to ecosystem restoration. Third parties may challenge agency decisions to move ahead with a project, using additional evidence that was not considered or, in its view, considered improperly. The suit may merely delay implementation of a project, or may be successful in stopping it, i.e., effectively reversing the agency decision. There are numerous cases where a third party has used the ESA as a tool to stop a project. Best known is the famous snail darter case, where a small, endangered fish forced the modification of a partially completed dam on the Little Tennessee River.[29] Conservationists hailed the case as a victory, but the same strategy could be used by an affected landowner to stop or delay Everglades restoration, to the dismay of most conservationists. At this time, the most obvious species used in this regard is the Cape Sable seaside sparrow (CSSS); but an earlier conflict involved the snail kite (see Chapter 21, "Modified Water Deliveries"). With the length of time and diversity of individual projects involved in the

Comprehensive Everglades Restoration Program (CERP, see Chapter 21), the use of the ESA for other listed species is likely. It has been determined that 36 listed species (federal and state) are likely to be affected by CERP.[612]

In 2003, the Department of the Interior engaged a team of scientists to discuss the potential effects that implementation of CERP might have on four listed species: the CSSS, wood stork, snail kite, and roseate spoonbill (the last state-listed only). The results were generally encouraging that overall restoration would ultimately benefit all of them, but it was recognized that intermediate steps of the restoration process could be problematic.[72] Legal constraints imposed by the ESA do not allow for interim reductions of a listed species even if in the long term, recovery should be enhanced.[60] The ESA was written before its implications on large ecosystem restoration programs were widely considered. The focus was stopping the alarming loss of individual species. In that regard, it should be noted that the USFWS "South Florida Multi-Species Recovery Plan" (MSRP)[614] is a monumental compilation all the recovery plans and species-specific information for the 68 federally listed species that occur in the South Florida ecosystem. The MSRP is intended to coordinate recovery programs with CERP. However, because of the limitations of the ESA, the MSRP does not provide guidance on multi-species strategies, as its name suggests, because it cannot resolve the problem of conflicts between listed species and what to do about them.

In the long run, the present ESA (and state equivalents) could cause more extinction than would occur if well-conceived ecosystem strategies were incorporated. To be blunt, single-species management is a trap. Protection of a species or population should be weighed against wider ecological values, i.e., general ecosystem functions including a suite of representative and keystone species that characterize the broader health of the ecosystem. In this context, triage might sometimes be required — not permissible in the present ESA. Most conservationists would look in horror at a decision that subjects a species, subspecies, or population to extinction or extirpation. However, the consequences of steadfast protection must be weighed against the potential demise of an ecosystem one species at a time, ending in ultimate failure of the original intent — to protect species from extinction. Clearly, the ESA needs revision — not elimination — to emphasize broad-based multi-species and ecosystem strategies as principal tools to prevent the extinction of species.

20 Synthesis — Ecological Relationships and Processes in the Everglades Region

The foregoing chapters have provided a basis for understanding the plant communities, flora, and fauna of the region. This chapter serves to put some important ecological interrelationships, widely scattered through the book, into one place. Two themes are provided: succession and food chains.

SUCCESSION [26, 128, 176]

Succession is the process of change from one plant community into another as a result of environmental circumstances. For example, a pineland in southern Florida changes into a tropical hardwood hammock without fire, a normal successional sequence. However, with regular fire, the process is set back and pineland is maintained. These examples of succession are mentioned in both Chapters 5 and 6. Many other successional relationships are mentioned in Chapters 2 through 11.

A generalized diagram showing the successional relationships of all South Florida plant communities is provided in Figure 20.1. Processes are not shown, only the relationships. At the bottom of the chart are permanently flooded habitats. Given stable sea level, these aquatic or marine habitats will accumulate sediments that lead to the establishment of wetland plant communities. Freshwater areas will lead to sawgrass and related shallow wetlands, and marine or estuarine areas will lead to mangrove swamps. At the top are hypothetical uplands without vegetation. Both examples of soil types and conditions lead to pinelands. Slash pine forests would dominate moist locations, while sand pine would dominate dry, well drained soils. The latter have not been covered because they are not related to Everglades habitats. They are dominated by sand pine (*Pinus clausa*) on peripheral habitats, including dunes of the Atlantic Coastal Ridge, from the Fort Lauderdale area and north. These habitats are now rare (mostly developed) but were likely to have been widespread over the peninsula when sea level was lower several thousand years ago, possibly even dominating the peninsula.

Succession of all communities, in the absence of fire and with stable sea level, leads to the ultimate or "climax" plant community of South Florida, the hardwood forest — more tropical in the southern portion of the Everglades region, and more temperate to the north. However, all sequences are reversible, mostly due to the effects of fire and changing water levels. Using this diagram as a framework, many scenarios can be developed for particular situations.

EVERGLADES PEATLAND SUCCESSION [252, 402]

To focus on the driving forces of the succession, Figure 20.2 shows known and hypothetical successional relationships in the Everglades peatlands, which comprise the vast majority of the ecosystem. Plant communities are shown in ovals and environmental factors, and processes that maintain or change plant communities are shown in boxes. Six drivers of change are identified:

- **Surface fire** that removes growing and dead vegetation but does not affect soil or roots
- **Soil fire** that reduces peat soil elevation and kills plant roots

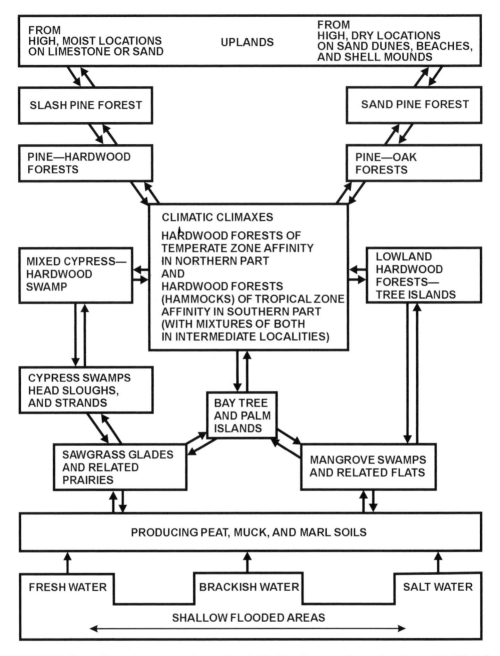

FIGURE 20.1 Generalized plant succession in South Florida. See text for explanations. (Modified from Alexander and Crook.)[24, 26]

- **Water flow** and its resulting erosion and transport of sediments
- **Alligator activity** that uproots vegetation and prevents soil accretion
- **Soil accretion** resulting mostly from accumulated plant material in saturated soils
- **Peat pop-up** (battery formation) in sloughs, providing a base for bayhead development, and likely perpetuating deep slough habitat

The two roles of fire are very different. Common surface fires burn standing vegetation without damaging roots or burning into soil. In the Everglades, such fires normally perpetuate the sawgrass

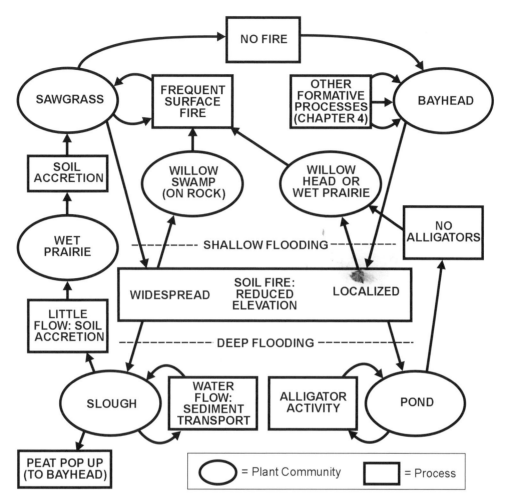

FIGURE 20.2 Plant community succession and its driving forces in Everglades peatlands. See text for explanations. (Modified from Ref. 252 with ideas from Refs. 7 and 40, and experience of the author.)

community (except temporarily when rising water follows soon after the fire: see the Sawgrass Marsh section, Chapter 3). Without such fire, shrubs would begin to invade with or without significant soil accretion, such as in the formation of strand tree islands (Chapter 4). The other role of fire is central to major or catastrophic changes in plant communities and occurs during severe droughts when higher elevation peat surfaces become dry enough to catch fire. Elevation of the peat surface is reduced, sometimes to bedrock where soils are shallow, creating very different plant communities — depending on the range of seasonal water depths — from sloughs or ponds (deeper) to wet prairies or willow swamps (shallower).

Opposing the effect of soil fire in wetlands is soil accretion. Accretion results mostly from the accumulation of plant material, particularly roots, but also from sediments trapped from moving water that is slowed in its passage through dense stands of marsh plants (see The Pre-Drainage Lake section, Chapter 11). Peat soil elevations in marsh communities can rise up to the level where dry-season aerobic decay balances wet-season anaerobic accretion. This balanced elevation relative to water levels is higher where trees and shrubs contribute woody peat, which is more resistant to decay (Chapter 4), so that bayheads are maintained at a higher elevation than sawgrass marsh.

Two factors that prevent soil accretion and serve to maintain deeper areas such as sloughs and ponds are water flow and alligator activity. The importance of water flow as a formative process has recently gained attention although the obvious orientation of plant communities with the direction of flow in the Everglades has been seen for more than a century. It is hypothesized that historic Everglades slough communities did not fill-in because water moving through them prevented accumulation of sediments and perhaps eroded the bottom. How sloughs became elongated is not understood. Figure 20.2 shows the origin of sloughs only in depressions created by fire, not any process that would result in elongation with the direction of water flow. Slough peat (*Loxahatchee peat*) is known to break loose because of its buoyancy, producing pop-ups (Chapter 4). Whether this process is important to maintaining and shaping sloughs is also not known. Obviously, there is more to be learned about the development and maintenance of slough landscape.

The influence of alligators in preventing the succession of ponds — many probably created as fire pits — is better understood and easily observed. Without the activity of alligators — especially larger individuals — ponds eventually fill in and become emergent wetland communities that would ultimately succeed to sawgrass marsh. This important, key role that large numbers of alligators have played in maintaining Everglades landscape and wildlife cannot be over emphasized (Chapter 17).

While Figure 20.2 shows the various processes of change, the end result in the natural environment is creation of, and then maintenance of, the ecosystem's features. Ecosystem steady-state maintenance, however, is based on relatively constant hydrology — within cyclic variations of rainfall that have recurred for the past century and stable sea level. The great changes that occurred through the 20th century to the present are the result of human impacts, especially altered hydrology — including flow patterns — but also departures from natural fire regimes, reduced numbers of adult alligators, the introduction of invasive exotic plants, and numerous other influences. By comparison, rising sea level has been a relatively minor influence. These impacts are covered in Chapter 21.

FOOD CHAINS AND FOOD WEBS [317, 459]

Figure 20.3 is a generalized food-chain diagram, which simplifies the transfer of biomass or energy among categories of living organisms (arranged in a "chain") and the environment. Green plants form the base of food chains. In that role, they are called *primary producers* by ecologists. Green plants use sunlight in photosynthesis, initially to split water into hydrogen and oxygen, a reaction that requires energy. "Reduced" hydrogen is incorporated into energy-containing substances that drive the biochemistry of plant growth and produce energy-storing substances such as sugars, which can also be linked together to form cellulose, a major plant structural material. Oxygen is a waste product of the photosynthetic process. In the food-chain steps, each called a *trophic* level, green plants (trophic level one) become the food of herbivores (trophic level two), also known as primary consumers. Herbivores consume plants for their growth and reproduction, and they themselves become the prey of carnivores (trophic level three), or secondary consumers. Lower carnivores become the prey of higher carnivores (trophic level four and higher) in the food chain that may have more levels ending in "top" carnivores. Dead material becomes *detritus* — energy-containing material available to decomposers: a bacteria and fungi. The preponderance of detritus is plant remains, either whole plants or plant parts such as twigs, leaves, flowers, dead seeds, and roots. Bacteria and fungi are the principal organisms that use detritus as a food source.

Detritus still contains abundant energy, much in the form of carbohydrates such as cellulose from plants. The microorganisms of decay attach to, or surround, detritus as they degrade it — thus the physical attachment between detritus and decomposers shown in Figure 20.3. Detritus is typically low in nitrogen, which is needed to make protein. But in association with decay microorganisms, detritus becomes enriched as these organisms fix nitrogen and uptake phosphorus and other nutrients from the surrounding medium. They use the detritus mostly as their source of energy and carbon.

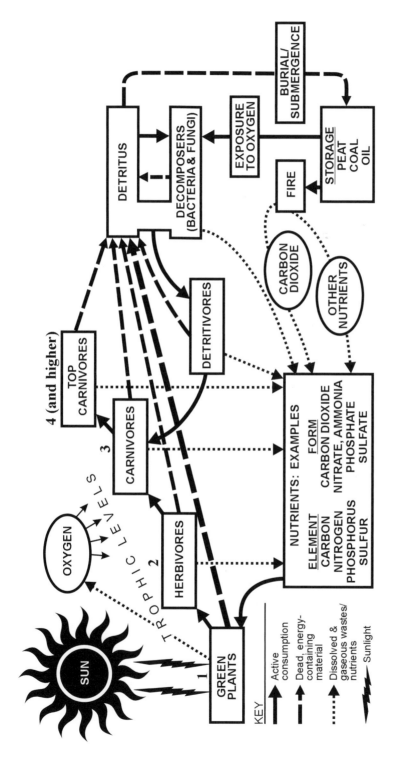

FIGURE 20.3 A generalized food chain and nutrient cycling diagram. See text for explanations. (Developed from the author's general knowledge.)

Detritus enriched by microorganisms is then exploited as food for another category of animals — the *detritivores*. Most detritivores are microscopic to small organisms — rotifers, small oligochaete worms related to earthworms, insect larvae, and small crustaceans such as amphipods. Their various means of feeding include direct consumption of detritus in sediments, filtering particles from the water, scraping surfaces of detritus and removing attached organisms, or shredding and consuming whole particles, such as tiny leaf fragments, together with the attached microorganisms. Larger organisms that consume detritus also eat a variety of other organisms in the process. In the Everglades, for example, the lake chubsucker is a detritivore, but it also consumes periphyton and small bottom-dwelling animals as it selects material in bottom sediments and scrapes surfaces. Detritivores, then, are eaten by various predators (carnivores) as merely another trophic link in the overall cycle of eat and be eaten.

As a generalization, only about ten percent of the energy of each trophic level is resident in the biomass (essentially the amount of living material) of the next level. About 90 percent is lost in the effort expended in finding food, digestion, growth, reproduction, and other work. Waste products are not shown proportionally, but the generalization is shown that plants produce relatively little waste material, but produce the most detritus of all trophic levels, proportional to their *biomass* (the total amount of energy in each tropic level). All consumer levels — herbivores through top carnivores — and decomposers produce wastes products containing nitrogen, phosphorus, sulfur, carbon dioxide and other wastes, all of which are reused by plants in their growth, and the cycle continues.

Not shown in the Figure 20.3 is the reality that animal species are seldom confined to a specific trophic level. Herbivores, for example, are often opportunistic carnivores. Adult Florida redbelly turtles are normally herbivores but will scavenge any available flesh or take an easy target such as a young apple snail. Some animals eat plant and animal matter as their regular course. The raccoon is a prime example in the Everglades, as it eats seeds and fruits, crayfish, fish, insects, and other animal life. Such a dietary habit is called *omnivory*, so the raccoon is termed an *omnivore* by ecologists — not assignable to a trophic level but bridging between herbivores the several carnivore levels.

Another feature of Figure 20.3 is storage — detritus that is protected from decomposers for long terms. Peat is the general storage medium in the Everglades, but on the scale of geologic time, coal and oil are more permanent storage products. The energy of each of these products can be released by fire, and peat is decomposed by bacteria and fungi if oxygen is present (see Marsh Soils, Chapter 3).

The understanding of food chains for specific ecosystems has important applications. One application is to understand the consequences of altering habitats so that prey species are affected. That change, in turn, affects predator species by altering their food base. Another application is in understanding the transfer of toxins within an ecosystem. It is known that many pesticides and mercury, for example, are passed through the food chain and concentrated with each step, a process called *biomagnification*. Knowing how many trophic steps are involved helps researchers predict the biomagnification consequences of different food chains. For example, if eagles in one environment regularly prey on ducks that have mostly eaten seeds of plants, then the food chain is short — only two steps from the plant producers. But if the eagles eat largemouth bass that have eaten spotted sunfish that have eaten mosquitofish that have eaten mosquito larvae that have grazed on periphyton, then the number of food chain steps is much larger — a total of five trophic steps above the plant producers. The potential difference in biomagnification can be huge. Understanding how toxins can be transferred through food chains obviously has important applications in human health, for example in defining risks of exposure for hunters and fishermen.

EVERGLADES FOOD WEB [252, 405, 477, 479, 480]

In spite of its apparent complexity, Figure 20.4 shows a greatly simplified food web for the freshwater Everglades marshes. Using the term *food web* instead of *food chain* emphasizes the

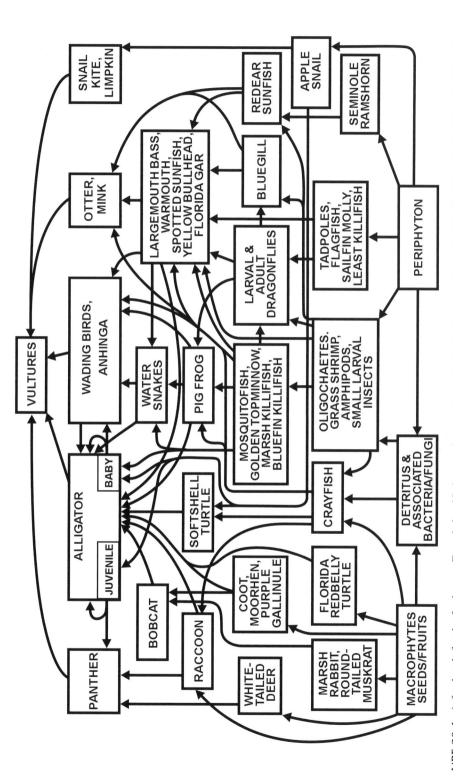

FIGURE 20.4 A food web for the freshwater Everglades. Native species mentioned in this book are emphasized. (Modified from Refs. 252 and 478, and using data from Ref. 292.)

multiple trophic pathways of real life. The tropic interrelationships of species mentioned in the text are shown and simplified to show the more common pathways. The figure shows that periphyton and detritus/microorganisms are the basis of most of the food web, with smaller numbers of species depending on live, rooted plants (macrophytes) such as leaves, roots, or fruits and seeds. In terms of numbers and biomass, the top predators in the Everglades are the alligator and wading birds, with much smaller roles for other predators, such as the northern river otter, Everglades mink, snail kite, and limpkin. The panther, shown for its publicized role as a top predator, has little presence in the Everglades but is important in the Big Cypress Swamp. The much more common bobcat has a larger role in the Everglades than shown in Figure 20.4. Bobcats are important predators of wading birds, such as white ibis, and sometimes even raid wading-bird and anhinga nests in rookeries. Although smaller in numbers, the food-web diagram also identifies vultures at the highest trophic level. Vultures are not predators; they avoid living prey. They are, though, in a potential trophic position to be top recipients of biomagnification of toxins, one of the important reasons for studying trophic relationships.

Omnivory is addressed by showing divergent food sources for some boxes. Crayfish, for example, are shown to consume plant material, detritus, and animals such as worms, larval insects, and other invertebrates. In fact, crayfish are opportunistic predators on any animal they can subdue. Such omnivorous tendencies are much more common than shown in Figure 20.4. In the box identifying tadpoles, flagfish, sailfin molly, and least killifish, the only species that is heavily dependent on periphyton is the sailfin molly. Anyone who has kept an aquarium of native fish has probably watched flagfish, which consume any algae offered, also attack small invertebrates such as grass shrimp without hesitation, and larval amphibians (tadpoles or pollywogs) are mostly herbivores but are known for cannibalism, given opportunity. Upon transformation to adult frogs they become strictly carnivorous so that food web diagrams need to show larval and adult frogs separately. Pig frogs, for example, eat any fish they can catch, applesnails, Seminole ramshorns, and nearly anything else that moves and fits in their large mouths (including baby water snakes, which become predators of pig frogs as adults). The diagram does not have enough space to show all the arrows.

Many other species, groups, and arrows could be shown in the Everglades food-web diagram. The important and diverse roles of many invertebrates are omitted in favor of emphasizing species that are more familiar. Examples of higher trophic level species that are also omitted include the red-shouldered hawk and barred owl. Many of these relationships are explored in the identified literature sources.

Lake Okeechobee Food Web [217, 263, 265, 459]

Figure 20.5 presents a very general, simplified food web for Lake Okeechobee's open-water pelagic zone. The simple, generalized food chain for the open waters of unenriched lakes is: phytoplankton to zooplankton to fish. The figure shows that Lake Okeechobee's pelagic food web is more complicated, with its base being phytoplankton, bacterioplankton, and suspended detritus particles (with their associated bacteria and fungi). Rooted plants and periphyton that prosper on sunlit surfaces of wetlands and of typical lake littoral zones also occur in Lake Okeechobee's pelagic zone as scattered rafts of floating material, adding more complexity (not shown).

Phytoplankton is composed of many species of algae that exist as single cells or small groups of cells, and most are so small that magnification of about 400X is required to effectively identify them, but some that live as large groups of cells can be seen by the naked eye. Water rich in phytoplankton is usually tinted green. Also suspended in the water are free-living bacteria, often called bacterioplankton. These organisms are considerably smaller than phytoplankton. The lake's pelagic herbivores are kinds that can harvest these microscopic foods. They include tiny organisms termed zooplankton, larger mostly bottom-dwelling macroinvertebrates such as clams and aquatic insects, and certain fishes that are capable of filtering out phytoplankton and detritus particles.

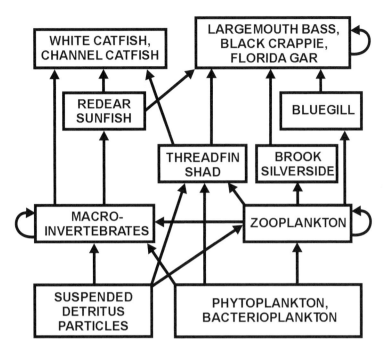

FIGURE 20.5 Lake Okeechobee pelagic food web, emphasizing native species and groups. (From Refs. 217 and 263.)

(With recent lake enrichment, detritus and bacterioplankton have become relatively more important, and phytoplankton has become dominated by cyanobacteria or blue-green algae, which are less used by zooplankton.* See Chapter 11.)

Zooplankton includes a range of small animals from microscopic single-celled forms, such as ciliates, to invertebrates visible to the unaided eye (perhaps requiring some near-sightedness). Zooplankton organisms swim and may effectively rise and descend in depth as they seek phytoplankton, but move slowly relative to currents and are thus displaced around the lake by wind-driven movement of water. Excluding single-celled forms, over 60 species of zooplankton have been identified in Lake Okeechobee. Abundant kinds include crustaceans such as tiny copepods (e.g., *Diaptomus dorsalis*) and "water fleas" (e.g., *Daphnia ambigua*). They are eaten by bluegills, specialists among sunfishes in harvesting small prey because of strainers associated with their gills. When not breeding in the littoral zone, adult bluegills roam in large schools through open water to feed on the larger zooplankton. The minnow-sized brook silverside is another fish that schools and feeds on zooplankton.

Macroinvertebrates include forms that are larger than zooplankton and are mostly associated with the bottom and its sediments — *benthic* macroinvertebrates. Lake Okeechobee's benthic environment harbors innumerable species of worms, snails, clams, and aquatic insects. Other macroinvertebrates live on the surface. Small snails can float by attaching their foot to the surface and many aquatic insects ride on the surface tension, such as whirligig beetles and spider-like bugs called water striders. Many aquatic insects and insect larvae swim in the water column, at least temporarily. Macroinvertebrates have many means of harvesting detritus, phytoplankton, bacterioplankton, and zooplankton. Feeding techniques include scraping surfaces and filtering water. Clams are mostly filter-feeders. Other macroinvertebrates including many aquatic insects are

---

* See Havens, K.E., and T.L. East. 1997. Carbon dynamics in the grazing food chain of a subtropical lake. *Journal of Plankton Research* 19: 1687-1711; and Havens, K.E., K.A. Work, and T.L. East. 2000. Relative efficiencies of carbon transfer from bacteria and algae to zooplankton in a subtropical lake. *J. Plankton Res.* 22: 1801-1809.

predatory on other macroinvertebrates and even small fish. Examples are dragonfly larvae and predaceous diving beetles. Lake Okeechobee's pelagic zones harbor about 120 species of macro-invertebrates.

Phytoplankton is also grazed directly by a few fish. The gizzard shad and juvenile threadfin shad are filter-feeders comparable to the anchovies of marine and estuarine habitats. They effectively harvest phytoplankton, zooplankton, and suspended detritus particles, which are often abundant in the lake's open waters as the result of sediment suspension. The gizzard shad is primarily an inhabitant of the littoral zone and remains a phytoplankton grazer and a "surface-scraper" its entire life. The threadfin shad, becoming much more dependent on zooplankton as it grows, is abundant in the pelagic zones. With its mixed diet, the threadfin shad is an omnivore.

Many fishes prey on macroinvertebrates. Redear sunfish are avid snail-eaters, but also consume many other invertebrates. Smaller catfish including a miniature species called the tadpole madtom (not shown in Figure 20.5) also prey mainly on macroinvertebrates. The tadpole madtom remains a macroinvertebrate specialist its entire life, feeding heavily on amphipods, but larger catfish species such as the white and channel catfishes grow to become important predators on other fishes as well as larger invertebrates. However, the lake's top predatory fishes that become mostly dependent on smaller fishes include the largemouth bass, the black crappie, and the Florida gar. Small numbers of the huge longnose gar, up to five feet long, are also top predators in the lake.

The pelagic zones of Lake Okeechobee are quite different from the littoral zone. The littoral zone is much like the Everglades marshes at the base of the food web, and includes the same top predators: alligators, wading birds, snail kites, otters, and so forth. The difference lies in the large number of juvenile fishes, of several species not found in the Everglades, that use the littoral zones as a nursery area and then move into the pelagic area for much of their adult life.

Mangrove Food Web [218, 266, 340, 447]

Figure 20.6 presents some important generalizations about the mangrove swamp food web. As an intertidal habitat dominated by trees that shade the waters below, detritus from the mangroves is the most important base of the food web, described in Chapter 8. There is a relatively less important but significant contribution by algae, much of which is attached to mangrove roots. The important contribution of the mangrove swamp, like the littoral zone of a lake, is to the adjacent open waters. Many important coastal species are tied to mangrove swamps during some phase of their life cycles, using the habitat as a nursery ground and/or the detritus as a food source. Important examples include many game fishes and crustaceans such as the pink shrimp.

The food-web diagrams presented in this chapter can serve as the basis for expanded models of how the various environments of the greater Everglades ecosystem function.

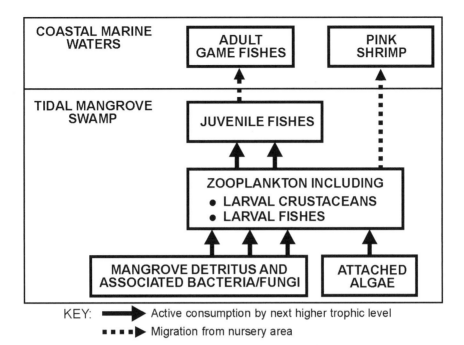

FIGURE 20.6 South Florida mangrove swamp food web showing relationship to coastal marine waters. (From Refs. 81, 266, and 340.)

Part IV

ENVIRONMENTAL IMPACTS

21 Man and the Everglades

The foregoing pages have dealt primarily with natural history, confining references to human impact on the region's environments to the introductory pages and the chapters on the Big Cypress and Lake Okeechobee. At this point, we are ready to explore the degradation of the Everglades — how it happened and the plans to fix it. Because the Everglades itself is linked to many other South Florida landscapes, the interrelated Everglades region (or "South Florida Ecosystem" as it is often called) is the framework of this examination. The impacts on southern Florida's natural systems are evident everywhere. They range from deliberate removal of plants and wildlife to the introduction of exotic plants and animals to the effects of pollutants and toxins to the extensive alterations of the region's natural hydrology.

Obviously, humans have caused the massive changes that have occurred to South Florida's natural environments since the 1880's, yet mankind has lived in the region for many millennia. Thus, it is initially prudent to regress in time and trace the region's human occupation from its beginning. This approach will provide a framework for understanding the recent changes within the context of man's overall activity. Subsequent sections will then cover modern impacts.

NATIVE AMERICANS AND THE EVERGLADES [94, 97, 98, 240, 248, 390, 413, 414, 590]

Numerous estimates place the arrival of mankind in Florida at least 13,000 years ago, nearly as early as the oldest known human evidence in the lower 48 states at Clovis, New Mexico. With sketchy evidence, there is a widespread assumption that humans entered North America from Asia by crossing a land bridge that existed between Siberia and Alaska during the last glaciation, which reduced sea level to its lowest about 20,000 years ago. The low sea level exposed marine shallows between the continents until their reflooding about 10,000 years ago. That region, including much of Alaska, was ice-free, but Canada was entirely covered by ice that would have blocked overland migration for thousands of years. However, there is the possibility that the early invaders were independent of contiguous land, using small boats, like modern Eskimos, for hunting marine mammals. Following a coastal route, their movement between continents and then down the west coast of Canada could have used a series of island encampments to reach the ice-free region of Washington State. Evidence of such a route would now lie under a few hundred feet of water. By whatever initial mode of invading North America, the newcomers wasted little time in crossing the continent to reach Florida.

Archaeologists have characterized Native American habitation in Florida as evolving through four general periods and numerous subperiods. Their ages and characteristics are shown in Table 21.1.

PALEO-INDIANS

Evidence shows that early human arrivals in Florida hunted large animals, the "megafauna," suggesting that these early Americans were influential in the extinction of numerous large mammal species and other groups in Florida, as well as throughout North America (see Chapter 18). The oldest evidence of human habitation in southern Florida is from Little Salt Spring in Sarasota County, immediately north of Charlotte Harbor (see Figure 2.2 and Color Figure 1). An unfortunate Native American had apparently fallen into Little Salt Spring, then an open sinkhole with a water level 87 feet below the level today. Today's level is only a few feet below the surrounding land surface so that an observer is unaware that Little Salt Spring is a deep

TABLE 21.1
Native American Cultural Periods in South Florida — emphasizing relationships to natural environments.[a]

Period	Dates[b]	Comments
Paleo-Indian	13,000? – 9000	Hunted megafauna with perfected stone projectile points, diet supplemented by smaller game and plants; sea level rising at about 6 feet per century (Chapter 10). Probable limited occupation of pre-Everglades terrain.
Archaic:		
Early	9000 – 7000	Extinction of megafauna; onset of dryer conditions; development of baskets and more versatile stone and wooden tools (e.g., chopper, boomerang, mortar); more diversity in hunting, gathering, and food preparation; applesnails and oysters first used; sea level continued rapid rise; difficult living conditions.
Middle	7000 – 5000	Dry climate persists in South Florida with xeric habitats in the interior but mesic habitats develop from increasing rainfall in central and northern Florida so that Lake Okeechobee begins forming from Kissimmee watershed; sea level rise slows, increasing use of shellfish and other coastal resources; cypress spreading into South Florida by the end, with Everglades area then coming to resemble modern Big Cypress.
Late	5000 – 3500	Rainfall increases, wetland plant communities become widespread; Everglades forms, displacing more diverse plant communities; more use of freshwater resources such as applesnails and turtles; estuaries become productive of shellfish e.g., oysters. Rapid cultural development and widespread human presence across Everglades area by the end, but abandonment of lower elevation sites probably due to rising water, while some low sites were perpetuated as tree islands by midden accumulation keeping ahead of rising water. Continued occupation and enlargement of the heads of higher tree islands in Everglades occurred.[97]
Formative:		
Transitional	3500 – 2750	Fired clay pottery comes into use; slow rate of sea level rise, estuaries stabilize and become very productive. Continued habitation of Everglades tree islands.
Glades I	2750 – A.D. 750	Settlements on the south shore of Lake Okeechobee and on Fisheating Creek, with ditch and canal construction, and possible local corn agriculture; possible mound construction in the Big Cypress. Long-term habitation at Bear Lake (near Flamingo, Everglades National Park) produced midden mounds associated with a canal for canoe travel. Extremely varied diet utilizing all edible animal life and many plants.
Glades II	A.D. 750 – 1200	Mound construction becomes common, indicating stratified society with ruling and/or priest class, intermediate levels, and labor class.
Glades III	A.D. 1200 – 1566	Calusa dominate South Florida. Main settlements around Charlotte Harbor and Marco Island, extending into the Ten Thousand Islands, up the Caloosahatchee River to Lake Okeechobee. The Tequesta occupy the southeast coast and use interior Everglades, with principal town at mouth of the Miami River. Shell mounds and canals constructed.
Glades-Historic	A.D. 1566 – 1763	Spanish contacted the Calusa capital near mouth of the Caloosahatchee in 1566: a starting point in the demise of the Calusa culture; abandonment of tree island habitations and exit of c. 300 remaining Calusa and Tequesta with Spanish to Cuba in 1763. Subsequent near absence of Native Americans in South Florida until Second Seminole War (1856–58).
Seminole/ Miccosukee	A.D. 1825 – present	Renewed habitation on the eastern Everglades rim, larger Everglades tree islands, and uplands in the Big Cypress. By the late 1800's, extensive reuse of the same Everglades tree islands abandoned by the Tequesta more than a century earlier. Travel in the Everglades by dugout canoes, with trails severed in 1928 by the Tamiami Trail. Subsequent diversion of tribal life to tourist trade, agricultural work, and other economic enterprises with decreasing dependence on natural environmental resources (see text).

[a] Source: adapted from references 97, 390, and 413.
[b] Unless indicated, dates are before present. (B.P.). Dates are variable among sources. Where possible, they follow Carr,[97] with deviations from McCally.[390]

sinkhole — it just looks like a round pond. Then trapped by steep, overhanging walls of the sinkhole cavern, this Paleo-Indian survived for a time on a ledge just above the water's surface and managed to start a fire with which he cooked a huge land tortoise — of a "megafaunal" species now extinct — that had also fallen into the sinkhole.[*] The ashes of his campfire have provided extraordinary evidence of the activity and date, between 12,000 and 13,000 years ago, very early in the Paleo-Indian period.

The Paleo-Indian period lasted until about 9000 years ago, and is characterized by a nomadic, hunting-gathering life style. The climate was cooler, windier, and much drier than today, perhaps initially even too dry to support scrub oaks like today's well-drained sandy ridges of central Florida. Shifting dunes may have been widespread and the terrain poor for human habitation. A lack of freshwater over large areas would have precluded survival in the interior of what was then a much wider peninsula due to lower sea level (see Chapters 1 and 10). Obviously, Paleo-Indians would have occupied riverbanks, but rivers were then only near the coast, not in the elevated, dry interior. Such sites would now lie in 100 or more feet of water, many far out in the Gulf of Mexico. Documented Paleo-Indian sites on the mainland are always associated with sinkholes or depressions where freshwater could be obtained, however precariously. In addition to Little Salt Spring, mentioned above, other South Florida sites having evidence of occupation include the nearby Warm Mineral Springs, also a sinkhole, and the 10,500-year-old[**] "Cutler Fossil site" near the modern shore of Biscayne Bay on the Atlantic Coastal Ridge south of Miami (see Color Figure 3). During the Paleo-Indian period, the Everglades area was almost entirely uplands, dotted with occasional wet depressions. One depression near modern Weston in Broward County (immediately south of the western part of Water Conservation Area 2B, see Color Figure 3) has revealed projectile points (Dalton points) similar to Paleo-Indian points of northern Florida and of the Cutler Fossil site (which contains the earliest evidence of marine exploitation in North America). However, due to sparse evidence, little is known about this ancient period in the Everglades region, and the human population throughout South Florida was probably sparse.

ARCHAIC PERIOD

The transition from Paleo-Indian to Archaic coincides with the loss of megafauna that the earlier Americans hunted. The Archaic Indians used tools that were less refined but more versatile in function, evidencing a corresponding change in subsistence. Sediment pollen studies have shown that Florida became even drier during early Archaic times, from 8000 to 5000 years ago, and southern Florida's small Native American population may have declined further. But by 6500 years ago, rising sea level and increasing rainfall, beginning sooner in northern and central Florida, began changing the region. With much of its water source from the Kissimmee watershed of central Florida, Lake Okeechobee began forming. Archaic culture in southern Florida showed an increasing harvest of shellfish, such as oysters, available in the expanding and more productive estuarine habitats, and hunting, fishing, and plant gathering in the wetter interior. By 5000 years ago, the Everglades region probably resembled today's Big Cypress Swamp with pockets of hammock forest, pinelands, marshes, and cypress in the deeper depressions. Marsh peatlands of the initial Everglades began to expand southward from the developing Lake Okeechobee. Many higher elevation sites in the Everglades area were inhabited, later to become tree islands as the Everglades formed. By the end of the Archaic Period, lower elevation habitation sites in and around the Everglades were abandoned due to rising water. Their evidence became buried beneath protective layers of peat in the evolving Everglades. Higher tree islands were continuously inhabited, so that evidence of Archaic culture remained buried beneath artifacts of more recent habitation. The increasingly diverse environment during the late Archaic time served to increase South Florida's human habitation greatly.

---

[*] *National Geographic* (reference 94) provides an excellent graphic depiction of this event; however, the reader should understand that this scenario is one of other possible explanations of the evidence.

[**] This date is revised from its previous carbon-14 estimate of 9800 years (Robert Carr, personal communication).

FORMATIVE PERIOD

This period began as a gradual transition from late Archaic culture but is distinguished by the advent of fired pottery. The main cultures within the Formative Period, called *Glades* cultures, are included in Table 21.1. These various subperiods are distinguished by decorations on pottery, generally increasing to maximum ornamentation, supplemented with carved shell and bone ornaments of the Glades III culture.

During the Glades times, increasing social organization culminated in a ruling/priest class, intermediate levels, and a lower labor class. Such organization enabled mound and canal construction. Burial mounds became common, with prominent examples along the south shore of Lake Okeechobee and on some tree islands in the Everglades during Glades III time. Large oyster shell mounds evidenced the extensive use of shellfish and probably represented a chieftain social order. Outstanding examples of mound complexes exist along the tidal rivers near Chokoloskee, now in Everglades National Park. There also was ditch construction, possibly drainage for corn agriculture, at a site on Fisheating Creek west of Lake Okeechobee, and canals were constructed apparently for canoe passage in the upper Caloosahatchee and near Bear Lake, linking interior lakes and Florida Bay at the southern tip of the peninsula near modern Flamingo in Everglades National Park. Except for the possible corn agriculture along Fisheating Creek, no other specific agriculture is documented in southern Florida through Glades III or earlier culture, but the use of fire may have been intended in some locations to enhance the survival and growth of plants with underground, edible tubers. The tubers of many plants are an adaptation to fire, and such use of fire represents a form of agriculture applied to natural plant communities.

Because of the growing populations in Glades times, middens (accumulated cultural wastes such as bones, shells, and pottery fragments of "kitchen" use) became widespread, and contributed to the growth and integrity of tree islands (see Chapter 4). Through Archaic and increasing through Glades cultures, an astonishing variety of foods were eaten without heavy dependence on any one species. Uplands, freshwater wetlands, and coastal marine and estuarine resources were all used. Animal foods included oysters, sharks, sea turtles, a wide variety of bony fishes, freshwater snails and clams, amphibians including frogs, toads, and siren, nine species of freshwater turtles, several water snakes including the poisonous cottonmouth, alligators, and innumerable mammals including opossum, round-tailed muskrat, cotton rat, fox squirrel, raccoon, gray fox, bobcat, rabbits, and deer. Among the relatively few birds taken were black vultures. The Calusa and Tequesta (Glades III culture) developed and used fishnets and even hunted manatees and the now extinct Caribbean monk seal with harpoons, although these marine mammals were not of high importance in diet. Plant resources were mostly fleshy fruits including those of mastic, cocoplum, cabbage palm, saw palmetto, seagrape, and hog plum, but also acorns and red mangrove sprouts. It is clear that by Glades times, little was neglected or wasted.

At the time of first European contact, the Calusa controlled the region of southwest Florida, centered around the prolific coastal resources from Charlotte Harbor through Marco Island and the Ten Thousand Islands, but also reaching up the Caloosahatchee to the Lake Okeechobee region of the interior (see Figures 2.5 and 7.1 for locations). The Calusa were also known to use sea-going canoes and traveled between Cuba and Florida. They dominated the also strong Tequesta, who occupied southeast Florida from modern Boca Raton through the Florida Keys and made transient — probably seasonal — use of Everglades tree islands. At the time of initial European (Spanish) contact, the total Calusa and Tequesta population was estimated at 20,000. Initial contact with Native Americans began in northeastern Florida (St. Augustine) about 1500 A.D. (500 B.P.) but effectively not until 1566 in southern Florida, when the Spanish, led by Pedro Menendez de Aviles, contacted the Calusa at their headquarter settlement in Charlotte Harbor near the mouth of the Caloosahatchee River. Subsequent Calusa and Tequesta decline in southern Florida resulted from fatal diseases, unwittingly transmitted by immune European carriers, and to slave raids mostly supported by British landowners in South Carolina during the early and mid-1700's. Tree island use in the Everglades was abandoned. The last confirmed Calusa and Tequesta presence in Florida was in 1763 when the remaining 300 banded together with the Spanish and fled to Cuba, escaping unfriendly British occupation of the territory.

SEMINOLE AND MICCOSUKEE TRIBES

Following the disappearance of the Calusa and Tequesta, there were few if any Native Americans in the Everglades and Big Cypress regions for at least 60 years. New settlement began about 1825 with the arrival of the Seminoles. These people had originally inhabited Alabama, Georgia, and northern Florida where they were part of a larger Indian confederation known to English-speaking settlers as "Creeks." The Creeks included Hitichiti-speaking Miccosukees of the lower Creek region, who began moving from Alabama and Georgia into northern Florida in the early 1700s in response to European settlers encroaching into their historic lands. The Upper Creeks spoke the closely related Muskogee language, had traded and cooperated with British, and thus came into conflict with settlers after the United States was created in 1784. They then began to move into Spanish Florida, sharing the land with the established Miccosukees. The "Seminole" name is of Muskogee origin and was applied to the combination of Upper Creeks and Miccosukees by European people, masking tribal distinctions to the public until the 1960s. Subsequent movement of the Seminoles into central and then southern Florida resulted from a series of three Seminole Wars, spanning the years from 1817 to 1858. The United States attempted to defeat and remove the Seminoles from Florida, but small numbers evaded. The Miccosukees, their numbers reduced to 50 individuals in Florida, took refuge deep in the Everglades. An estimated 300 Muskogee-speaking Seminoles sought refuge in several locations, including lands northwest of Lake Okeechobee, portions of the Big Cypress Swamp, and lands on the eastern edge of the Everglades.

In southern Florida, the Seminoles (including Miccosukees) resumed life styles that partially emulated the earlier Calusa and Tequesta. By the late 1800s, they were using many of the same Everglades tree islands abandoned by the Tequesta more than a century earlier. However, the new occupation made more specific use of islands, some for agriculture, some for hunting camps, and some for burial purposes. Other differences involved trade with the South Florida's growing Anglo population. In the 1890s, for example, that trade included plume-feathers of wading birds, which were essentially unused by the earlier inhabitants.

Significant changes in the Seminoles' use of the Everglades began while the Tamiami Trail was constructed, with completion in 1928. The road blocked north-south canoe travel, altering traditional use that had become culturally entrenched for more than a half century. The Trail also altered Seminole life style by opening the interior to tourist trade along the road and enabled canoe travel to Miami via the Tamiami Canal that paralleled the road. Settlement began moving from tree islands to the road. Following dedication of Everglades National Park in 1947, access to park lands south of the Tamiami Trail was curtailed.

In 1957, the Seminole Tribe legally incorporated, and in 1962, the Miccosukee Tribe — earlier concealed in the public's eye under the Seminole name — became an officially recognized sovereign government. It was the Miccosukees who had lived deep in the southern Everglades, in addition to parts of the Big Cypress. The Seminoles in South Florida lived in the Big Cypress north of the Miccosukee areas and also occupied lands east of the Everglades near Fort Lauderdale and lands northwest of Lake Okeechobee. Both tribes' South Florida lands today generally follow this history, with the Miccosukees occupying a narrow 666-acre strip along the Tamiami Trail on the west side of the Everglades, operating a casino/resort on historic easterly Everglades land on the Tamiami Trail at Krome Avenue, and having a 75,000-acre reservation straddling Alligator Alley and a 189,000-acre lease of Everglades lands in Water Conservation Area 3A. The latter area, accessed from the Tamiami Trail, is used for tourist airboat rides and for tribal activities on tree islands. The three Seminole Indian reservations are in the Big Cypress Swamp north of Alligator Alley, in Hollywood Florida, and on the northwest side of Lake Okeechobee (see Figures 7.1 and 11.1). They also have communities along the Tamiami Trail and in Immokalee, Fort Pierce, and Tampa.

OVERVIEW OF NATIVE AMERICAN IMPACTS

Following the initial ice-age intrusion of humans into North America, with large impacts as an introduced, invasive species, Native American cultures adjusted to sustained use of the environment — the easy game animals were gone. Obviously, the subsequent generations of Native Americans through the early Seminole and Miccosukee cultures in southern Florida also altered the environment due to hunting and gathering and due to the use of fire. However, their effect on the natural environment was small relative to impacts that have occurred since the late 1800's. The historical presence of Native Americans from Archaic times into the early 20th century must be regarded as part of the natural ecosystem. The entire evolution of Lake Okeechobee, the Big Cypress Swamp, and the Everglades occurred in their presence and under their influence. Obviously, modern Native Americans do not emulate their ancestors' impacts for numerous reasons. Paramount among them are modern modes of travel — such as airboats and automobiles — and the fixed geography of legal land ownership and land-use rights. No longer are Native Americans free to shift encampments and hunting/gathering locations between the Everglades and neighboring lands in response to changing conditions, such as seasonal water levels, to follow sustainable natural resources. Today's Seminoles and Miccosukees have responded by pursuing alternate available professions and now have little dependence on resources of the natural environment.

HYDROLOGY, LAND USE, AND THE C&SF PROJECT [232, 252, 343, 564]

Lowering the natural water levels has had by far the greatest initial impact on the integrity of the vast expanses of wetlands of the Everglades region. The general history of alterations of the hydrology, from the Kissimmee River through Lake Okeechobee and the Everglades, is outlined in the introduction by Marjory Stoneman Douglas and in Chapter 11. Early drainage efforts sought to provide for navigation and to reduce water levels for agriculture in marsh and swamplands. There was little regard for conserving water for irrigation and potable uses. Freshwater was thought to be inexhaustible. Flood control efforts focused on directing interior waters through canals to coastal tidal waters as quickly as feasible. Canals from Lake Okeechobee were constructed through the Everglades to the Atlantic (see Chapter 11), which not only lowered Lake Okeechobee and partially drained the Everglades, but also interrupted the original flow pattern of the Everglades. Within a few decades, uncontrolled drainage brought many problems. In dry years, most notably during an intense drought in 1944–1945, it endangered the regional water supply due to saltwater intrusion into former freshwater aquifers and caused uncontrollable soil fires with attendant, choking smoke. Alternatively, the system had insufficient capacity and design to prevent flooding in very wet years such as 1947. Therefore, flood control progressed to a higher level of technology on a massive scale: water management, embodied in the Central and Southern Florida Project for Flood Control and Other Purposes (C&SF Project), authorized by Congress in 1948. (Major features of the C&SF Project can be seen on Color Figures 3 and 6.) Implementation began in the early 1950's and was essentially complete by 1973.

The C&SF Project had four main components. First, it established a perimeter levee through the eastern portion of the Everglades, blocking sheet flow so that lands farther east, on and adjacent to the Atlantic Coastal Ridge, would be protected from direct Everglades flooding. The levee was about 100 miles long. It defined a new, artificial eastern edge of the Everglades and became the westward limit of agricultural, residential, and other land development for Florida's lower east coast from West Palm Beach to Homestead (Figure 21.1). Only a few areas were subsequently developed west of it, most notably the controversial "8.5-Square-Mile Area," a rural residential area in Miami-Dade County south of the Tamiami Trail, north of Homestead.[339] The perimeter levee severed the eastern 16 percent of the Everglades from its interior (see Color Figures 2 and 3).

Second, the C&SF Project designated a large area of the northern Everglades, south of Lake Okeechobee, to be managed for agriculture (Color Figure 6). Only a portion near the lake had

FIGURE 21.1 Historic Everglades: looking northeast over developed residential land in southern Broward County, Florida, just east of Water Conservation Area 3B. Development here is typical for the region, involving removal of about three feet of surficial Everglades muck (peat) and excavation of the underlying limestone (usually requiring blasting) to make lakes that provide crushed limestone fill for land areas and for now-required stormwater storage for flood protection. In some areas, lakes exceed 50 feet due to the need to obtain sufficient fill material of these low-lying areas. (Photo by T. Lodge.)

previously been developed, leaving much room for agricultural expansion in what had been a vast, nearly unbroken expanse of sawgrass. Except for two large tracts used for wildlife management — the Rotenberger and Holey Land* Wildlife Management Areas — all of this area eventually was used for agriculture, primarily sugarcane. Called the Everglades Agricultural Area (EAA), it encompassed about 27 percent of the historic Everglades and was a major factor in the economic justification of the C&SF Project (see following section).

Third, water conservation became the primary designated use for most of the remaining Everglades between the EAA and the Tamiami Trail at the northern border of Everglades National Park. The water conservation areas were limited on the east by the eastern perimeter levee, and on the west by the EAA and an incomplete levee bordering the Big Cypress Swamp — the L-28, broken to allow flows from Mullet Slough. The area was divided into three units, essentially wetland impoundments, called Water Conservation Areas (WCAs). The northernmost (WCA-1), also designated for wildlife management, eventually became the Arthur R. Marshall Loxahatchee National Wildlife Refuge and included most of the historic Hillsborough Lake area of the Everglades, noted for its abundance of tree islands and sloughs. WCA-2 and WCA-3 (the largest) were divided into A and B units. The WCAs are separated from one another and from Everglades National Park by levees and are regulated by canals and interconnections by water control structures. Huge pump stations move water into or out of the WCAs, giving protection to the surrounding lands by receiving excess water at times of impending flood and by storing and selectively releasing water to compensate for drought conditions (Figure 21.2). WCA-3 also delivers water to Everglades National

---

* Named for its bomb craters from Air Force target practice during World War II.[170, 339]

FIGURE 21.2 The S-9 pump station in Broward County, looking westward. It sits astride the eastern perimeter levee of Water Conservation Area 3 at Holiday Park off of U.S. 27. The L-67 levee/canal (angling from left) and other canals meet the Everglades eastern levee at this point, about 5 miles south of Alligator Alley. The huge pumps here can pump water in either direction — "backpumping" into WCA-3 (background) from the C-11 Canal (left foreground), which drains the southwestern part of Broward County's developed area, or from WCA-3 into the C-11 to be sent to the Atlantic Ocean at Fort Lauderdale. S-9 backpumping has become a legal pollution issue because nutrient-rich water from developed areas is often sent into WCA-3, part of the Everglades Protection Area. (Photo by T. Lodge.)

Park through control gates along the Tamiami Trail (Figure 21.3). In total, the WCAs encompass 32 percent of the historic Everglades (see Color Figure 3).

The fourth feature of the C&SF Project was the enlargement of the overall canal system into a more interconnecting network and the installation of the pumps and control gates, such as those for the WCAs mentioned above, to completely regulate the waters of the system including the Kissimmee, Caloosahatchee, St. Lucie, Lake Okeechobee, and the Everglades. The primary system now includes about 1000 miles each of canals and levees, 150 water control structures, and 16 major pump stations.[612]

Everglades National Park, established in 1947, contained lands only of the southern Everglades, originally amounting to about 21 percent of the historic freshwater ecosystem. Recently, another 4 percent was added — an area called the Northeast Shark Slough located west of the eastern perimeter levee and south of WCA-3B. This area had been over-drained but undeveloped. Its addition raised the amount of historic freshwater Everglades in the park to 25 percent. Except for local rainfall, Everglades National Park is dependent on operation of the C&SF Project for delivery of water under the Tamiami Trail from WCA-3A (and more recently from WCA-3B) through control structures that can be opened and closed.

From the ecological standpoint, the C&SF Project divided the central and northern Everglades into uncoordinated pieces, but it had a profound effect on human existence in southeastern Florida, making land development safe and inviting. While it controlled nature, its design was thought to protect huge areas in a natural state in the water conservation areas, widely perceived to be the engineering solution to saving the dwindling Everglades. There was no intent to ruin an ecosystem.*

FIGURE 21.3 Looking west along Tamiami Trail and Tamiami Canal as seen from the L-67 Canal (entering the Tamiami Canal from the right foreground). The original Tamiami Trail and Canal lie parallel to the left (south) of the modern ones. The structure at bottom center is S-333, a gate that regulates the flow and level of water moving east in the Tamiami Canal. In the upper center is S-12D, one of the four gates (S-12 structures) that regulate the release of water from Water Conservation Area 3A (right) under the Tamiami Trail into Everglades National Park (left). S-12D is in about the middle of the historic Shark River Slough. (Photo by T. Lodge.)

THE EVERGLADES AGRICULTURAL AREA (EAA) [564, 612]

The EAA is about 700,000 acres or over 1000 square miles (see Color Figure 6). Together with the drained Everglades lands along the Atlantic Coastal Ridge, it represents extremely altered Everglades terrain yet has some important wildlife values that are little known. For this reason, it is important to understand the general nature of the EAA and how it is farmed.

Sugarcane is by far the largest crop, recently covering about 65 percent of the EAA (Figure 21.4). Other crops are sod (lawn grass), vegetables such as beans, lettuce, celery, corn, radishes, and rice. EAA soils are naturally rich in nitrogen but low in phosphorus. Soil subsidence (see Chapter 3) has been a major concern since initial drainage and development.

A typical field in the EAA is 40 acres and configured in a rectangle a half-mile long by an eighth-mile wide. The long sides of each field are bordered by ditches connected by adjustable gates through a low dike to 15-foot-wide canals that border the ends of the fields. The canals connect to larger canals of the regional system, from which water can be added or removed by pumps. This configuration allows growers to control the groundwater level or flooding in each field. Four such fields are usually planted and maintained together, making a square measuring a half mile on each side — a "quarter section" of 160 acres.

For sugarcane, a tall tropical grass, the water level is controlled about 20 inches below the surface for the year that the crop is growing, including one summer rainy season. Sugarcane is planted in the fall or winter and is harvested the first time about a year later, most harvesting

---

FIGURE 21.4 Historic Everglades: a view of sugarcane fields in the Everglades Agricultural Area southeast of Lake Okeechobee in August. Black sawgrass peat of recently prepared fields shows at left. Actively growing sugarcane, probably about 8 months old, is at right. It will be harvested in the next dry season. A fallow, flooded field is in the center background, a typical condition for the mid-rainy season done to minimize soil subsidence. (Photo by T. Lodge.)

done from October into March. The mature field is burned to remove unwanted foliage, and the stalks are then cut near ground level and removed for processing. Second and third crops, called ratoons, are normally grown from the rooted base of the stalks remaining after harvest so that a typical sugarcane field is in continuous use for three years. During the summer rainy season following the final harvest, the field normally lies dormant and flooded to reduce soil subsidence, and then it is planted in a rotation crop the next dry season. Sugarcane requires relatively little phosphorous, although early practices applied far more than necessary. Recent practice has been more conservative.

Water control for vegetable crops is similar to sugarcane except that they are not grown during the summer due to the seasonal market for South Florida produce. Thus, vegetable fields can be flooded in the rainy season, reducing soil subsidence and stormwater treatment needs. Vegetable crops require more fertilizer than sugarcane. Rice differs in being grown in flooded conditions, and it requires the least fertilizer of all EAA crops. Rice agriculture most closely emulates Everglades hydrology, not requiring dewatering during the wet season and tolerating flooding through the rest of the year.

Wildlife of the EAA has had little attention, but the few studies indicate some important values. Rice, and especially an experimental program where crayfish were cultured with rice, supported a broad spectrum of Everglades wildlife. Other values include late summer foraging habitat for huge numbers of southward-migrating shorebirds, using fallow, flooded sugarcane, vegetable, and rice fields, helping these species compensate for lost (developed) coastal mudflats. Many larger wading birds also prosper in these flooded fields, including wood storks, which cannot find such a plentiful food-base in native South Florida habitats during the summer and early fall seasons. Two species of ducks, the mottled duck and the fulvous whistling duck, are prospering in southern Florida in large part due to EAA habitats.[56, 247, 278, 279, 301, 585, 662]

Although important wildlife values have been identified in the EAA, there have been relatively few cooperative ventures between wildlife agencies and EAA farmers. Through the rest of the country, wildlife managers and farmers regularly cooperate in wildlife programs. But with a few notable exceptions, there is not a beneficial relationship between EAA farmers and wildlife agencies. Minor adjustments of flooding regimes of fallow fields and an increase of rice acreage are obvious areas where wildlife benefits could be fruitful.

The largest concern regarding the EAA is its future once agriculture is no longer feasible — either from soil subsidence or changes in economics that sustain agricultural production. The land use that replaces the EAA will be critical to the Everglades as well as to the water supply for South Florida.

COASTAL WATERS [611, 612]

Alterations of freshwater quality and of flows to the coasts have been excessive. Modifications of the natural system began in 1881 (see Chapter 11). Historically, freshwaters were stored by increased flooding of the interior and were released relatively slowly and evenly through a myriad of small rivers and creeks, and diffuse groundwater flows to coastal waters, where they nurtured estuarine habitats. Early drainage efforts, greatly increased by the C&SF Project, directed releases to specific coastal locations where canals "dumped" their flows. Key issues involve the Caloosahatchee River, the St. Lucie Canal, Biscayne Bay, and Florida Bay. These environmental impacts are important considerations in overall Everglades restoration. The nature of the impacts is described here.

Charlotte Harbor

The connection of the Kissimmee River/Lake Okeechobee watershed to the Caloosahatchee River, initially in 1881, directed much larger, often pulsed, releases to Charlotte Harbor, an extraordinary natural estuary of the Myakka, Peace, and Caloosahatchee rivers (see Figure 2.5). Charlotte Harbor's productivity was naturally enhanced by phosphorus from the Peace River, which drained the phosphate region of west-central Florida (see Chapter 1). The natural, low-level enrichment had nurtured the Calusa Indian cultures. Especially since the 1960's, with C&SF "improvements" in South Florida's water management system, pulsed releases of fresh-water from Lake Okeechobee have substantially damaged estuarine functions of the lower Charlotte Harbor area.

St. Lucie/Indian River

Mirroring the situation in Charlotte Harbor, the connection of Lake Okeechobee through the entirely artificial St. Lucie Canal to the South Fork of the natural St. Lucie River severely damaged the estuarine ecology of the St. Lucie River as well as the connecting Indian River, a coastal lagoon. The St. Lucie River connects to the Atlantic Ocean at Stuart, Florida (for location, see Figure 2.5), and the inlet itself has complex history, having been artificially opened in the late 1890s. Prior to that time, the St. Lucie only connected to the Indian River.

Biscayne Bay

The C&SF Project also overhauled how freshwater from the interior reached Biscayne Bay (see Color Figure 3). Historically, numerous transverse glades carried seasonal overflows from the Everglades across the Atlantic Coastal Ridge (see Chapter 2 and Color Figure 2), and groundwater from the Biscayne Aquifer leaked into Biscayne Bay in vast but diffuse flows. There are accounts of being able to dip freshwater from "boils" in the bay using a bucket from a boat, and springs were common along the mainland shore.[191, 227, 402] Mangrove swamps, most extensive along the southern part of the bay, were nourished by freshwater inputs mixing into tidal water. Widespread groundwater seepage was observed on the exposed bottom at low tide. Drainage works constructed in the second decade of the 1900's, later expanded by the C&SF Project, changed all of that, with discharges of the interior's freshwater directed into the bay at points. The Miami River was once a short watercourse beginning as rapids where surface water of the Everglades spilled over the coastal ridge. The rapids location is now in the heart of the city, over a mile and a half east of Miami International Airport next to Miami's N.W. 27th Avenue and just south of the river's massive replacement, the Miami Canal, initially completed in 1913. The Miami Canal, carrying water all the way from Lake Okeechobee, is now the headwaters of the Miami River, and its flows to Biscayne Bay are excessive compared to the original relative trickle. Restoring an ecologically functional shoreline in Biscayne National Park has become an important element of Everglades restoration, directed to reestablishing diffuse flows through the coastal mangroves.[232, 343, 612]

Florida Bay

Florida Bay (see Figure 10.1) was historically renowned for its excellent commercial and sport fishing and for its abundance of wading birds, particularly roseate spoonbills. Beginning in the 1970s, however, declines were observed, which greatly intensified from the mid-1980's into the 1990s. About 15 square miles of seagrass beds died in the western half of the bay between 1987 and 1990, and about 90 square miles were damaged.[502] Sporadic planktonic algal blooms and extensive losses of wading birds, fish, spiny lobsters, shrimp, sponges, and even mangroves of the islands were documented.[68, 619]

Initial investigations focused on hypersalinity from diverted freshwater flows.[397] A historic maximum salinity of 75 ppt (over twice the salinity of sea water) was recorded in May 1990, and a particularly interesting study involving a type of coral found in the bay added credibility. The coral was found to record the presence of humic compounds of freshwater origin in its growth bands, which indicated decreased freshwater inputs to the bay beginning in 1912; this correlates with the initial Everglades drainage works.[560] There was much public criticism of reduced flows through Taylor Slough and adjacent eastern portions of Everglades National Park. In the mid-1960's, a major drainage canal called C-111 (Figure 21.5) was constructed near the eastern boundary of Everglades National Park.* Together with earlier canals and other drainage modifications, C-111 further interrupted and diverted flows from Taylor Slough — water that would have naturally moved through the eastern area of the park into Florida Bay through a broad area of sheet flow. The new release of these waters, through C-111, was into Barnes Sound where it created problems comparable to the Caloosahatchee River and St. Lucie Canal because of the concentrated release point.[343]

---

* When C-111 was under construction (1964–66), citizens observed that it was larger and deeper than comparable drainage canals and that a drawbridge was being provided for the U.S. 1 highway crossing. The design had been intended — without the public participation required after 1969 by the National Environmental Policy Act (NEPA) — to serve as a drainage canal *and* to allow barge traffic to carry rocket assemblies between the Kennedy Space Center and a proposed Aerojet General Corporation testing facility located near the main entrance to Everglades National Park. The testing project was never funded, but public outrage over the construction of a canal that might seriously affect the waters of a national park was enormous (even the National Park Service voiced a strenuous but unheeded objection) and precipitated a lawsuit by the National Audubon Society against the U.S. Army Corps of Engineers. An even larger concern than the drainage aspect of the project was saltwater intrusion. It was resolved that an earthen plug would remain near the mouth of the canal (later replaced by a water control structure) to prevent the inland movement of salt water.[17, 99, 343] The water control structure remains today, as does the alternate name for C-111, the "Aerojet Canal." (The previous reputation of the Corps of Engineers now stands in contrast to its role as a steward of wetland protection under the Clean Water Act, and as the lead federal agency in Everglades restoration.)

Even the National Park Service has been responsible for controversial canal projects, notably the Buttonwood Canal in Everglades National Park. When that canal was completed in 1957, it allowed salt water from Florida Bay to enter Coot and eastern Whitewater bays, which were rapidly and drastically altered from their historic freshwater/brackish condition.[128, 619] Subsequent installation of a barrier in the canal at the Flamingo marina in July 1982, was an admirable action to correct the historic environment of Coot and Whitewater bays.

FIGURE 21.5 The extreme southeastern Everglades, looking northwestward over Canal C-111, with the U.S. 1 drawbridge at the center. Although part of the C&SF Project, the C-111 history has been controversial, from its design to accommodate private industry to its potential to carry saltwater to the edge of Everglades National Park, to its rerouting of water that would have followed sheet flow and creeks to upper Florida Bay, to its berm (seen with dirt road to the left of the canal) that impounded the saline tide of Hurricane Betsy in the right, background area, to its periodic massive point discharges to Barnes Sound. Note that the canal and highway cut across the north-south orientation of the tree island corridors, which indicate the historic direction of natural flows. See text for more details. (Photo by T. Lodge)

Because of the identified degradation of Florida Bay, extensive research efforts were launched, with much of it published in 2002.[75, 173, 181, 203, 293, 335, 341, 368, 439, 500, 559] Significant findings were the following:

- The freshwater inputs to Florida Bay from the Everglades are so small relative to direct rainfall and to exchanges of tidal waters through the passages in the Middle Keys and especially across the broad western connection of Florida Bay with the Gulf of Mexico that nutrient sources from the Everglades are negligible.
- Nutrients and other water quality problems, such as imported algal blooms from currents coming down the southwest coast of Florida into Florida Bay are a significant problem. They implicate inputs to the eastern Gulf of Mexico from Florida's west coast rivers such as the Caloosahatchee, which contributes enriched water from Lake Okeechobee.
- The limiting nutrient for western Florida Bay is nitrogen, not phosphorus. An historic, natural source of phosphorus to Gulf coastal waters is from rivers such as the Peace River. Nitrogen from South Florida agriculture has enriched eastern and central Florida Bay, where phosphorus is limiting. Nitrogen combined with phosphorus from the west have caused problems in the central bay and the Florida Keys. The Everglades Forever Act's focus on lowering phosphorus would likely not have a beneficial result in Florida Bay because negligible phosphorus currently enters Florida Bay from the Everglades.

- Except for pulsed releases over intolerant biota, salinity by itself, high or low, is seldom a problem, but merely identifies the source of the water. The larger problem is other contaminants carried with the water.
- Roseate spoonbills are a good indicator of the health of Florida Bay and adjacent coastal habitats because they respond to the production of juvenile fish.
- Florida Bay is extremely complex, having a range of characteristics over its geographic extent. No data are available to establish a reasonable, pre-drainage era baseline that could serve as a target for restoration at this time.

WATER QUALITY — INTERIOR FRESH WATER

Water quality has many aspects including nutrients — primarily nitrogen and phosphorus compounds used by plants for growth — dissolved substances that affect pH (acid/base scale), hardness, salinity, and toxins that may inhibit or poison living processes. Degraded water quality has created widespread changes in the Everglades. The effects have been more recent than the changes due to altered hydrology, and the visible extent of the damage, while growing, has not yet reached the magnitude of the hydrologic changes. Water quality concerns in the freshwater Everglades have focused on two issues: nutrient enrichment involving phosphorus and contamination by mercury. This focus does not mean that these are the only concerns. Another concern, for example, involves the amount of calcium dissolved in the water, because periphyton in soft water (low calcium) enters food chains more efficiently. The calcium issue in Everglades restoration involves water stored underground in aquifer storage and recovery where the water becomes unnaturally saturated in calcium. Also, in the coastal marine waters of Florida Bay, nitrogen enrichment is probably more important than phosphorus, addressed in the coastal waters section. The freshwater issues of phosphorus and mercury are covered in the following sections.

PHOSPHORUS [108, 149, 393, 568]

The basis for understanding water quality in the Everglades is set in Chapter 3, and for Lake Okeechobee in Chapter 11. Phosphorus is known to be the limiting nutrient in the Everglades. Natural levels between four and ten parts per billion (ppb, the same measure as micrograms per liter) characterized the original system, although it is thought that incoming waters from Lake Okeechobee, prior to agricultural development of its watershed, were naturally higher — at least 20 ppb. The northern Everglades acted as a natural nutrient-removal system as it grew and laid down peat soils. Considerable research has attempted to quantify the level of phosphorus that the natural Everglades could assimilate without causing an *imbalance of flora and fauna* — the legal term for allowable change. Consensus among most scientists is that the level is about 10 ppb, but some work has indicated that the concentration is as high as 16 ppb.[497],*

With its conversion into the EAA, the northern Everglades became a nutrient source instead of its historical role as a "sink." Without treatment, EAA stormwater contained high phosphorus levels from the fertilizer applications, often in excess of 500 ppb. As the entire EAA became developed in the 1970's, the problem of where to put excess stormwater was increasingly solved by backpumping to Lake Okeechobee. It was recognized, however, that Lake Okeechobee was suffering from eutrophication (see Figure 11.8) from all of the intensely used agricultural lands in its watershed. A water management decision in 1979 protected the lake from EAA nutrients, redirecting EAA stormwater to the water conservation areas, including WCA-1 (the Loxahatchee National Wildlife Refuge).

---

*The reader should be cautioned that phosphorus chemistry is complex. A simple measure of concentration may not represent what is available to plants.

Soon, there was a proliferation of southern cattails along the Hillsboro Canal inside WCA-1 and in the immediate vicinity of the release gates (S-10 structures) where water leaves WCA-1 and enters WCA-2A. In WCA-2A there is no perimeter canal so that the water is forced to follow sheet flow through the area, and the cattail proliferation continued expanding into WCA-2A (see Color Figure 3). Cattail infestations also began to appear along the North New River Canal on the west side of WCA-2A and into WCA-3A where that water is released under U.S. 27 through the S-11 gates. Armed with this evidence and very high phosphorus levels in water and soils, the issue entered legal status. On October 27, 1988, the federal government sued the State of Florida (specifically the Department of Environmental Regulation* and the South Florida Water Management District) for violations of Florida's own standards in waters entering the Arthur R. Marshall Loxahatchee National Wildlife Refuge and waters directed to Everglades National Park.[337]

The legal action stimulated research on phosphorus levels and treatment technology and prompted improved agricultural practices in the EAA. The lawsuit was settled by an agreement embodied in the Everglades Forever Act of 1994, which required the state to develop a numerical standard for phosphorus entering the "Everglades Protection Area" (basically the WCAs and Everglades National Park) and to set a timetable for conformance. The South Florida Water Management District constructed a demonstration project for wetland treatment, which consisted of 4000 acres in the EAA abutting the northwest side of the Loxahatchee National Wildlife Refuge. The formation of peat soil was the intended long-term means of phosphorus removal. The goal was to reduce the average influent concentration of phosphorus from 175 ppb to less than 50 ppb. It was more successful than anticipated, and a total of over 41,000 acres of such Stormwater Treatment Areas (STAs) were designed and constructed (see Color Figure 6).[241, 306, 568] The standard has been set at 10 ppb phosphorus (but the timetable for when this level will be required may be delayed). Treatment technologies are still being developed, and the standard of 10 ppb will be very difficult to achieve, although levels of about 25 ppb have been easily obtained in STAs.

The EAA is not the only source of degraded water quality entering the Everglades Protection Area. Inputs from the huge S-9 pump station (see Figure 21.2), which back-pumps stormwater into WCA-3A from developed areas of southwestern Broward County when flooding may occur. This discharge entered litigation that reached the U.S. Supreme Court.**

MERCURY[209, 215, 256, 319, 353, 482, 568]

A major issue in water quality has been the high levels of mercury in Everglades fauna — a new discovery in the late 1980's. Even in remote areas of Everglades National Park, levels in largemouth bass and other predatory fish were so high that consumption warnings were posted for fishermen. The Florida standard for mercury in fish (maximum allowable) is now 0.5 parts per million (ppm, equal to micrograms per gram of tissue) and a lower standard of 0.1 ppm is being considered. Edible tissue levels in largemouth bass were commonly at 1.0 ppm in the central and southern Everglades. In mid-1989, following growing concern over the mercury issue, a Florida panther was found dead in the Everglades. Liver samples (typically having the highest level among body tissues) from this individual contained 110 ppm mercury, which indicated that mercury poisoning was the probable cause of death. This particular panther had frequented the Shark River Slough, where deer (which have low mercury levels) were scarce, and it is known that raccoons and alligators formed a large part of its diet. In mid-1991, the last two female panthers inhabiting Everglades National Park died. Their mercury levels were lower than in the earlier case, but mercury poisoning

---

* Now known as the Florida Department of Environmental Protection.

** The case involved the question of the South Florida Water Management District's responsibility. The District contended that the pump does not cause pollution but merely transports water as authorized under the District's responsibility to prevent flooding of developed lands. Adversaries contended that the District knowingly polluted the Everglades Protection Area by back-pumping water that caused an *imbalance of flora and fauna*, therefore violating state water quality standards. The case was remanded to a lower court for technical clarifications, its status at this writing.

was still regarded as a probable contributor to the death of at least one of these individuals. It became clear that mercury was bioaccumulating in the region's fauna, and subsequent research showed that alligators, raccoons, and wading birds all had high levels, far higher than the largemouth bass.[304, 355, 353, 639] Concern about mercury stimulated an avalanche of research and regulatory reform. That effort has resulted in an encouraging achievement in the understanding and abatement of this critical toxic material.

Mercury is a naturally occurring element that is familiar as a silvery metallic liquid. It has no known biological function but forms chemical compounds of various toxicities. The most toxic is the simple, organically bound form, methylmercury, which readily enters living organisms, and is biomagnified through food chains. Like other mercury compounds, it can cause mutations, abnormal growths, and neurologic/behavioral disorders leading to death at higher concentrations. In humans, embryonic development and infancy are most susceptible to mercury damage because of interference with tissue differentiation, even at levels that are not dangerous to adults.

Key findings of the research were the following:

1. The main sources of mercury to the Everglades have been regional, not global, and nearly all the mercury has been deposited from the atmosphere by rainfall and "dryfall" (e.g., with dust). The sources were primarily waste incineration and electric power generation from burning fossil fuels, particularly coal (which is used more in central and northern Florida).

2. The central and southern Everglades, with large areas prone to seasonal drying, are well suited to forming methylmercury. It forms in sediments at the interface between aerobic and anaerobic conditions, such as under a periphyton mat. Sulfur-reducing bacteria are instrumental in causing the "methylation" of mercury as a byproduct while they reduce sulfate to hydrogen sulfide, the chemical that gives some anaerobic sediments and water the rotten-egg odor. Far less methylation occurs in areas of continuous flooding such as the Loxahatchee area of the northern Everglades.

3. Everglades soil analyses have shown that mercury accumulation rose slowly until about 1900. Between 1900 and 1990, the rate of accumulation increased at least five fold, with most of it after 1950.

4. Mercury was found in animal samples from all trophic levels, but is notably less in lower trophic levels. Applesnails, for example contained only about 0.02 ppm.

5. As would be expected from biomagnification, nearly all higher trophic levels harbored greater mercury concentrations — often near or above the level of human health concern and of negative wildlife impacts. In addition to the panther, very high levels have been found in raccoons and wading birds, with an example high level of 55 ppm in a great blue heron (liver tissue). Examples of commonly eaten fish with high levels are yellow bullhead, warmouth, bluegill, and largemouth bass. Among fish, the highest levels have been found in Florida gar and bowfin. Fish consumption restrictions have been posted for Water Conservation Areas 2 and 3, and for the Everglades National Park areas in and near the Shark River Slough, namely north and west of the park's road to Flamingo.

6. Mercury is deposited in the growing feathers of birds, and nestling wading birds are protected from high tissue levels while their feathers are growing. However, at the time of fledging, their feathers stop growing. At that time, levels of mercury in Everglades fledglings are high enough that any additional mercury accumulated from feeding on Everglades fish could affect survival. Problems might be poor feeding success and slow reactions to predators. It is surmised that mercury has negatively affected wading birds, but verification is lacking and would be difficult to be determined.

Mercury source controls have greatly reduced inputs to the Everglades. Reduced emissions from power plants and incinerators since the late 1980s have been dramatic, with incinerator releases lowered by about 99 percent. Deposition into the Everglades has been reduced accordingly, and tissue mercury levels have decreased throughout trophic levels — 60 percent lower in fish tissue and 70 percent lower in bird feathers. At the time of this writing, levels in fish in the central and southern Everglades are still high enough to be of concern for human and wildlife health, but projections are favorable for lifting restrictions in the near future. Stormwater treatment areas (STAs), when kept flooded, have not contributed to mercury problems, but drying did cause STA-2 (see Color Figure 6) to produce substantial methylmercury, a condition that is expected to recur with droughts but at sequentially lower levels. Full recovery from mercury effects in the Everglades is expected to occur within 30 years, unless global sources increase significantly outside of our control ability.

CHANGES IN WILDLIFE [61, 224, 449, 507, 604]

Except for wading birds, changes in Everglades wildlife populations are poorly documented. Accounts of biologists, hunters, fishermen, and others who explored the region in the early to middle decades of the 1900's are helpful but not quantitative (see Recollections). The factors that have led to wildlife changes range from direct hunting and collecting ("plume" birds, alligator hunting/poaching, and specimen collecting, e.g., tree snails and orchids) to automobile road kills to direct loss of habitat (land development, introduced species) to large changes in habitat character. Some of these will be discussed below.

WADING BIRDS

There is only sketchy evidence of the early populations of wading birds (herons, egrets and their relatives) in the Everglades. A crude estimate for the 1870s, prior to any significant human intervention, is about 2.5 million birds. Plume hunting cut these numbers to perhaps a half million by 1910. Wood storks were relatively unaffected, but the snowy egret, reddish egret, and roseate spoonbill populations were decimated. Their decorative breeding feathers — plumes — were the most desired for the fashion market in women's hats. Laws protecting these birds were partially effective, but it was the success of a public awareness campaign during the early years of this century, rendering the fashion socially unacceptable, that finally removed the economic incentive for plume hunters. Populations of most of these birds then rebounded during the next few decades, to a high point for the century in about 1935.[507]

Beginning in the 1930's, relatively accurate counts have been kept (Table 21.2). The deterioration of wading bird populations was extensive, from highs in the 1930's to lows in the late 1980's

TABLE 21.2

Estimated breeding populations of selected wading birds since the 1930's in the central and southern Everglades[a]

Species	1931–1946	1974–1981	1982–1989	1996–1998	1999–2001	2001–2003
great egret	6,500	6,500	4,200	8,000	14,500	16,900
small herons[b]	25,000	8,000	2,500	2,700	17,200	16,100
white ibis	200,000	29,000	12,500	4,500	48,000	41,500
wood stork	6,500	2,600	750	450	3,700	3,200

[a] Source: modified from Ogden[449] (Table 22.2 therein) for numbers through 1989 and from reference 224 thereafter. Where ranges are given in these sources, averages are used to simplify this table.

[b] Small heron counts are almost entirely snowy egrets and tricolored herons, species that utilize rookeries and would have been taken by plume hunters.

or early 1990's, depending on the species. Concurrent with these losses were movements of rookeries from historic locations on mangrove islands along tidal creeks and rivers of the southern Everglades* to new locations on tree islands in the WCAs. The basis of the change in rookery locations is not entirely clear, but deteriorating nesting success in the southern Everglades was dramatic, with fewer and fewer years supporting the development of the "super colonies," which were a barometer of favorable conditions for the birds. Between 1931 and 1946, 90 percent of wading bird nesting in the Everglades was in the mangrove-marsh ecotone (transition) rookeries at the south end of the Shark River Slough. By the 1974–1989 period, only 15 percent of the nesting remained there, and the total number of nesting birds was far smaller. The remaining nesting birds shifted to the WCAs.

Several conditions favored the natural occurrence of the historic, pre-drainage rookeries at the southern end of the Everglades. The location provided many mangrove islands along the upper tidal creeks of the great mangrove forests. Such islands are favored by wading birds as an obvious behavioral adaptation to avoiding nest predators such as raccoons. The location also provided ready access to a greater range of hydroperiods and habitats than any other place in the Everglades. The Shark River Slough, with its very long hydroperiod and respective production of aquatic prey, occupied a path about eight miles wide through the southern Everglades. It was flanked by a gradation of freshwater habitats extending to the short-hydroperiod (six to seven months) marl prairie with abundant solution holes — the rocky glades. As typical dry seasons progressed, "drying fronts" where prey was concentrated moved toward the rookeries from all directions — inward toward the edges of the Shark River Slough and generally southward through the expanses of Everglades. This progression normally left a large, shallow, freshwater pool above the buttonwood embankment at the end of the dry season. In addition, the great flow of freshwater through the historic ecosystem supported a highly productive estuarine mangrove zone, which provided habitat and prey base alternatives to the freshwater Everglades throughout the year on regular tidal cycles. Thus, within short range, the birds nesting in the mangrove zone could take advantage of rich habitat diversity and probably the best fish production in the region, but this favorable condition was all based on freshwater flows from the north — dependable flows that changed only gradually in the transition from wet to dry season.[57, 153, 208, 358, 449, 628]

The initial drainage of the Everglades from 1910 into the 1940's diminished the amounts of water that moved southward through the Shark River Slough, Taylor Slough, and into the estuaries.[560] The southern Everglades rookeries persisted through this stage, however. It was only after the 1950's, with implementation of the C&SF Project, that their numbers plummeted. While the WCAs slowly began to support rookeries with the demise of the southern Everglades locations, conditions of the WCAs were thought not to be favorable, an idea supported by the small, unstable rookeries. It was surmised that the birds needed to follow long-lasting drying fronts through the Everglades on a predictable day-to-day basis. The problem with the water conservation areas was that they broke the continuity of the Everglades into uncoordinated compartments so that drying fronts were greatly reduced in spatial extent, often meeting levees where the favorable conditions ended (Figure 21.6). Furthermore, the WCAs were used for stormwater storage, so that conditions sometimes changed quickly, reversing drying fronts with negative effects on wading birds. Such problematic hydrology also occurred as pulsed releases through Everglades National Park to the southern Everglades. Through the 1970s and 80s, there were numerous changes in water management aimed at relieving some of these problems, without apparent success.[57, 153, 155, 208, 374, 449, 454]

---

*Much smaller rookeries occurred elsewhere in the Everglades and along the shores of Lake Okeechobee, but the available records indicate that they were minor and unstable compared to the giant southern Everglades rookeries. Also, comparable to the southern Everglades location, large rookeries of the Usumacinta-Grijalva system are located at the freshwater marsh-mangrove interface.

FIGURE 21.6 Abrupt transition caused by the division of Water Conservation Area (WCA) 3. WCA-3A is to the left. The L-67A canal and levee (with dirt road on top) run through the center, and a transitional area between WCA-3A and 3B lies to the right. The differences in hydrology are seen in the wetter conditions in WCA-3A. Slough habitat in WCA-3A is dark, evidencing open water and lily pads. To the right, remnant slough habitat is covered with periphyton, making it light colored, and sawgrass has encroached, reducing the amount of slough habitat.

The recent improvement in nesting success of wading birds, primarily in the WCAs, has been encouraging (see Table 21.2) and has spurred reevaluation of wading bird science. Some recent years nearly emulated the 1930's (although displaced from the southern Everglades). The basis of the success is not well understood, but gradual adjustment by the birds to the water conservation areas may be important. Other important insight into wading bird success was gained when it was identified that super-colony development nearly always occurred one to two years after severe droughts. Other years usually have mediocre to poor success even when most circumstances appear favorable. Speculation on this correlation is that severe droughts mobilize nutrients into improved productivity of marsh habitats after reflooding, which results in larger prey populations. This hypothesis is testable and future research will undoubtedly focus in this area.[212]

Whether the reduction in mercury contamination has helped in the recent success of wading birds is not known but fits the circumstances. Despite the success in the WCAs, it is hoped that Everglades restoration will foster reestablishment of the southern Everglades rookeries. Hampering such reestablishment, however, is the fact that the historic mangrove islands that supported the southern Everglades rookeries have grown together — coalesced — for reasons not well understood but probably related to decreased flows from the freshwater Everglades.

The Alligator

Alligators were once extremely abundant, but estimating the historical Everglades population is not possible. Anecdotal accounts such as by Glen Simmons[537] and Totch Brown,[85] who made much of their living hunting and later poaching alligators beginning in the 1920's and 30's, respectively, indicate that alligators were abundant in peripheral landscapes of the Everglades where they are rare or absent today. Alligator hunting for their hides, and habitat drainage reduced populations through the 1930's and 40's, but they rebounded to local abundance in the 1950's with protection afforded by Everglades National Park following its creation in 1947. Renewed declines due to alligator poaching occurred in the early 1960's but subsequent effective legal protection spawned a rebound (see Chapter 17). However, alligator populations today are undoubtedly smaller than historically, and significant numbers no longer occupy historic peripheral Everglades habitats where they once provided important wildlife benefits there. In these shorter hydroperiod marsh habitats that become seasonally dry, alligator holes were highly valuable as aquatic refugia. Slough habitats and canals are now the mainstay of alligators in the Everglades. The alligators' role there in assisting other wildlife is relatively small because these habitats produce fish and aquatic invertebrates without the help of alligators.[387]

Specimen Collecting

Of the huge changes in South Florida, taking specimens from the wild might seem minor. Yet collectors have taken a large toll on wildlife, here used in the broad sense including all native plants and animals — invertebrates, fish, reptiles, etc. Many of the more unusual orchids and ferns of the region are now all but extinct.[128, 131] The intricately colored tree snails (Color Figure 10), once found in a myriad of varieties throughout the tropical hardwood hammocks of the area, have been collected so copiously for their beautiful shells that numerous varieties no longer exist. In fact, tree snail collectors in the past were even known to burn hammocks in order to facilitate collecting the local variety of tree snail, damaging both the hammocks as well as the snail populations.[61] Reptile collectors can also damage populations, and the rarity of the large and beautiful indigo snake (an important predator of rattlesnakes) is partially their doing. Collecting — much of it now illegal — will continue to be a problem for wildlife managers in the future.

Most of the specimen-collecting problem springs from the urge that many people feel to possess wildlife, either caught personally or purchased from wildlife collectors. I recall my craving, as a youngster interested in aquatic wildlife, to have my own snakes, turtles, frogs, and fish in cages and aquariums. The first such trophy often evokes a stronger desire to have more. Peer pressure stimulates competition, aggravating the problem. For me, it was discovering nature photography while in my mid-20's that quelled my drive to possess wild plants and animals, and

I became more satisfied by observing wildlife in its natural environment. Education is critical to overcoming the urge to possess wildlife and to instill an ethical regard for the natural environment, including not supporting illegal wildlife collectors — poachers. Wildlife regulations and enforcement for those who do not get the message should be a last resort — it is nearly impossible to police the wilderness.

INTRODUCED SPECIES [92, 120, 168, 533, 535, 536]

When I arrived in South Florida in 1966, there were few exotic species in the region's natural landscapes. Having an interest in freshwater fishes, I was intrigued that several tropical aquarium species such as the firemouth (*Cichlasoma meeki*) were reportedly established, and I searched for them, initially without success. The giant or marine toad (*Bufo marinus*) and the brown anole (*Anolis sagrei*), a small lizard of shrubbery-level habitat, were common on the University of Miami campus, but both were then restricted to urban/suburban areas. Water hyacinth, a floating plant, had been established in Florida since the 1880's and was a big problem clogging lakes and rivers farther north but was only a minor concern in the Everglades region south of Lake Okeechobee. Brazilian pepper and melaleuca, both invasive trees, were well established but not considered a threat to natural environments except by a few "alarmists," who were generally ignored. Since those days, the number of naturalized nonindigenous species and the extent of damage done has risen dramatically. Scientists as well as the public are now genuinely alarmed at the magnitude of the problem, whether looking at the extent of damage to natural habitats or to the various high costs of control — in the wild, in waterways, in agriculture, and in the backyard. The interchangeable terms, *exotic, alien, introduced,* and *nonindigenous* used to describe non-native plants and animals have become everyday vocabulary.

The introduction of a plant or animal to a new environment may have various outcomes. Some fail or survive only in agriculture, horticulture, or home aquaria. Others escape to disturbed areas where natural communities have been stressed or removed. Roadsides, for example are often dominated by smutgrass (*Sporobolus indicus* var. *indicus*) to the extent that many of us call it "DOT grass" (for the Department of Transportation). The extreme behavior of an introduced species is called "invasive," the condition where it moves into man-made and/or natural environment causing economic hardship and/or displacing native species, often radically altering environments and their ecological functions. South Florida has a large share of introduced species that fit the "invasive" label.

Plants [167, 243, 521]

Plants are the basis of ecosystem productivity and provide most of the non-geological structure of an ecosystem. Thus, alterations of plant communities can cause extensive ecological changes. The largest threat of invasive plants in natural landscapes is damaging the functions of natural communities. The number of exotic plants now established in South Florida is shown in Figure 21.7 in proportion to the native flora. In total, 779 introduced plants (species or varieties of species) have successfully invaded natural environments. Of the introduced flora, 106 kinds (nearly 14 percent) of the South Florida "escapees" have become so pervasive that they are listed by the Florida Exotic Plant Pest Council as the most invasive, or "Category I" exotic pest plants. Two-thirds of these have escaped from ornamental horticulture. Only 16 percent are related to agriculture, and the remainder have accidental and unknown origins. Selected examples of the Category I invasive exotic pest plants are provided in Table 21.3. Three of the most damaging to South Florida terrestrial ecosystems are melaleuca, Brazilian pepper, and Old World climbing fern.

Melaleuca

The greatest invasive threat to short-hydroperiod marsh and swamp habitats of the Everglades and Big Cypress Swamp is this tree from Australia (Figure 21.8). Most often called "melaleuca" after

TABLE 21.3
Selected nonindigenous invasive biota in South Florida and their effects

Common name	Scientific name	General character/origin/Florida introduction date/impacts
plants[a]		
Old World climbing fern	*Lygodium microphyllum*	Vine/tropics of SE Asia, Australia, and Africa/1958/forms dense growths over ground, shrubs, and trees, shading out natural vegetation; most prolific in shallow freshwater wetlands such as drained cypress, Everglades tree islands, and low pinelands[76, 234, 521]
Hydrilla	*Hydrilla verticilata*	Submerged aquatic plant/Sri Lanka/1950's/infests lakes and canals; seriously competes with natural submerged plants; deleterious to water quality and fish populations; impedes water flow in canals requiring expensive periodic removal.[167, 521]
Burma reed	*Neyraudia reynaudiana*	Giant grass/southern Asia/1916/invades dunes and pine rocklands, there supporting hot fires that often kill pines.[167, 521]
water-hyacinth	*Eichhornia crassipes*	Floating plant/South America/1884/covers surface water shading submerged plants; obstructs navigation and water control structures.[536]
air-potato	*Dioscorea bulbifera*	Vine/Old World tropics/1905/most damaging in hammocks where it shades understory and inhibits tree recovery after disturbances such as hurricanes.[47, 521]
Australian pine	*Casuarina equisetifolia*	Tree/Australia/1887/mostly invades disturbed areas, shading recovery and smothering herbaceous plants with leaf litter; damaging in beach/dune associations especially where roots can obstruct turtle and crocodile nesting on low energy (narrow) beaches.[168, 521]
Brazilian pepper	*Schinus terebinthifolius*	Shrub or low tree/Central America/1840s/quickly invades disturbed areas, including pinelands, hammocks, and upper edge of mangrove swamps, especially after hurricanes; supports very limited bird and reptile populations compared to native communities.[168, 521]
Carrotwood	*Cupaniopsis anacardioides*	Tree/Australia/1980/invades spoil islands and other disturbed areas including Australian pine areas. Spreading rapidly and little is known about its potential impacts.[521]
lather leaf	*Colubrina asiatica*	Vinelike shrub/tropical Asia/1950's/invades coastal hammocks, mangroves, overgrowing natural plant communities. Most prevalent in the Florida Keys, potential still unknown.[47, 521]
Melaleuca, paper bark	*Melaleuca quinquenervia*	Tree/tropical Australia/1906/invades shallow marshes and cypress/pine transitional wetlands, excessively altering native habitats; provides poor animal habitat.[521]
animals		
lobate lac scale	*Paratachardina lobata*	Scale insect/India and Sri Lanka/1999/a tiny x-shaped dark red scale insect that infests a wide range of native and exotic trees and shrubs, mostly fastening to twigs and small branches. Waste products cause growth of black sooty mold on infected plants.[b, 348, 568]

Common name	Scientific name	Description
Bromeliad or Mexican weevil	*Metamasius callizona*	Beetle/Mexico and Central America/1989/adults and larvae feed on air plants (*Tillandsia* spp.), killing most species; spreading through Florida from an initial infestation in Broward County; probable introduction on infected air plants imported from beetle's native range.[205, c]
walking catfish	*Clarias batrachus*	Fish/SE Asia/1960's/predator with wide diet; survives poor water quality by breathing air; crosses land in wet weather reaching new habitat, sometimes where amphibians formerly reproduced with isolation from fish predation; eaten by numerous birds.[92, 118]
pike killifish	*Belonesox belizanus*	Fish/Central America/1957/small estuarine to freshwater predator of smaller fishes, e.g., mosquitofish.[118] Occupies a food-chain position that would increase mercury biomagnification to wading birds.
Mayan cichlid	*Cichlasoma urophthalmus*	Fish/Mexico and Central America/1983/most commonly seen nonindigenous fish in South Florida freshwaters; voracious predator of smaller fish; wide salinity tolerance — survives in seawater. May interfere with nursery function of estuaries by preying on juvenile marine species. Potential impacts little known.[118, 354, 360]
blue tilapia	*Oreochromis aureus*	Fish/Africa/1961/large freshwater herbivore, alters aquatic plant communities, gets too large for wading birds to eat. Introduced accidentally from test for control of nonindigenous aquatic plants that concluded it should not be released.[118, 123]
Cuban treefrog	*Osteopilus septentrionalis*	Treefrog/Cuba/1931/a large treefrog, feeds on insects and other amphibians; may be responsible for eliminating native treefrogs in residential areas.[92]
Burmese python	*Python molurus bivittatus*	Huge constrictor (snake)/SE Asia/1980's/established in mangroves and possibly Royal Palm area of Everglades National Park. Typically eats mammals such as raccoons, opossums, and dogs; could possibly take young endangered American crocodiles as well as humans. Potential impacts little understood.[408, 409]
black rat	*Rattus rattus*	Rodent/Europe/late 1800's or earlier/inadvertently introduced by ships, possibly as early as the 1500s; widely established and known to compete with native rodents including the endangered Key Largo woodrat.[291, 332, 614]
feral pig	*Sus scrofa*	Hoofed mammal/Eurasia/1500's/occurs throughout Florida with population estimated at a half million; common in Big Cypress Swamp; rooting habit is the most destructive impact of any introduced mammal in Florida; also destroys nests of sea turtles, gopher tortoises, indigo snakes, and shore birds; important prey for Florida panther.[332, 378, 614]

[a] These examples were selected by the author from the of the of the Florida Exotic Pest Plant Council's 2003 list of 61 most invasive (Category I) exotic species that occur in southern Florida.[199]

[b] At the time of this writing little literature was available on this species. See Univ. of Florida web site: http://creatures.ifas.ufl.edu/orn/scales/lobate_lac.htm.

[c] Also see Creel, Olan Ray. The evil weevil: what will Florida lose? *The Palmetto*, the Quarterly Magazine of the Florida Native Plant Society, Winter 99-2000.

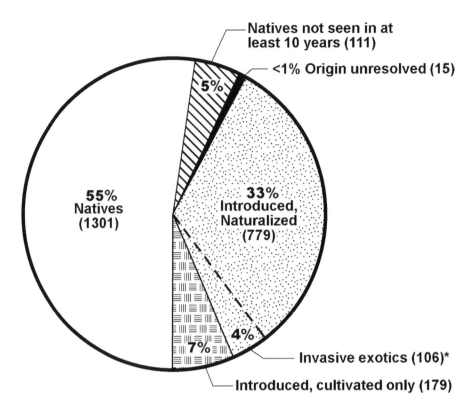

FIGURE 21.7 The composition of South Florida's flora, from the latitude of Lake Okeechobee southward. The total flora now includes 2385 taxa (kinds of plants: species and named subspecies/varieties), up from an original count of just over 1400 (extant natives plus taxa that have not been seen in recent years, now possibly extirpated). At least 958 taxa have been successfully introduced, primarily for horticulture and agriculture, but only 179 have remained confined to cultivation. About 779 nonindigenous taxa have "escaped" to become naturalized among wild native plants. Of the naturalized flora, 106 taxa (4 percent of the total flora and 11 percent of all naturalized nonindigenous plants) have become so successful — encroaching into, and altering natural habitats — that they have been listed by the Florida Exotic Pest Plant Council* as "invasive species." Sixty-one of these are known to be damaging to native plant communities and are recommended for control or eradication. (Pie Chart primarily compiled from Refs. 199* and 222 (website)).

its scientific genus name (*Melaleuca*), it is also known as paper-bark, punk tree, cajuput, and white bottlebrush tree. In Florida, melaleuca may grow to 100 feet in a straight stature, unlike its shorter branching stature in Australia, giving evidence that biological controls are not present in Florida. Its sensitivity to freezing weather has restricted it to the southern half of the Florida peninsula, where it grows in uplands and wetlands, invading sawgrass marshes, pine flatwoods, and cypress swamps, especially where partially drained. Melaleuca forms dense forests that exclude native vegetation and obstruct habitat use by wading birds. It was promoted in horticulture and agricultural for windbreaks and honey production and was intentionally seeded into the Everglades for the purpose of land development — trees transpire more water than marsh vegetation, so it was thought it would dewater the Everglades marshes. The spread of melaleuca was enhanced by its extreme tolerance of fire. Huge melaleuca forests spread from fine, wind-dispersed seeds into the Everglades and Big Cypress from both coasts.

After melaleuca's ecologically damaging characteristics were realized, labor-intensive and costly efforts to reduce its coverage were launched. The efforts have involved hand-applied herbicides and helicopter applications in dense patches in the Everglades, Big Cypress Swamp, and

FIGURE 21.8 A dense melaleuca forest in historic Everglades of Broward County, Florida. This cut-away view was provided by use of the foreground as a new school site. It is easy to see why most wading birds cannot use mature melaleuca forests as foraging locations. Their wings are far too long to land and there is no protection from lurking predators such as bobcats. (Photo by T. Lodge.)

Lake Okeechobee's littoral zones. They have been highly successful but do not offer the permanent solutions that biological control would. Biological control has been started with two extensively tested insects from Australia: the melaleuca snout beetle (*Oxyops vitiosa*) and a psyllid fly (*Boreioglycaspis melaleucae*). They show promising results, but more biological control agents will probably be required to make melaleuca behave like a native plant — lower height, a more branching stature, and no dense monocultures. Whether or not such control can be achieved depends on continued funding of ongoing physical/chemical control and of additional biological control agents, currently under research.[67, 283, 329, 568]

Brazilian Pepper

More versatile than melaleuca, especially in its invasion of upland communities, Brazilian pepper was initially called Florida holly for marketing of this large shrub as an ornamental in South Florida. Its seeds, in small decorative red berry-like clusters, have been widely spread by birds and mammals after passing through the digestive tracts. Raccoons spread them locally, and migrating robins, for example, have dispersed Brazilian pepper to the northern peninsula where it meets its temperature limit. With a noticeable proliferation after Hurricane Donna (1960), Brazilian pepper has widely invaded many plant communities in the Everglades region, especially pinelands and a variety of coastal lowland habitats including shallow mangrove swamps. Particularly troublesome is its ability to form closed forests in abandoned agricultural land and shallow coastal marsh habitats and its prevention of the recovery of pinelands after hurricane damage. Extensive areas of the western part of Everglades National Park are now Brazilian pepper forests and are effectively ruined as feeding habitat for wading birds. All pine rocklands damaged by Hurricane Andrew in South Florida would be Brazilian pepper thickets if it were not for huge efforts by government biologists and countless volunteers. Biological controls are now under quarantine development but not yet released.[62, 168, 169, 568]

Old World Climbing Fern

Relative to melaleuca and Brazilian pepper, the Old World climbing fern is a new invader in South Florida. It became established in the wild in 1965 in Martin County, where it soon became a problem in the headwater cypress wetlands of the Loxahatchee River system. By 1989 it had spread into Loxahatchee National Wildlife Refuge of the northern Everglades where it has completely covered tree islands in large areas of the refuge. Its growth into the forest canopy provides "fire ladders" by which surface fires ascend and kill trees that would ordinarily survive and prosper from natural fires. With several years' growth, successive layers become as thick as three feet — impossible to hike through (Figure 21.9). Recently, it has become established in the Big Cypress Swamp and western Everglades National Park, and there is no feasible means of combating it yet in sight. The Old World climbing fern now appears to be the most difficult problem in all of Everglades restoration.[76, 234]

Animals

In addition to introduced plants, South Florida also supports innumerable introduced animals (see Table 21.3). Compared to the basic role played by plants in providing habitats and serving as the base of food chains, nonindigenous animals do not cause such large changes in ecosystems except, of course, where animals directly alter plant communities.

Invertebrates

Two prominent examples that may alter plant communities are the lobate lac scale insect and the bromeliad beetle. These two recent arrivals may cause enormous ecosystem changes. The scale insect has infected over 39 species of native shrubs and trees and a larger number of ornamental

FIGURE 21.9 An infestation of Old World climbing fern in pinelands on State Road 711, Martin County, Florida. The blanket covering the natural habitat is an obvious conservation concern, smothering the area's native plants and animals. But imagine also that you are a hunter with no particular interest in nature compared to your quest for doves, rabbits, or deer. How will you ever find your quarry? The mat of stiff fern roots and "stems" over the ground can be up to three feet thick. How will you ever walk through it? (Photo by T. Lodge.)

plants. Wax myrtle, a common native shrub that is used extensively in landscaping, is often killed by lobate lac scale infestations. The bromeliad beetle potentially affects over 20 species of native air plants, killing many of them. Bromeliads are not only an asset of southern Florida's natural beauty, but provide important habitat for many small animals such as treefrogs. To keep the lobate lac scale and the bromeliad beetle in perspective, nearly 1000 nonindigenous insects now occur in Florida. Among them are three tiny wasp species that are the specific pollinators of ornamental fig trees, namely the lofty fig (*Ficus altissima*), the banyan (*F. beghalensis*), and the laurel fig (*F. microcarpa*). These three figs were previously unable to reproduce for the lack of pollination and thus remained confined to their planted locations. Now they have become invasive in natural communities and potentially damaging to structures such as buildings and bridges.[205]

Many other invertebrates have been introduced by the aquarium and ornamental-plant trade, often inadvertently as hitchhikers among plants and in soil. Several terrestrial and freshwater snails are examples, some of which have become garden pests while others compete with native snails and are potential human-disease hosts. Two such species are now common, the melanias (*Melanoides* spp.), inch-long freshwater snails with elongated shells, which are intermediate hosts of the human lung fluke, a worm-like parasite.[640]

Fishes

The warm temperatures of South Florida's waters have fostered the successful escape of many tropical freshwater aquarium fishes and the intentional introduction of a few. The key role that fish play in Everglades food chains makes introduced species an important concern. The pike killifish (Figure 21.10) was an early release: a dumped research project. This small voracious predator, adept at catching mosquitofish and other similarly sized fishes, remained in ditches and canals of southeastern Miami-Dade County for years but suddenly spread across the Everglades beginning in the late 1980's. It is one of the relatively few introduced fishes that has successfully invaded

FIGURE 21.10 A pike killifish measuring about 4 inches long, collected by the author from its initial infestation location in coastal agricultural ditches near Cutler, Miami-Dade County, about 1972. Since then it has spread to Key Largo, the Big Cypress Swamp, and the Everglades northward into Broward County. (Photo by T. Lodge.)

shallow marsh habitat distant from artificial waters such as canals. The walking catfish was a later aquarium introduction, now commonly seen in the beaks of cormorants, larger wading birds, and Florida gar, apparently displacing the native yellow bullhead in that role. Seventeen species of the cichlid (pronounced *sĭ′-klĭd*) family are now permanently established in Florida, seven specifically in Everglades National Park. Two abundant examples are the Mayan cichlid and blue tilapia (see Table 21.3). The Mayan cichlid has become a regional game fish, but its occurrence in upper estuarine habitats may have serious consequences as a predator of juvenile marine species such as snook and redfish. The blue tilapia, a large herbivorous fish (thus not a valued game species), has become abundant nearly throughout the Florida peninsula, where it has visibly altered habitats by construction of two-foot wide nest "craters" on shallow, littoral zones of lakes and canals. It has also altered native aquatic plant populations. Beginning in the late 1970's, the blue tilapia became abundant at the Anhinga Trail in Everglades National Park, where it — possibly with help from the introduced spotted tilapia (*Tilapia mariae*) — apparently eliminated growths of naiad (*Najas* sp.), a submerged aquatic plant. Naiad had been an important cover for smaller fish, as well as a food supply supporting native golden shiners, which in turn were important prey for numerous water birds (see front cover). An intentional introduction was the peacock bass (*Cichla ocellaris*), a beautifully colored, South American cichlid that is a game fish equivalent to the native largemouth bass (and considerably larger than the Mayan cichlid). The Florida Game and Fresh Water Fish Commission introduced it, without extensive review, both as a game fish and to control the abundance of other introduced cichlids in southeastern Florida. It has become a popular game fish but has not made a noticeable dent in the populations of other cichlids.[118, 123, 348, 354, 360, 605]

Reptiles and Amphibians

Relatively few introduced amphibians have succeeded in the wild in southern Florida although common in residential areas are the muted, cricket-like calls of the tiny West Indian greenhouse frog (*Eleutherodactylus planirostris*) and the giant "marine toad" (mentioned above), intentionally introduced from Mexico. Among reptiles, introduced lizards now outnumber native species in Florida. Nearly all lizards now seen in South Florida residential areas are introduced species. They include eight daytime-active species of anoles, such as the brown anole (mentioned above) now seen on boardwalks in Everglades National Park as well as in residential areas and any of several geckos, such as the tropical house gecko (*Hemidactylus mabouia*) seen inside houses at night. The largest of the introduced lizards is the great green iguana (*Iguana iguana*), an herbivorous lizard established from the pet trade, originally from localities in Central America and northern South America. Adult males may reach six feet, including their long tails. In an unusual event near Homestead, Florida, construction workers modifying a canal had seen several iguanas use a floating silt curtain as a convenient catwalk to cross the canal even though great green iguanas are adept swimmers as well as climbers. A six-foot spectacled caiman (also introduced, see Chapter 17) had similarly noticed the activity and quietly approached a three-foot iguana using the curtain floats. When it was within a few feet, the caiman suddenly lunged and caught the lizard. Following a short struggle before an astonished audience, the caiman ate the iguana.* Much more ominous from the human viewpoint is the now confirmed presence of the huge Burmese python in Everglades National Park. Its potential threat to dogs and children is an obvious concern, in addition to whatever wildlife impacts it may bring (see Table 21.3).[59, 92, 408, 409]

Birds

With partial exceptions such as the widely and long-established European starling and the recent explosive population growth of the Eurasian collared dove, few introduced birds have invaded southern Florida's wilderness. Residential areas, however, abound with them. The Miami/Fort Lauderdale area harbors at least ten parrot species, established from escaped pets. The monk

---

* Observation reported by Dr. Michael J. Andrejko, Natural Resources Program Manager, Homestead Air Reserve Base, summer, 2003.

parakeet, originally from temperate (southern) South America, has become particularly abundant and occurs in the eastern edge of the Everglades in Broward County. A recent introduction that shows an inclination to invade the Everglades is the purple swamp hen from the Middle East. It looks like a gigantic purple gallinule, being similarly colored but several times heavier. The purple swamp hen is a vegetarian and is strong enough to uproot tubers of marsh plants such as arrowhead. It is established in the Silver Lakes and Weston developments of Broward County, adjacent to Everglades water conservation areas.[296, 348, 427]

Mammals and the House Cat Dilemma

Twenty-six nonindigenous mammals have become established in Florida. Examples include the black rat and feral pig, both abundant throughout the state (see Chapter 18).[332] Although only marginally established in the wild, the domestic or house cat (*Felis catus*, originally derived from the wildcat, *F. silvestris*, of Europe and Africa) is abundant as outdoor pets and the "homeless" equivalent, *feral* cats. Domestic cats have been documented as the greatest single cause of bird mortality across the United States. Bird kills by other human causes, including hunting, collisions against windows and communication towers (telephone, radio, and TV), and kills by motor vehicles — all combined — are less than a third of cat predation. Domestic cats of U.S. urban/suburban areas kill on the order of a billion birds annually.[120, 231] Cat predation has also been documented as a leading cause of further jeopardizing some endangered mammals such as the lower keys marsh rabbit.[614] Free-ranging house cat abundance far exceeds equivalent predator populations that would occur in a wild, natural environment. For perspective, studies of the bobcat in the Big Cypress Swamp indicate a population of about one individual per two square miles (0.5/sq. mi.).[331] Based on the 2.2 times greater weight of bobcats (14 to 29 pounds) compared to domestic cats (6.5 to 13 pounds),[647] the *equivalent* predator population of domestic cats would be 1.1 individuals per square mile. However, where people support domestic cats in residential areas (owned pets and feral cat colonies), there may easily be 640 free-ranging cats per square mile (crudely estimated at one per acre).[348] This level is nearly 600 times greater than the natural level of equivalent wild predators. It does not take exact numbers to see that cities and their residential sprawl have become death traps for migratory birds because of uncontrolled outdoor pet and feral cats. Kept indoors, house cats present no such problem.

Controlling the Introduction of Plants and Animals

The dilemma of the constantly increasing numbers of nonindigenous plants and animals stems from society's reluctance to require demonstration of compatibility should a species escape and become established. Resistance to regulation stems largely from economic concerns of the ornamental landscape and pet-trade industries, where restrictions are viewed as damaged profits. The few regulations in place govern two concerns: 1) species that are known to have already caused problems — an after-the-fact situation; and 2) obviously dangerous species such as all large crocodiles, poisonous snakes, and piranhas.[123] The sparse regulations are in spite of enormous costs incurred in controlling introduced biota. On the other hand, the introduction of species for the purpose of biological control, such as those developed for melaleuca, requires years of testing in costly quarantine facilities, with testing results scrutinized in extensive reviews prior to approval, or rejection. Modern biological controls have a proven track record. Alligatorweed, once a severe navigation problem in waterways of the southeast, has been effectively controlled by a beetle that research identified as being an alligatorweed-specific herbivore. Clearly, similar rigorous proof should be required of all proposed introductions so that ecosystems can be spared further damage, and taxpayers, agriculture, and backyard gardeners can avoid huge control costs.[120, 167, 536]

OFF-ROAD VEHICLES [175, 176, 522, 644]

A variety of off-road vehicles (ORVs) has evolved for traveling in wetlands where the water is too shallow or obstructed for conventional boats and soils are too soft for cars or motorcycles. ORVs have greatly enhanced our ability to enter remote areas where few would otherwise venture. Wilderness recreation and hunting are the main uses. Four general categories of ORVs are used in the variable terrain of the Big Cypress Swamp, including the small, three-wheeled all-terrain cycle or ATC; the four-wheeled all-terrain vehicle (ATV); a plethora of improvised versions of the "swamp buggy" (typically four-wheel drive vehicles with large aircraft or tractor tires); various tracked vehicles, either half-track with front steering tires or full-track; and airboats. In the open, flat, more flooded Everglades terrain, the use of airboats (flat-bottom boats driven by an aircraft or automotive engine and pusher propeller) far surpasses other types, with small numbers of tracked vehicles also used.

The direct impact of ORVs includes soil rutting, damage to vegetation, spread of introduced species, noise, aesthetic (visual) alteration of wilderness, specimen collecting (especially in remote areas), and intrusive interference with wildlife. Secondary impacts include alteration of water flow and interruption of fire, with attendant functional changes to the ecosystem. In the Big Cypress, swamp buggies and tracked vehicles are used and cause extensive soil rutting. Major ORV "mud highways" have developed (Figure 21.11), especially near entry points. Areas of marl soils are particularly damaged because the ruts are very slippery when wet. Subsequent ORVs avoid the earlier ruts, making additional tracks. One such area became a quarter of a mile wide where almost all vegetation was eliminated. Aside from some major ORV trails and their substantial visual impacts (especially from aerial views), damage in the Big Cypress Swamp is considered to be small, although impact to wildlife due to human intrusion is little understood. The National Park Service has recently installed crushed limerock trails in the Big Cypress National Preserve in an effort to minimize ORV damage. In the Everglades water conservation areas, tracked vehicles have substantially damaged tree islands.[175]

The impact of airboats in the Everglades has been considered relatively minor. Their operation is limited to official use in Everglades National Park and to strictly regulated use in the Loxahatchee National Wildlife Refuge, but there is extensive commercial and private recreational use in the water conservation areas and the Northeast Shark Slough, south of WCA-3B. Relative to other ORVs, airboats cause the least damage to soils and vegetation, but repeated use of the same trails results in uprooting vegetation and soil displacement, producing channels (Figure 21.12). Some areas are strewn with such trails (Figure 21.13). The hydrological and biological impacts of soil disturbance and altered flow patterns (reducing sheet flow) are not well understood. However, cattail establishment is commonly associated with areas of heavy airboat traffic, and cattails often invade abandoned trails, even those lightly traveled.

Airboat impacts on wildlife is also little known, but airboat harassment may be problematic. Noise effects would seem obvious, as most airboats are excessively loud, but noise by itself has been extensively studied (mostly with respect to aircraft) and results indicate that wildlife easily habituate to constant or repeated noise and sudden, surprise noises cause only startle reactions. Physical intrusion and noise combined are far more damaging. Because airboats often travel 40 to 60 miles per hour (even higher speeds are sometimes achieved), significant adverse wildlife reactions seem inevitable. Further evaluation is warranted.

SOLVING THE PROBLEMS: EVERGLADES RESTORATION [612]

An important purpose of this book is to provide enough background on the Everglades and related ecosystems of southern Florida so that the reader can understand and participate in efforts to restore the region's natural functions. Everglades restoration is a multi-dimensional, moving target, with planning and implementation that will continue for nearly three decades. Thus, only a brief exposure to restoration plans is provided here. Other publications and web sites abound with current information.

FIGURE 21.11 Off-road vehicle (ORV) trails in the Big Cypress National Preserve. ORV operators trying to avoid ruts of earlier ORVs have proliferated extensive damage in many areas. To control this problem, the National Park Service has recently installed crushed limerock trails and requires ORV operators to use them. (Photo from Ref. 644, Figure 9.6c, used by permission from CRC Press.)

FIGURE 21.12 Well-traveled airboat trail through moderately dense sawgrass, as seen from the airboat. Water Conservation Area 3A. (Photo by T. Lodge.)

FIGURE 21.13 An aerial view of abundant airboat trails, Northeast Shark Slough south of the Tamiami Trail, looking southwest. (Photo by T. Lodge.)

Many initiatives to fix problems began in the 1970's after it was recognized that the Everglades and many adjacent parts of the "South Florida ecosystem" were seriously deteriorating. Recognized problems included: declining wading bird populations; melaleuca infestations; deteriorating health of Lake Okeechobee; the region's estuaries and Florida Bay; and damaging nutrient-related changes in the Everglades water conservation areas. Numerous separate programs were initiated to solve these unwanted changes. Melaleuca control and mercury reduction, both described earlier in this chapter, are promising successes. Other key initiatives are Modified Water Deliveries, projects under the 1994 Everglades Forever Act, and the joint state/federal Comprehensive Everglades Restoration Program.

Modified Water Deliveries, "Mod Waters"

Of the efforts to rehabilitate specific damaged portions of the Everglades, this program stands out. It was envisioned to re-hydrate parched and over-drained parts of the original Shark River Slough as part of the 1989 federal Everglades Expansion Act. The design would breach the L-67 levees (see Color Figures 3 and 4 and Figure 21.6) and allow flows from over-flooded WCA-3A to enter water-deprived WCA-3B and then continue through new passages under the Tamiami Trail into "Northeast Shark Slough" (NESS) south of the trail. This last area was to be annexed to Everglades National Park under the act.

Mod Waters immediately entered controversy with a jeopardy opinion by the USFWS concerning the snail kite population in WCA-3A where over-flooding had promoted a large applesnail population on which the kites were prospering. Further research and mediation arranged by the National Audubon Society[454] solved concerns over the snail kite, and the jeopardy opinion was reversed. However, another problem was the existence of a residential area that had been allowed to develop by Miami-Dade County, on the west side of the Everglades eastern protective levee that separated flood-prone lands from developing areas to the east (see Color Figure 3). With renewed flows through NESS, Mod Waters would have flooded this community. After many iterations, a plan was selected to buy-out the western portion of this community and protect the eastern portion with an additional levee. The final hurdles have apparently been solved and Mod Waters implementation should occur by 2007. It includes a 3000-foot bridge on the Tamiami Trail and numerous culverts to reestablish flow that the trail blocked for over 85 years.[339, 612]

EVERGLADES FOREVER ACT

A massive effort that is part of ecosystem restoration is a State of Florida program to improve water quality in the Everglades Protection Area. The 1994 Everglades Forever Act, which settled the federal lawsuit regarding water quality standards and Everglades pollution (see phosphorus section, this chapter) stipulates that a phosphorus criterion be developed, now set at 10 ppb for waters entering the Everglades. Treatment areas — the Everglades Construction Project — were designed and constructed for compliance with this act (see Color Figure 6).

COMPREHENSIVE EVERGLADES RESTORATION PLAN (CERP)

Out of rising concern for the entire Everglades ecosystem, an overall revision of the C&SF Project was officially born in the federal Water Resources Development Act of 1992. Initially it was called the "Restudy," short for the Central and Southern Florida Project Comprehensive Review Study. The Restudy document was submitted to Congress in 1999 and upon approval became the Comprehensive Everglades Restoration Plan (CERP). It is a joint effort of the federal government and the State of Florida. Important guiding principles of CERP are to:

- Restore, preserve, and protect the South Florida ecosystem while providing required flood protection and water supply. Emulating the original system's water storage is the largest general element of the plan. The overall generalization for CERP's purpose is to ensure the "right quantity, quality, timing, and distribution" of water.
- Incorporate water quality considerations and criteria (under Section 401 of the Clean Water Act). Water quality is the greatest area where the modern plan deviates from the original C&SF Project as there were no water quality standards or regulations to be considered then except for recognition that freshwater supplies required protection from encroaching saline coastal waters.
- Base the plan on the best available science (see references 429-435).
- Incorporate adaptive monitoring, assessment, and management as the program progresses to allow for corrections to the plan and its implementation as more is learned about performance of the system (see RECOVER, below).

CERP was developed by examining the hypothetical performance of 21 parts of the South Florida ecosystem. Among these are Lake Okeechobee, the Caloosahatchee and the St. Lucie estuaries, each of the three water conservation areas, two freshwater physiographic regions in Everglades National Park, Florida Bay, and numerous others. Two teams were responsible for plan formulation, an Alternative Development Team (ADT) and an Alternative Evaluation Team (AET). The ADT developed alternatives and the AET judged their performance. Through an iterative process between these teams, they developed and examined six alternatives, four of which (alternatives A-D) were explored in detail. Alternative D was selected and further refined through 13 versions to become the selected plan, called Alternative D13R.

Various computer models were used as tools in the selection process. The hub was the South Florida Water Management Model, a computer model that integrates surface and groundwater over the region from Lake Okeechobee to Florida Bay. It incorporated data from the 31-year period between 1965 and 1995 and produced results of water depths (including drydown levels) for scenarios of routing water through the region based on the alternatives. These hydrologic results (in the form of tables, graphs, and maps) were then used in other models including water quality models (a Lake Okeechobee eutrophication model, an Everglades phosphorus transport model, and analyses of phosphorus-reduction performance by the non-CERP Everglades Construction Project and proposed reservoirs), the River of Grass Evaluation Model (ROGEM), and the Across Trophic Level System Simulation model (ATLSS). The South Florida ecosystem's ecological integrity and the region's water supply and flood control were examined with these models for the future with CERP Alternative D13R and without CERP. The "without CERP" alternative looked at the year 2050, extrapolating only projects that were in place or would be done independent of CERP, such as phosphorus reduction by the Everglades Construction Project (see Color Figure 6).

For ecologists, the ATLSS model is of particular interest. The model runs were done for the Cape Sable seaside sparrow, white-tailed deer, snail kite, overall Everglades landscape fish production, and wading bird foraging conditions. It was used throughout the selection process to compare alternatives for a variety of conditions, including wet, normal, and dry years. ATLSS graphically displays the with- and without-project scenarios and the difference between them for the study area or any selected geographic subregion. An example output for long-legged wading birds is shown in Color Figure 7.

Alternative D13R is based on 49 operational and structural features called "components." Many of these components require that water-control structures be constructed and/or that existing structures be removed. Other components are strictly operational — defining how an existing feature will be operated, such as Lake Okeechobee's water levels. While the overall, general plan (Alternative D13R) has been selected, the details of exactly where facilities will be located and configured have yet to be determined. The following four example components explain how CERP will become a reality.

Component A6 (North of Lake Okeechobee Storage Reservoir)

This component involves reservoir storage and water quality treatment for stormwater in a three-county area (Glades, Highlands, and Okeechobee counties) north of Lake Okeechobee. It is intended to retain wet-season water that would otherwise cause flooding and require discharge through the lake and existing outlet canals to coastal waters. The benefits of Component A6 are reduced flooding (including avoiding high lake levels that damage the littoral zone), reduced pulses of water to the Caloosahatchee and St. Lucie estuaries, water retention for dry-season ecosystem benefits, and enhanced water supply for humans. The reservoir storage capacity and the size of associated water quality treatment area(s) were selected based on modeling the contribution to the overall D13R alternative but with only the regional definition of their locations in the three-county area. The criteria, set through the modeling process, were as follows:

Reservoir:

- 17,500 acres at 11.5 feet maximum depth (to hold up to about 200,000 acre-feet)
- Inflow pump capacity of 4800 cubic feet per second (cfs)
- Outflow release structure(s) that can carry 4800 cfs

Stormwater treatment area:

- 2500 acres at 4.0 feet maximum depth
- Gravity flow outlet

It is up to subsequent detailed work — conducted by the Corps of Engineers, the South Florida Water Management District, or a consultant working under one of these agencies — to select the actual location/configuration of the reservoir (or reservoirs) and associated stormwater treatment areas. The work involves: 1) identifying suitable land that minimizes impacts to important wildlife habitat (especially where listed species occur), to wetlands, and to land uses such as towns, highways, schools, parks, unique agricultures land, etc.; and 2) determining whether it is feasible to obtain the land or the use of the land through legal easements. As required by law, the process is open to the public. Only after selecting and obtaining sufficient land can final design and construction occur. For the many components like A6 where features are only conceptually defined in D13R, it is easy for the reader to understand that a great deal of work and time is involved.

Component GG4 (Lake Okeechobee Aquifer Storage and Recovery)

This component involves the construction of 200 huge wells to store excess water from Lake Okeechobee in the regionally saline Upper Floridan Aquifer, some 800 to 1000 feet underground in the vicinity of the lake (see Figure 1.2). The D13R-specified capacity for all of these wells totals about a billion gallons per day (3000 acre-feet per day). The hydrologic benefits mimic Component A6 (above), but the unique features of aquifer storage and recovery (ASR) that make it desirable are: 1) almost no land is required; and 2) it avoids the loss of water to evaporation, which is excessive for reservoirs. About 70 percent of water stored in a saline aquifer can be recovered (pumped out) before water that has mixed with salt precludes further withdrawal. This level of recovery far exceeds surface storage, which may be as low as 30 percent due to evaporation. ASR has been used in many applications in Florida for drinking-water supply but never on a scale approaching what is proposed in this component of CERP. Pilot projects are under investigation.

Component G6 (EAA Storage Reservoir)

Since its development, the EAA depended directly on Lake Okeechobee and the WCAs (through the regional canal network) for its water supply and flood relief, accomplished by pumping water in the needed direction. This component is to correct that deficiency, specifying three 20,000-acre (31-square mile) storage reservoirs with six-foot depth variation to receive or deliver stormwater. These reservoirs were conceptually designed to receive excess water from Lake Okeechobee and the EAA. The coupling to Lake Okeechobee helps prevent damaging releases to the estuaries like the other two components discussed above. They also relieve hydrologic stresses on the WCAs from EAA releases. The general specification designates use of EAA stormwater treatment areas — developed independently as the state's Everglades Construction Project — for treated releases to the WCAs. Like Component A6 above, only the performance requirements of these three reservoirs, not the exact locations, were specified in D13R. Thus, they require site selection, land purchase, and final design prior to construction.

Component QQ6 (Decompartmentalization of WCA-3)

This component specifies modifications of WCA-3 so that unrestricted, passive flow can occur between WCA-3A, 3B, and Everglades National Park. Some of its details were the final changes that established Alternative D13R. It involves both operational and structural modifications. The operational modifications have to do with how water is delivered into WCA-3 from upstream sources. The main structural modifications are (refer to Color Figures 3 and 4):

- Fill-in the Miami Canal in most of WCA-3.
- Degrade the L-67B levee and install eight overflow structures along the length of L-67A.
- Fill in 7.5 miles of the south end of the L-67 canal (along the L-67A levee), from the Tamiami Trail northward.
- Remove the L-28 levees on the west side of WCA-3.

- Remove the L-29 levee (that forms the south end of WCA-3A and 3B) and the S-12 gates that currently regulate flows from WCA-3A into Everglades National Park.
- Elevate the Tamiami Trail on a new levee (to replace the L-29) but provide a series of bridges to allow sheet flow into Everglades National Park.

For components like QQ6, the geographic location of structures is defined — they exist or were aligned to replace existing structures. Site selection and land purchases, therefore, are minimal or not involved — only final design and construction. Nevertheless, the construction phasing of components so that they operate in hydrologic coordination will require many years. Thus, it is easy to understand that Everglades restoration under CERP will be ongoing long past the life of this book to bring CERP's Alternative D13R to reality.

Restoration, Coordination, and Verification (RECOVER)

How will we know whether CERP is effective? It is the responsibility of RECOVER to develop overall CERP performance measures and to oversee development of special performance measures that pertain to the individual components. Respective monitoring programs will then provide performance data for RECOVER's adaptive assessment process in order to recommend adaptive-management strategies for "course corrections." In this way, unforeseen circumstances and unanticipated consequences can be authorized within CERP, pending availability of funds, instead of requiring new legislation.

Restoring the Everglades is obviously an enormous undertaking for both CERP and separate state and federal programs. CERP alone carries a price tag of over $8 billion and is projected to take nearly 30 years. What is the certainty of its success? Some comfort can be taken in the built-in adaptive assessment and management strategy. However, an important part of the plan is water supply for a projected population. The initial C&SF Project was initiated in 1950 when the population in the 16-county planning area encompassed by CERP was about 900,000. That area has grown to over 6 million, far in excess of the population envisioned by the water supply planned in the original C&SF Project. The population expectation used in CERP for the year 2050 is 11.6 million. Population growth and associated environmental impacts are highly unpredictable thus represent an unavoidable weakness in CERP. Lower growth could obviously be accommodated by the plan, but substantially higher growth and attendant water demand, and other impacts, could require major adjustments, an area emphasized by restoration critics.[*]

THE SHORT-TERM PROGNOSIS: THE NEXT FIFTEEN YEARS

With the recent improvement in wading bird reproductive success, the effective reduction of mercury contamination, progress in phosphorus reduction, and the beginning of implementation of CERP, there is a basis for optimism in the short-term future of the Everglades. Important ongoing efforts include curtailing phosphorus enrichment, solving technical listed-species constraints, and controlling nonindigenous species. Other difficult constraints are inevitable lawsuits. Listed species (see inset box, Chapter 19), water quality, and land acquisition are obvious areas of existing and future legal concern, but legal tactics can be expected from innovative sources. Through all of these challenges, maintaining the momentum of CERP, state programs, and their respective funding is critical. Cooperation among environmental interests, as emphasized by The Everglades Coalition, is the key to success here.

---

[*] See Weisskoff, R. 2003. A tale of two models: IMPLAN and REMI on the economics of Everglades restoration. In: D.J. Rapport and others (eds.). *Managing for Healthy Ecosystems*. Lewis Publishers, Boca Raton, Florida; Weiskoff, R. In press, 2004. *The Economics of Everglades Restoration: Missing Pieces in the Future of South Florida*. Edward Elgar Press, Northampton, MA.; and reference 643.

Phosphorus enrichment has dominated Everglades science and politics since 1989. Under state programs, great progress has been made in phosphorus treatment technology and on-farm best management practices. However, a level of 10 ppb phosphorus will be difficult to attain and appears to be critical to the Loxahatchee National Wildlife Refuge as well as the central and southern Everglades where slough and wet prairie habitats occur. Based on the pre-drainage Lake Okeechobee overflows probably having been 20 ppb or somewhat higher, northern WCA-2A and northern WCA-3A could likely assimilate those levels without damaging the remainder of the Everglades Protection Area, provided that such waters are forced to follow sheet flow and are not routed deep into the Everglades through canals. Recent performance data for STA 2 support this contention (see SFWMD 2004 Consolidated Report).[568]

Of major concern in the first edition of this book (1994), melaleuca biological control combined with intensive mechanical and herbicidal eradication programs appear to be solving the previously ominous impact of this species. Continued funding and success, however, are required. Brazilian pepper and especially Old World climbing fern will require effective controls not yet available. If Old World climbing fern cannot be stopped and it comes to dominate the Everglades, there would be just reason to abandon having a national park named "Everglades."

The forecast made in 1938 by biologist Daniel Beard[61] (later to become the first Superintendent of Everglades National Park) was all too prophetic:

Practically without exception, areas that have been turned over to the [National Park] Service as national parks have been of superlative value with existing features so outstanding that if the Service were able to merely retain the status quo, the job was a success. This will not be true of the Everglades National Park. The reasons for even considering the lower tip of Florida as a national park are 90 percent biological ones, and hence highly perishable. Primitive conditions have been changed by the hand of man, abundant wildlife resources exploited, woodland and prairie burned and reburned, water levels altered, and all the attendant, less-obvious ecological conditions disturbed.

Director Cammerer recently said "I would much rather have a national park created that might not measure up to all everybody thinks of it at the present time, but which, 50 or 100 years from now, with all the protection we could give it, would have attained a natural condition comparable to primitive conditions...." If the National Park Service is prepared to follow the strategy thus expressed, the Everglades National Park seems justified. If it is not ready to do this, the writer wishes to state emphatically, the Everglades is *not* justified.

THE LONG-TERM VIEW

What is the future beyond the next few decades? Major questions are the following:

- The continuing pernicious problem of introduced plants and animals.
- Effectiveness of extremes in maintaining Everglades landscapes and functions, including severe drought and fire, and excessive flooding with rapid flow. Striving to duplicate the average year by manipulation of water controls will likely "lose the battle." At issue are the importance of water flow in maintaining Everglades ridge-and-slough landscape and the related subtle problem of sawgrass overtaking slough habitat.
- The performance of aquifer storage and recovery, and how to replace water storage if the consequences of ASR are not acceptable.
- Underestimated population growth and its environmental impacts, importantly water demand.
- The fate of the EAA after farming — rock pits and residential development versus water storage, treatment, and wildlife habitat. What ultimately happens in the EAA is critical.
- Effective use of the best science ahead of special-interest politics.
- Long-term continued funding for Everglades restoration.

- Rising sea level.
- Problems such as endocrine disruptors — which influence living processes including development and reproduction in wildlife species and humans — that we have either not addressed due to insufficient information or not yet recognized.

The most optimistic part of CERP is its built-in adaptive assessment and management strategy under RECOVER. Most of the above problems can at least be addressed and their solutions recommended through CERP, with the ultimate problem then being funds to implement corrections. Problems that essentially cannot be addressed through the process are the loss of required science to special-interest politics, the loss of funding, and rising sea level. Hopefully, international cooperation will begin to curb global warming and attendant rising sea level — the problem is far larger than southern Florida. Without correction and a continuing foot-per-century rise, saltwater will obliterate Everglades National Park's freshwater habitats within 500 years, leaving only the central and northern Everglades (currently in the water conservation areas). However, the loss of effective science and the loss of funding for CERP and other programs would cause demise of the ecosystem at an orders-of-magnitude faster rate. As in the short term, cooperation among environmental interests will continue to be the key.

An expensive but manageable correction that may be demonstrated necessary early in the long term is accommodating much more water flow than anticipated in the current plan. The possibility that the Tamiami Trail and perhaps even Alligator Alley will need to be raised as bridges would be a significant cost — estimated at $150 million for the Tamiami Trail, possibly more than the tax-paying public would accept. With luck, such expensive corrections will not be necessary to produce a functional Everglades restoration. In this regard, the insight of David Fairchild (see Chapter 2) on the Tamiami Trial, recalled from a field excursion in about 1917, makes a fitting conclusion:[191]

Engineers had begun their work on the Tamiami Trail, and rock was being scooped out to form the roadbed of that long, ugly, straight line of glaring white, which was slowly penetrating the magnificent water landscape that was the Everglades. I have photographs taken that year showing Safford [a U.S. Dept. of Agriculture scientist, see reference 572] botanizing in one of the fascinating hammocks, which, like beautiful islands, were surrounded by water and made a unique and enchanting wonderland, full of mystery. There was a romance in knowing it was the home of great snakes and lizards and turtles and of those incomparably beautiful tree snails, the famous liguus, and that in the algae-covered waters below were being produced the countless myriads of fresh-water mollusks, which formed the food for the millions of beautiful plume birds, flocks of which were to be seen feeding everywhere in the shallow waters of the 'Glades.

How little we dreamed, those of us who thought then only of the prospect of being given access to those 'Glades, and of being able to travel across them to the west coast of Florida, what the Trail would one day become; that trail through a superb landscape which it had taken perhaps many thousands of years to produce and the like of which was to be found nowhere else on the planet.

References

1. Abrahamson, Warren G. and David C. Hartnett. 1990. Pine flatwoods and dry prairies. *In* Ronald L. Myers and John J. Ewel (editors). *Ecosystems of Florida*. University of Central Florida Press, Orlando.
2. Acosta, Charles A. and Sue A. Perry. 2001. Impact of hydropattern disturbance on crayfish population dynamics in the seasonal wetlands of Everglades National Park, USA. *Aquatic Conservation: Marine and Freshwater Ecosystems* 11: 45-57.
3. Acosta, Charles A. and Sue A. Perry. 2002. Spatio-temporal variation in crayfish production in disturbed marl prairie marshes of the Florida Everglades. *Journal of Freshwater Ecology* 17(4): 641-650.
4. Acosta, Charles A. and Sue A. Perry. 2002. Spatially explicit populations responses of crayfish *Procambarus alleni* to potential shifts in vegetation distribution in the marl marshes of Everglades National Park, U.S.A. *Hydrobiologia* 477: 221-230.
5. Ager, Lothian. 1971. The fishes of Lake Okeechobee, Florida. *Quarterly Journal of the Florida Academy of Sciences* 34(1): 53-62.
6. Alden, Peter, Richard B. Cech, Richard Keen, Amy Leventer, Gil Nelson, and Wendy B. Zomlefer. 1998. *National Audubon Society Field Guide to Florida*. Alfred A. Knopf, New York.
7. Alexander, Taylor R. Personal communications with the author, 1968-2003.
8. Alexander, Taylor R. 1953. Plant succession on Key Largo, Florida, involving *Pinus caribaea* and *Quercus virginiana*. *Quarterly Journal of the Florida Academy of Sciences* 16(3): 133-38.
9. Alexander, Taylor R. 1953. The largest mahogany tree. *Everglades Natural History* 1(1).
10. Alexander, Taylor R. 1954. Paradise Key on fire. *Everglades Natural History* 2(4).
11. Alexander, Taylor R. 1954. Trees against the sea. *Everglades Natural History* 2(4).
12. Alexander, Taylor R. 1955. Observations on the ecology of the low hammocks of southern Florida. *Q. Journ. Fla. Acad. Sci*. 18(1): 21-7.
13. Alexander, Taylor R. 1958. Ecology of the Pompano Beach Hammock. *Q. Journ. Fla. Acad. Sci*. 21(4): 299-304.
14. Alexander, Taylor R. 1958. High hammock vegetation of the southern Florida mainland. *Quarterly Journal of the Florida Academy of Sciences* 21(4): 293-8.
15. Alexander, Taylor R. 1958. Temperature variance in microclimates of southern Florida. *Proceedings of the Florida State Horticultural Society* 71: 356-8.
16. Alexander, Taylor R. 1967. A tropical hammock on the Miami (Florida) Limestone – a twenty-five year study. *Ecology* 48(5): 863-7.
17. Alexander, Taylor R. 1967. Effect of Hurricane Betsy on the southeastern Everglades. *Quarterly Journal of the Florida Academy of Sciences* 30(1): 10-24.
18. Alexander, Taylor R. 1968. Acacia choriophylla, a tree new to Florida. Quarterly Journal of the Florida Academy of Sciences 31(3): 197-8.
19. Alexander, Taylor R. 1971. Sawgrass biology related to the future of the Everglades ecosystem. *Soil and Crop Science Society of Florida Proceedings* 31: 72-4.
20. Alexander, Taylor R. 1974. *Schizaea germanii* rediscovered in Florida. *American Fern Journal* 64.
21. Alexander, Taylor R. 1974. Evidence of recent sea level rise derived from ecological studies on Key Largo, Florida. *In* P.J. Gleason (editor). *Environments of South Florida: Present and Past*. Memoir 2: Miami Geological Society, Miami, Florida.
22. Alexander, Taylor R. 1975. Four alien plants no longer welcome. *Museum* 7(5): 5-8.
23. Alexander, Taylor R. and Alan G. Crook. 1973. Recent and long-term vegetation changes and patterns in South Florida. Final Report, Part 1. Mimeo Rep. (EVER-N-51). U.S. Department of the Interior, National Park Service, 215 p., NTIS No. PB 231939.
24. Alexander, Taylor R. and Alan G. Crook. 1974. Recent vegetational changes in southern Florida. *In* P.J. Gleason (editor). *Environments of South Florida: Present and Past*. Memoir 2: Miami Geological Society, Miami, Florida.

25. Alexander, Taylor R. and Alan G. Crook. 1975. Recent and long-term vegetation changes and patterns in South Florida. Final Report, Part 2. Mimeo. Rep., U.S. Department of the Interior, National Park Service, 865 p. NTIS No. PB 264462.

26. Alexander, Taylor R. and Alan G. Crook. 1984. Recent vegetational changes in southern Florida. *In* P. J. Gleason (editor). *Environments of South Florida: Present and Past II*. Miami Geological Society, Miami. (republication of the 1974 article)

27. Alexander, Taylor R. and John H. Dickson III. 1970. Vegetational changes in the National Key Deer Refuge. *Quarterly Journal of the Florida Academy of Sciences* 33(2): 81-9.

28. Alexander, Taylor R. and John D. Dickson III. 1972. Vegetational changes in the National Key Deer Refuge – II. *Quarterly Journal of the Florida Academy of Sciences* 35(2): 85-96.

29. Alvarez, Ken. 1993. Twilight of the Panther: Biology, Bureaucracy and Failure in an Endangered Species Program. Myakka River Publishing, Sarasota, Florida.

30. Alvear Rodriguez, Elsa M. 2001. The use of nesting initiation dates of roseate spoonbills (*Ajaia ajaja*) in northeastern Florida Bay as an ecosystem indicator for water management practices, 1935-1999. Masters thesis, Florida Atlantic University, Boca Raton, Florida.

31. Anderson, D.L. and P.C. Rosendahl. 1998. Development and management of land/water resources: the Everglades, agriculture, and South Florida. *Journal of the American Water Resources Association* 34(2): 235-249.

32. Anonymous. 1997. Record 14-foot gator judged nuisance, killed. *The Miami Herald*, Thursday, October 2, 1997, page 5B (with photograph).

33. Armentano, Thomas V., Robert F. Doren, William J. Platt, and Troy Mullins. 1995. Effects of Hurricane Andrew on coastal and interior forests of southern Florida: overview and synthesis. *Journal of Coastal Research* 21: 111-144.

34. Armentano, Thomas V., David T. Jones, Michael S. Ross, and Brandon W. Gamble. 2002. Vegetation pattern and process in tree islands of the southern Everglades and adjacent areas. *In* Fred H. Sklar and Arnold van der Valk (editors). *Tree Islands of the Everglades*. Kluwer Academic Publishers, Boston, Massachusetts.

35. Ashton, Ray E., Jr. and Patricia Sawyer Ashton. 1988. *Handbook of Reptiles and Amphibians of Florida, part one: The Snakes*. Windward Publishing, Inc., Miami, Florida.

36. Ashton, Ray E., Jr. and Patricia Sawyer Ashton. 1988. *Handbook of Reptiles and Amphibians of Florida, part three: The Amphibians*. Windward Publishing, Inc., Miami, Florida.

37. Ashton, Ray E., Jr. and Patricia Sawyer Ashton. 1991. *Handbook of Reptiles and Amphibians of Florida, part two: Lizards, Turtles & Crocodilians* (revised second edition). Windward Publishing, Inc., Miami, Florida.

38. ASR Systems web site: *www.asrforum.com*.

39. Aumen, Nicholas G. 1995. The history of human impacts, lake management, and limnological research on Lake Okeechobee, Florida (USA). *In* N.G. Aumen and R.G. Wetzel (editors). Ecological studies on the littoral and pelagic systems of lake Okeechobee, Florida (USA). *Archiv für Hydrobiologie*, special issue: Advances in Limnology, 45: 1-356.

40. Aumen, Nicholas G. (editor). 2003. The role of flow in the Everglades ridge and slough landscape. South Florida Ecosystem Restoration Working Group, Science Coordination Team. 62 pp.

41. Aumen, N.G. and Susan Gray. 1995. Research synthesis and management recommendations from a five-year, ecosystem-level study of Lake Okeechobee, Florida (USA). *In* N.G. Aumen and R.G. Wetzel (editors). Ecological studies on the littoral and pelagic systems of lake Okeechobee, Florida (USA). *Archiv für Hydrobiologie*, special issue: Advances in Limnology, 45: 1-356.

42. Aumen, N.G. and R.G. Wetzel (editors). 1995. Ecological studies on the littoral and pelagic systems of Lake Okeechobee, Florida (USA). *Archiv für Hydrobiologie*, special issue: Advances in Limnology, 45: 1-356.

43. Austin, D. F. 1980. Historically important plants of southeastern Florida. *Florida Anthropologist* 33: 17-31.

44. Austin, D. F. 1997. Glades Indians and the plants they used. *The Palmetto* 17: 7-10.

45. Austin, D. F. and D.M. McJunkin. 1978. An ethnoflora of Chokoloskee Island, Collier County, Florida. *Journal of the Arnold Arboretum* 59: 50-67.

46. Austin, Daniel. (undated, c. 1995). *Vascular Plant List of Fakahatchee Strand State Park*. Funding and printing by the Naples Chapter of the Florida Native Plant Society and Bonita Bay Properties.

47. Austin, Daniel F. 1998. Invasive exotic climbers in Florida: biogeography, ecology, and problems. *Florida Scientist* 61(2): 106-117.

48. Austin, Daniel F. 2001 (update). Florida's Lost Ethnoflora? URL: *http://www.fau.edu/divdept/science/envsci/ethnoflora.html*.

49. Austin, Daniel F. and Elizabeth Smith. (undated, c. 1998). *Pine Rockland Plant Guide: a field guide to the plants of South Florida's pine rockland community.* Miami-Dade Dept. of Environmental Resources Management. Miami.

50. Avery, George N. 1980. Plants of Everglades National Park, a preliminary checklist of vascular plants. Report T-574, U.S. National Park Service, South Florida Research Center, Everglades National Park, Homestead.

51. Avery, George N. and Lloyd L. Loope. 1980. Endemic taxa in the flora of South Florida. Report T-558, U.S. National Park Service, South Florida Research Center, Everglades National Park, Homestead.

52. Bacchus, Sidney T. 2002. The "ostrich" component of multiple stressor model: undermining South Florida. *In* James W. Porter and Karen G. Porter (editors). *The Everglades, Florida Bay, and Coral Reefs of the Florida Keys: an Ecosystem Source Book.* CRC Press, Boca Raton, Florida.

53. Balciunas, Joseph K. and Ted D. Center. 1991. Biological control of *Melaleuca quinquenervia*: prospects and conflicts. *In* Ted D. Center, Robert F. Doren, Ronald H. Hofstetter, Ronald L. Myers, and Louis D. Whiteaker (editors). Proceedings of the Symposium on Exotic Pest Plants. U.S. Department of the Interior, National Park Service, Washington, D.C.

54. Baldwin, Andrew H., William J. Platt, Kari L. Gathen, Jeannine M. Lessmann, and Thomas J. Rauch. 1995. Hurricane damage and regeneration in fringe mangrove forests of southeast Florida, USA. *Journal of Coastal Research* 21: 169-183.

55. Bancroft, G. Thomas and Reed Bowman. 2001. White-crowned pigeon (*Columba leucocephala*). *In* A. Poole and F. Gill (editors). *The Birds of North America*, No. 596. The Birds of North America, Inc., Philadelphia, PA.

56. Bancroft, G. Thomas and R.J. Sawicki. 1995. The distribution and abundance of wading birds relative to hydrologic patterns in the water conservation areas of the Everglades. Final report to the South Florida Water Management District, Contract No. C-3137, 198 pages.

57. Bancroft, G. Thomas, Allan M. Strong, Richard J. Sawicki, Wayne Hoffman, and Susan D. Jewell. 1994. Relationships among wading bird foraging patterns, colony locations, and hydrology in the Everglades. *In* Steven M. Davis and John C. Ogden (editors). *Everglades: the Ecosystem and Its Restoration.* St. Lucie Press, Delray Beach, Florida.

58. Barnes, Jay. 1998. *Florida's Hurricane History.* University of North Carolina Press, Chapel Hill, North Carolina.

59. Bartlett, Richard D. and Patricia P. Bartlett. 1999. *A Field Guide to Florida Reptiles and Amphibians.* Gulf Publishing Company, Houston, Texas.

60. Baur, Donald C. and Wm. Robert Irvin. 2002. Endangered Species Act: Law, Policy, and Perspectives. American Bar Association Publishing, Chicago, Illinois.

61. Beard, Daniel B. 1938. Wildlife Reconnaissance, Everglades National Park Project. U.S. Department of the Interior, National Park Service. Washington, D.C.

62. Bennett, Fred D. and Dale H. Habeck. 1991. Brazilian peppertree — prospects for biological control in Florida. *In* Ted D. Center, Robert F. Doren, Ronald H. Hofstetter, Ronald L. Myers, and Louis D. Whiteaker (editors). Proceedings of the Symposium on Exotic Pest Plants. U.S. Department of the Interior, National Park Service, Washington, D.C.

63. Bennetts, Robert E. 1993. The snail kite, a wanderer and its habitat. *The Florida Naturalist* 66(1): 12-15.

64. Bennetts, Robert E., Michael W. Collopy, and James A. Rodgers, Jr. 1994. The snail kite in the Florida Everglades: a food specialist in a changing environment. *In* Steven M. Davis and John C. Ogden (editors). *Everglades: the Ecosystem and Its Restoration.* St. Lucie Press, Delray Beach, Florida.

65. Blake, N.M. 1980. Land into Water — Water into Land: A History of Water Management in Florida. University Presses of Florida, Tallahassee.

66. Blount, Robert S., III. 1993. *Spirits of Turpentine.* Florida Agricultural Museum, Tallhassee, Florida.

67. Bodle, Michael J., Amy P. Ferriter, and Daniel D. Thayer. 1994. The biology, distribution, and ecological consequences of *Melaleuca quinquenervia* in the Everglades. *In* Steven M. Davis and John C. Ogden (editors). *Everglades: the Ecosystem and Its Restoration.* St. Lucie Press, Delray Beach, Florida.

68. Boesch, Donald F., Neal E. Armstrong, Christopher F. D'Elia, Nancy G. Maynard, Hans W. Paerl, and Susan L. Williams. 1993. Deterioration of the Florida Bay ecosystem: an evaluation of the scientific evidence. Report dated September 15, 1993, to the Interagency Working Group on Florida Bay sponsored by the National Fish and Wildlife Foundation, the National Park Service, and the South Florida Water Management District.

69. Bosence, Daniel. 1989. Biogenic Carbonate Production in Florida Bay. *Bulletin of Marine Science* 44(1): 419-433.

70. Bottcher, A.B. and F.T. Izuno (editors). 1994. *Everglades Agricultural Area (EAA): Water, Soil, Crop, and Environmental Management.* University of Florida Press, Gainesville, Florida.

71. Bouton, S.N., P.C. Frederick, M.G. Spalding, and H. Lynch. 1999. The effects of chronic, low concentrations of dietary methylmercury on appetite and hunting behavior in juvenile great egrets (*Ardea albus*). *Environmental Toxicology and Chemistry* 18: 1934-1939.

72. Bowman, Reed, Martin Cody, Peter Frederick, Randall Hunt, Barry Noon, and Jeffrey Walters. 2003. South Florida Ecosystem Restoration: Multi-Species Avian Workshop Scientific Panel Report. Sustainable Ecosystems Institute, Portland, Oregon.

73. Boyer, J.N. and R.D. Jones. 1999. Effects of freshwater inputs and loading of phosphorus and nitrogen on the water quality of eastern Florida Bay. *In* K.R. Reddy, G.A. O'Connor, and C.L. Schelske (editors). *Phosphorus Biogeochemistry in Subtropical Ecosystems.* Lewis Publishers, Boca Raton, Florida.

74. Boyer, Joseph N. and Ronald D. Jones. 2002. A view from the bridge: external and internal forces affecting the ambient water quality of the Florida Keys National Marine Sanctuary (FKNMS). *In* James W. Porter and Karen G. Porter (editors). *The Everglades, Florida Bay, and Coral Reefs of the Florida Keys: an Ecosystem Source Book.* CRC Press, Boca Raton, Florida.

75. Brand, Larry E. 2002. The transport of terrestrial nutrients to South Florida coastal waters. *In* James W. Porter and Karen G. Porter (editors). *The Everglades, Florida Bay, and Coral Reefs of the Florida Keys: an Ecosystem Source Book.* CRC Press, Boca Raton, Florida.

76. Brandt, Laura A. and David W. Black. 2001. Impacts of the introduced fern, *Lygodium microphyllum*, on the native vegetation of tree islands in the Arthur R. Marshall Loxahatchee National Wildlife Refuge. *Florida Scientist* 64(3): 191-196.

77. Brandt, Laura A., Jennifer E Silveira, and Wiley M Kitchens. 2002. Tree islands of the Arthur R. Marshall Loxahatchee National Wildlife Refuge. *In* Fred H. Sklar and Arnold van der Valk (editors). *Tree Islands of the Everglades.* Kluwer Academic Publishers, Boston, Massachusetts.

78. Brezonik, P.L. and C.D. Pollman. 1999. Phosphorus chemistry and cycling in Florida lakes: global issues and local perspectives. *In* K.R. Reddy, G.A. O'Connor, and C.L. Schelske (editors). *Phosphorus Biogeochemistry in Subtropical Ecosystems.* Lewis Publishers, Boca Raton, Florida.

79. Briggs, John C. 1958. A list of Florida fishes and their distribution. *Bulletin of the Florida State Museum, Biological Sciences* 2(8): 224-318.

80. Brooks, H.K. 1984. Lake Okeechobee. *In* P.J. Gleason (editor). *Environments of South Florida: Present and Past II.* Miami Geological Society, Miami, Florida.

81. Browder, Joan A. 1985. Relationship between pink shrimp production on the tortugas grounds and water flow patterns in the Florida Everglades. *Bulletin of Marine Science* 37(3): 839-856.

82. Browder, Joan A., Patrick J. Gleason, and David R. Swift. 1994. Periphyton in the Everglades: spatial variation, environmental correlates, and ecological implications. *In* Steven M. Davis and John C. Ogden (editors). *Everglades: the Ecosystem and Its Restoration.* St. Lucie Press, Delray Beach, Florida.

83. Browder, Joan A., Victor R. Restrepo, Jason K Rice, Michael B. Robblee, and Zoula Zein-Eldin. 1999. Environmental influences on potential recruitment of pink shrimp, *Farfantepenaeus duorarum*, from Florida Bay nursery grounds. *Estuaries* 22(2B): 484-499.

84. Brown, Larry N. 1997. *A Guide to the Mammals of the Southeastern United States.* The University of Tennessee Press, Knoxville, Tennessee.

85. Brown, Loren G. "Totch." 1993. *Totch: a life in the Everglades*. University Press of Florida, Gainesville, Florida.

86. Brown, Paul Martin. 2002. *Wild Orchids of Florida*. University Press of Florida, Gainesville.

87. Brown, Randall B., Earl L. Stone, and Victor W. Carlisle. 1990. Soils. *In* Ronald L. Myers and John J. Ewel (editors). *Ecosystems of Florida*. University of Central Florida Press, Orlando, Florida.

88. Bruno, Maria Cristina, William F. Loftus, Janet W. Reid, and Sue A. Perry. 2001. Diapause in copepods (Crustacea) from ephemeral habitats with different hydroperiods in Everglades National Park (Florida, U.S.A.). *Hydrobiologia* 453/454: 295-308.

89. Buker, George E. 1997. *Swamp Sailors in the Second Seminole War*. University Press of Florida, Gainesville, Florida.

90. Bull, L.A., D.D. Fox, D.W. Brown, L.J. Davis, S.J. Miller, and J.G. Wullschleger. 1995. Fish distribution in limnetic areas of Lake Okeechobee, Florida. *In* N.G. Aumen and R.G. Wetzel (editors). Ecological studies on the littoral and pelagic systems of Lake Okeechobee, Florida (USA). *Archiv für Hydrobiologie*, special issue: Advances in Limnology, 45: 1-356.

91. Busch, David E., William F. Loftus, and Oron L. Bass, Jr. 1998. Long-term hydrologic effects on marsh plant community structure in the southern Everglades. *Wetlands* 18(2): 230-241.

92. Butterfield, Brian P., Walter E. Meshaka, Jr., and Craig Guyer. 1997. Nonindigenous amphibians and reptiles. *In* Daniel Simberloff, Don C. Schmitz, and Tom C. Brown (editors). *Strangers in Paradise: Impact and Management of Nonindigenous Species in Florida*. Island Press, Washington, D.C.

93. Campbell, Mark R. and Frank J. Mazzotti. 2001. Mapping Everglades alligator holes using color infrared aerial photographs. *Florida Scientist* 64(2): 148-158.

94. Canby, Thomas Y. 1979. The search for the first Americans. *National Geographic* 156(3): 330-363.

95. Carleton, James T. and Mary H. Ruckelshaus. 1997. Nonindigenous marine invertebrates and algae. *In* Daniel Simberloff, Don C. Schmitz, and Tom C. Brown (editors). *Strangers in Paradise: Impact and Management of Nonindigenous Species in Florida*. Island Press, Washington, D.C.

96. Carr, Archie and Coleman J. Goin. 1959. *Guide to the Reptiles, Amphibians and Fresh-water Fishes of Florida*. University of Florida Press, Gainesville, Florida.

97. Carr, Robert S. 2002. The archaeology of Everglades tree islands. *In* Fred H. Sklar and Arnold van der Valk (editors). *Tree Islands of the Everglades*. Kluwer Academic Publishers, Boston, Massachusetts.

98. Carr, Robert S. and John G. Beriault. 1984. Prehistoric man in southern Florida. *In* P.J. Gleason (editor). *Environments of South Florida: Present and Past II*. Miami Geological Society, Miami, Florida.

99. Carter, Luther J. 1974. *The Florida Experience*. Resources for the Future, Inc., The Johns Hopkins University Press, Baltimore, Maryland.

100. Causey, Billy D. 2002. The role of the Florida Keys National Marine Sanctuary in the South Florida ecosystem restoration initiative. *In* James W. Porter and Karen G. Porter (editors). *The Everglades, Florida Bay, and Coral Reefs of the Florida Keys: an Ecosystem Source Book*. CRC Press, Boca Raton, Florida.

101. Center, Ted D., J. Howard Frank, and F. Allen Dray, Jr. Biological control. *In* Daniel Simberloff, Don C. Schmitz, and Tom C. Brown (editors). *Strangers in Paradise: Impact and Management of Nonindigenous Species in Florida*. Island Press, Washington, D.C.

102. Cerulean, Susan and Ann Morrow. 1998. *Florida Wildlife Viewing Guide* (second edition). Falcon, Helena, Montana.

103. Chafin, Linda G. 2000. *Field Guide to the Rare Plants of Florida*. Florida Natural Areas Inventory, Tallahassee, Florida.

104. Chapman, Tim (photographer). 2003. *The Miami Herald*, Wednesday, September 3, 2003, page 1B (photograph).

105. Chávez, Miguel (editor). 1988. *Ecología y Conservación del Delta de los ríos Usumacinta y Grijalva* (Memorias). Instituto Nacional de Investigacíon sobre Recursos Bióticos (INIREB) – División Regional Tabasco, 720p. (This hard-to-find document is the proceedings of a symposium held in Tabasco, Mexico, in 1987. It includes papers on the Everglades. A copy is in the South Florida Water Management District library, West Palm Beach.)

106. Chen, Ellen and John F. Gerber. 1990. Climate. *In* Ronald L. Myers and John J. Ewel (editors). *Ecosystems of Florida*. University of Central Florida Press, Orlando, Florida.

107. Chick, John H. and Carole C. McIvor. 1994. Patterns in the abundance and composition of fishes among beds of different macrophytes: viewing a littoral zone as a landscape. *Can. J. of Fish. Aquat. Sci.* 51: 2873-2882.

108. Childers, Daniel L., Ronald D. Jones, Joel C. Trexler, Chris Buzzelli, Susan Dailey, Adrienne L. Edwards, Evelyn E. Gaiser, Krish Jayachandaran, Anna Keene, David Lee, John F. Meeder, Joseph H.K. Pechmann, Amy Renshaw, Jennifer Richards, Michael Rugge, Leonard J. Scinto, Pierre Sterling, and Will Van Gelder. 2002. Quantifying the effects of low-level phosphorus additions on unenriched Everglades wetlands with *in situ* flumes and phosphorus dosing. *In* James W. Porter and Karen G. Porter (editors). *The Everglades, Florida Bay, and Coral Reefs of the Florida Keys: an Ecosystem Source Book*. CRC Press, Boca Raton, Florida.

109. Clausen, C, J., A.D. Cohen, C. Emiliani, J.A. Holman, and J.J. Stipp. 1979. Little Salt Spring, Florida: a unique underwater site. *Science* 203: 609-614.

110. Cohen, Arthur D. 1984. Evidence of fires in the ancient Everglades and coastal swamps of southern Florida. *In* P.J. Gleason (editor). *Environments of South Florida: Present and Past II*. Miami Geological Society, Miami, Florida.

111. Cohen, Arthur D. and William Spackman. 1984. The petrology of peats from the Everglades and coastal swamps of southern Florida. *In* P.J. Gleason (editor). *Environments of South Florida: Present and Past II*. Miami Geological Society, Miami, Florida.

112. Comp, G.S., and W. Seaman, Jr. 1985. Estuarine habitat and fishery resources of Florida. *In* William Seaman, Jr. (editor) *Florida Aquatic Habitat and Fishery Resources*. Florida Chapter of the American Fisheries Society, Kissimmee, Florida.

113. Compton, John S. 1997. Origin and paleoceanographic significance of Florida's phosphate deposits. *In* Anthony F. Randazzo and Douglas S. Jones (editors). *The Geology of Florida*. University Press of Florida, Gainesville, Florida.

114. Conner, William H., Thomas W. Doyle, and Daniel Mason. 2002. Water depth tolerances of dominant tree island species: what do we know? *In* Fred H. Sklar and Arnold van der Valk (editors). *Tree Islands of the Everglades*. Kluwer Academic Publishers, Boston, Massachusetts.

115. Conover, Michael. 1972. A study of *Procambarus alleni* in the Everglades. Unpublished paper, under direction of Durbin Tabb, Institute of Marine Sciences, University of Miami, and George K. Reid, Florida Presbyterian College (source: Everglades National Park reference library, Pam. box #67, Pam. #6-C).

116. Cook, Clayton B., Erich M. Mueller, M. Drew Ferrier, and Eric Annis. 2002. The influence of nearshore waters on corals of the Florida reef tract. *In* James W. Porter and Karen G. Porter (editors). *The Everglades, Florida Bay, and Coral Reefs of the Florida Keys: an Ecosystem Source Book*. CRC Press, Boca Raton, Florida.

117. Correll, Donovan S., and Helen B. Correll. 1982. *Flora of the Bahama Archipelago*. A.R. Gantner Verlag KG, FL-9490 Vaduz.

118. Courtenay, Walter R., Jr. 1997. Nonindigenous fishes. *In* Daniel Simberloff, Don C. Schmitz, and Tom C. Brown (editors). *Strangers in Paradise: Impact and Management of Nonindigenous Species in Florida*. Island Press, Washington, D.C.

119. Cowen, R.K. 2002. Oceanographic influences on larval dispersal and retention and their consequences for population connectivity. *In* P.F. Sale (editor). *Coral Reef Fishes*. Academic Press.

120. Cox, George W. 1999. *Alien Species in North America and Hawaii: Impacts on Natural Ecosystems*. Island Press, Washington, D.C.

121. Cox, James A. and Randy S. Kautz. 2000. *Habitat Conservation Needs of Rare and Imperiled Wildlife in Florida*. Office of Environmental Services, Florida Fish and Wildlife Conservation Commission, Tallahassee, Florida.

122. Cox, James, Randy Kautz, Marueen MacLaughlin, and Terry Gilbert. 1994. *Closing the Gaps in Florida's Wildlife Habitat Conservation* System. Office of Environmental Services, Florida Game and Fresh Water Fish Commission, Tallahassee, Florida.

123. Cox, James A., Lt. Thomas G. Quinn, and H. Hugh Boyter, Jr. 1997. Management by Florida's Game and Fresh Water Fish Commission. *In* Daniel Simberloff, Don C. Schmitz, and Tom C. Brown (editors). *Strangers in Paradise: Impact and Management of Nonindigenous Species in Florida*. Island Press, Washington, D.C.

124. Craft, C.B. and C.J. Richardson. 1993. Peat accretion and N, P, and organic C accumulations in nutrient-enriched and unenriched Everglades peatlands. *Ecological Applications* 3(3): 446-458.
125. Craighead, Frank C., Sr. 1963. *Orchids and Other Air Plants of the Everglades National Park*. Everglades Natural History Association, University of Miami Press, Coral Gables, Florida.
126. Craighead, Frank C., Sr. 1964. Land, mangroves, and hurricanes. *The Fairchild Tropical Garden Bulletin* 19(4): 1-28.
127. Craighead, Frank C., Sr. 1968. The role of the alligator in shaping plant communities and maintaining wildlife in the southern Everglades. *The Florida Naturalist* 41: 2-7, 69-74.
128. Craighead, Frank C., Sr. 1971. *The Trees of Southern Florida*. Volume 1. The natural environments and their succession. Univ. of Miami Press, Coral Gable, Floridas.
129. Craighead, Frank C., Sr. 1971. Is man destroying South Florida? *In* William Ross McCluney (editor). *The Environmental Destruction of South Florida*. University of Miami Press, Coral Gables, Florida.
130. Craighead, Frank C., Sr. 1973. The effects of natural forces on the development and maintenance of the Everglades, Florida. National Geographic Society Research Reports, 1966 Projects, pages 49-67.
131. Craighead, Frank C., Sr. 1984. Hammocks of South Florida. *In* P.J. Gleason (editor). *Environments of South Florida: Present and Past II*. Miami Geological Society, Miami, Florida.
132. Crumpacker, David W., Elgene O. Box, and E. Dennis Hardin. 2001. Potential breakup of Florida plant communities as a result of climatic warming. *Florida Scientist* 64(1): 29-43.
133. Cunningham, Kevin J., David Bukry, Tokiyuki Sato, John A. Barron, Laura A. Guertin, and Ronald S. Reese. 2001. Sequence stratigraphy of a South Florida carbonate ramp and bounding siliciclastics (late Miocene-Pliocene). *In* Thomas M. Missimer and Thomas M. Scott (editors). Geology and Hydrology of Lee County, Florida, Durward H. Boggess Memorial Symposium. Special Publication No. 49, Florida Geological Survey, Tallahassee, Florida.
134. Cunningham, Kevin J., Stanley D. Locker, Albert C. Hine, David Bukry, John A. Barron, and Laura A. Guertin. 2003. Interplay of late cenozoic siliciclastic supply and carbonate response on the southeast Florida Platform. *Journal of Sedimentary Research* 73(1): 31-46.
135. Cunningham, Kevin J., Donald F. McNeill, Laura A. Guertin, Paul F. Ciesielski, Thomas M. Scott, and Laurent de Verteuil. 1998. New tertiary stratigraphy for the Florida Keys and southern peninsula of Florida. *GSA Bulletin* 110(2): 231-258.
136. Curnutt, J.L., A.L. Mayer, T.M. Brooks, L. Manne, O.L. Bass Jr., D.M. Fleming, M.P. Nott, and S.L. Pimm. 1998. Population dynamics of the endangered Cape Sable seaside sparrow. *Animal Conservation* 1(1): 11-21.
137. Darby, Philip C., and Robert E. Bennetts. (in preparation, 2004). Apple snail life history in relation to hydrologic regimes in South Florida wetlands (working title).
138. Darby, Philip C., Robert E. Bennetts, Steven J. Miller, and H. Franklin Percival. 2002. Movements of Florida apple snails in relation to water levels and drying events. *Wetlands* 22(3): 489-498.
139. Darby, Philip C., Patricia L. Valentine-Darby, and H. Franklin Percival. 2003. Dry season survival in a Florida apple snail (*Pomacea paludosa* Say) population. *Malacologia* 45: 179-184.
140. D'Avanzo, Charlene. 1990. Long-term evaluation of wetland creation projects. *In* Jon A. Kusler and Mary E. Kentula (editors). *Wetland Creation and Restoration: the Status of the Science*. Island Press, Washington, D.C.
141. Davies, Thomas D. and Arthur D. Cohen. 1989. Composition and significance of the peat deposits of Florida Bay. *Bulletin of Marine Science* 44(1): 387-398.
142. Davis, Gary E., 1992. Assessment of Hurricane Andrew impacts on natural and archaeological resources of Big Cypress National Preserve, Biscayne National Park, and Everglades National Park (executive summary). U.S. National Park Service, University of California, Davis, Cooperative Park Study Unit, California.
143. Davis, Gary E. and Jon W. Dodrill. 1989. Recreational fishery and population dynamics of spiny lobsters, *Panulirus argus*, in Florida Bay, Everglades National Park, 1977-1980. *Bulletin of Marine Science* 44(1): 78-88.
144. Davis, John H., Jr. 1940. The ecology and geologic role of mangroves in Florida. Carnegie Institution, publication No. 517. Washington, D.C.
145. Davis, John H., Jr. 1943. The natural features of southern Florida, especially the vegetation, and the Everglades. The Florida Geological Survey, Geological Bulletin No. 25, Tallahassee, Florida.

146. Davis, John H., Jr. 1946. Peat deposits of Florida, their occurrence, development, and uses. The Florida Geological Survey, Geological Bulletin No. 30, Tallahassee, Florida.
147. Davis, John H. 1967. General map of natural vegetation of Florida. Agricultural Experimentations, Institute of Food and Agricultural Sciences, University of Florida, Gainesville, Florida.
148. Davis, Richard A., Jr. 1997. Geology of the Florida coast. *In* Anthony F. Randazzo and Douglas S. Jones (editors). *The Geology of Florida.* University Press of Florida, Gainesville, Florida.
149. Davis, Steven M. 1994. Phosphorus inputs and vegetation sensitivity in the Everglades. *In* Steven M. Davis and John C. Ogden (editors). *Everglades: the Ecosystem and Its Restoration.* St. Lucie Press, Delray Beach, Florida.
150. Davis, Steven M., Lance H. Gunderson, Winifred A. Park, John R. Richardson, and Jennifer E. Mattson. 1994. Landscape dimension, composition, and function in a changing Everglades ecosystem. *In* Steven M. Davis and John C. Ogden (editors). *Everglades: the Ecosystem and Its Restoration.* St. Lucie Press, Delray Beach, Florida.
151. Davis, Steven M. and John C. Ogden (editors). 1994. *Everglades: the Ecosystem and Its Restoration.* St. Lucie Press, Delray Beach, Florida.
152. Davis, Steven M. and John C. Ogden. 1994. Introduction. *In* Steven M. Davis and John C. Ogden (editors). *Everglades: the Ecosystem and Its Restoration.* St. Lucie Press, Delray Beach, Florida.
153. Davis, Steven M. and John C. Ogden. 1994. Toward ecosystem restoration. *In* Steven M. Davis and John C. Ogden (editors). *Everglades: the Ecosystem and Its Restoration.* St. Lucie Press, Delray Beach, Florida.
154. Dawes, C.J., S.S. Bell, R.A. Davis, Jr., E.D. McCoy, H.R. Mushinsky, and J.L. Simon. 1995. Initial effects of Hurricane Andrew on the shoreline habitats of southwestern Florida. *Journal of Coastal Research* 21: 103-110.
155. DeAngelis, Donald L. 1994. Synthesis: spatial and temporal characteristics of the environment. *In* Steven M. Davis and John C. Ogden (editors). *Everglades: the Ecosystem and Its Restoration.* St. Lucie Press, Delray Beach, Florida.
156. DeAngelis, Donald L. 2003. Overview of the Across Trophic Level System Simulation (ATLSS) program: model development, field study support, validation, documentation, and application (abstract). Joint Conference on the Science and Restoration of the Greater Everglades and Florida Bay Ecosystem, April 13-18, 2003, Palm Harbor, Florida.
157. DeAngelis, Donald L., Sara Bellmund, Wolf M. Mooij, M. Philip Nott, E. Jane Comiskey, Louis J. Gross, Michael A. Huston, and Wilfried F. Wolff. 2002. Modeling ecosystem and population dynamics on the South Florida hydroscape. *In* James W. Porter and Karen G. Porter (editors). *The Everglades, Florida Bay, and Coral Reefs of the Florida Keys: an Ecosystem Source Book.* CRC Press, Boca Raton, Florida.
158. DeAngelis, Donald L. and Peter S. White. 1994. Ecosystems as products of spatially and temporally varying driving forces, ecological processes, and landscapes: a theoretical perspective. *In* Steven M. Davis and John C. Ogden (editors). *Everglades: the Ecosystem and Its Restoration.* St. Lucie Press, Delray Beach, Florida.
159. Deuerling, Richard. 1983. Class I Injection Well Inventory. Florida Department of Environmental Regulation, Groundwater Section.
160. De Vorsey, Louis. 2002. Navigating Fisheating Creek. *Exploring Mercator's World* 7(5). (on line at: *www.mercatormag.com/article.php3?i=141*).
161. Deyrup, Mark and Richard Franz (eds.). 1994. *Volume IV. Invertebrates. In* Ray E. Ashton, Jr. (series editor). *Rare and Endangered Biota of Florida.* University Presses of Florida, Gainesville, Florida.
162. Diamond, Craig, Darrell Davis, and Don C. Schmitz. 1991. Economic Impact Statement: the addition of *Melaleuca quinquenervia* to the Florida Prohibited Aquatic Plant List. *In* Ted D. Center, Robert F. Doren, Ronald H. Hofstetter, Ronald L. Myers, and Louis D. Whiteaker (editors). Proceedings of the Symposium on Exotic Pest Plants. U.S. Department of the Interior, National Park Service, Washington, D.C.
163. Dickson, John D. III, Roy O. Woodbury, and Taylor R. Alexander. 1953. Check list of flora of Big Pine Key, Florida and surrounding keys. *Quarterly Journal of the Florida Academy of Sciences* 16(3): 181-97.
164. Diersing, Nancy. 2002. Ice age forest discovered in sanctuary waters. *Sounding Line*, Winter 2002 (newsletter of the Florida Keys National Marine Sanctuary).

165. Dineen, J. Walter. 1984. Fishes of the Everglades. *In* P.J. Gleason (editor). *Environments of South Florida: Present and Past II*. Miami Geological Society, Miami, Florida.

166. Dix, E.A., and J.N. MacDonigle. 1905. The Everglades of Florida: a region of mystery. *Century Magazine* 47: 512-527.

167. Doren, Robert F., Amy Ferriter, and Harriet Hastings (editors). Undated (2001). Weeds Won't Wait: the strategic plan for managing Florida's invasive exotic plants. Part One: an assessment of the most invasive plants in Florida; Part Two: The strategy for managing invasive plants in Florida. A report to the South Florida Ecosystem Restoration Task Force, Florida International University, Miami, Florida.

168. Doren, Robert F. and David T. Jones. 1997. Management in Everglades National Park. *In* Daniel Simberloff, Don C. Schmitz, and Tom C. Brown (editors). *Strangers in Paradise: Impact and Management of Nonindigenous Species in Florida*. Island Press, Washington, D.C.

169. Doren, Robert F., Louis D. Whiteaker, and Anne Marie LaRosa. 1991. Evaluation of fire as a management tool for controlling *Schinus terebinthifolius* as secondary successional growth on abandoned agricultural land. *Environmental Management* 15(1): 121-129.

170. Douglas, Marjory Stoneman. 1997. *The Everglades: River of Grass*. 50th anniversary edition, with updates by R. Loftis and C. Zaneski. Pineapple Press, Inc., Sarasota, Florida.

171. Douglas, Marjory Stoneman (personal communications). Coconut Grove, Florida, 1989–1997.

172. Doyle, Thomas W., Thomas J. Smith, III, and Michael B. Robblee. 1995. Wind damage effects of Hurricane Andrew on mangrove communities along the southwest coast of Florida, USA. *Journal of Coastal Research* 21: 159-168.

173. D'Sa, Eurico J., James B. Zaitzeff, Charles S. Yentsch, Jerry L. Miller, and Russell Ives. 2002. Rapid remote assessments of salinity and ocean color in Florida Bay. *In* James W. Porter and Karen G. Porter (editors). *The Everglades, Florida Bay, and Coral Reefs of the Florida Keys: an Ecosystem Source Book*. CRC Press, Boca Raton, Florida.

174. Duever, Michael J. 1984. Environmental factors controlling plant communities of the Big Cypress Swamp. *In* P.J. Gleason (editor). *Environments of South Florida: Present and Past II*. Miami Geological Society, Miami, Florida.

175. Duever, Michael J., John E. Carlson, and Lawrence A Riopelle. 1981. Off-road vehicles and their impacts in the Big Cypress National Preserve. Report T-614, National Park Service, South Florida Research Center, Everglades National Park, Homestead, Florida.

176. Duever, Michael J., John E. Carlson, John F. Meeder, Linda C. Duever, Lance H. Gunderson, Lawrence A. Riopelle, Taylor R. Alexander, Ronald L. Myers, and Daniel P. Spangler. 1986. The Big Cypress National Preserve. National Audubon Society Research Report No. 8, National Audubon Society, New York, New York. This document contains a separate vegetation map in a back-cover pocket: see McPherson, 1973.

177. Duever, M.J., J.F. Meeder, L.C. Meeder, and J.M. McCollom. 1994. The climate of South Florida and its role in shaping the Everglades ecosystem. *In* Steven M. Davis and John C. Ogden (editors). *Everglades: the Ecosystem and Its Restoration*. St. Lucie Press, Delray Beach, Florida.

178. Dumas, Jeannette V. 2000. Roseate spoonbill (*Ajaia ahaja*). *In* A. Poole and F. Gill (editors). *The Birds of North America*, No. 490. The Birds of North America, Inc., Philadelphia, PA.

179. Dunkle, Sidney W. 1989. *Dragonflies of the Florida Peninsula, Bermuda, and the Bahamas*. Scientific Publishers, Gainesville, Florida.

180. Dunkle, Sidney W. 2000. *Dragonflies through Binoculars: a Field Guide to Dragonflies of North America*. Oxford University Press, New York, New York.

181. Durako, Michael J., Margaret O. Hall, and Manuel Merello. 2002. Patterns of change in the seagrass-dominated Florida Bay hydroscape. *In* James W. Porter and Karen G. Porter (editors). *The Everglades, Florida Bay, and Coral Reefs of the Florida Keys: an Ecosystem Source Book*. CRC Press, Boca Raton.

182. Egler, Frank E. 1952. Southeast saline Everglades vegetation in Florida, and its management. *Vegetatio Acta Botanica III* (Fasc. 4-5): 213-265.

183. Eisenberg, John F. 1989. *Mammals of the Neotropics: Volume 1, the Northern Neotropics*. University of Chicago Press, Chicago, Illinois.

184. Enos, Paul. 1989. Islands in the bay — a key habitat of Florida Bay. *Bulletin of Marine Science* 44(1): 365-386.

208. Frederick, Peter C. 1993. Wading bird nesting success studies in Water Conservation Areas of the Everglades. Final report to South Florida Water Management District, West Palm Beach..

209. Frederick, P.C. 2000. Mercury and its effects in the Everglades ecosystem. *Reviews in Toxicology* 3: 213-255.

210. Frederick, Peter C. and Michael W. Collopy. 1988. Reproductive ecology of wading birds in relation to water conditions in the Florida Everglades. Florida Cooperative Fish and Wildlife Research Unit, School of Forestry Research and Conservation, University of Florida. Technical Report No. 30.

211. Frederick, Peter C. and William F. Loftus. 1993. Responses of marsh fishes and breeding wading birds to low temperatures: a possible behavioral link between predator and prey. *Estuaries* 16(2): 216-222.

212. Frederick, Peter C. and John C. Ogden. 2001. Pulsed breeding of long-legged wading birds and the importance of infrequent severe drought conditions in the Florida Everglades. *Wetlands* 21(4): 484-491.

213. Frederick, Peter C. and George V.N. Powell. 1994. Nutrient transport by wading birds in the Everglades. *In* Steven M. Davis and John C. Ogden (editors). *Everglades: the Ecosystem and Its Restoration*. St. Lucie Press, Delray Beach, Florida.

214. Frederick, Peter C. and Marilyn G. Spalding. 1994. Factors affecting reproductive success of wading birds (Ciconiiformes) in the Everglades ecosystem. *In* Steven M. Davis and John C. Ogden (editors). *Everglades: the Ecosystem and Its Restoration*. St. Lucie Press, Delray Beach, Florida.

215. Frederick, P.C., M.G. Spalding, and R. Dusek. 2002. Wading birds as bioindicators of mercury contamination in Florida: annual and geographic variation. *Environmental Toxicology and Chemistry* 21: 262-64.

216. Frederick, P.C., M.G. Spalding, M.S. Sepulveda, G.E. Williams, L. Nico, and R. Robins. 1999. Exposure of great egret (*Ardea albus*) nestlings to mercury through diet in the Everglades ecosystem. *Environmental Contamination and Toxicology* 18(9): 1940-1947.

217. Fry, Brian, Patricia L. Mumford, Franklin Tam, Don D. Fox, Gary L. Warren, Karl E. Havens, and Alan D. Steinman. 1999. Trophic position and individual feeding histories of fish from Lake Okeechobee, Florida. *Can. J. of Fish. Aquat. Sci.* 56: 590–600.

218. Fry, Brian and Thomas J. Smith III. 2002. Stable isotope studies of red mangroves and filter feeders from the Shark River estuary, Florida. *Bulletin of Marine Science* 70(3): 871-890.

219. Furse, J.Beacham and Donald D. Fox. 1994. Economic fishery valuation of five vegetation communities in Lake Okeechobee, Florida. *Proceedings of the Annual Conference of the Southeastern Association of Fish and Wildlife Agencies* 48: 575-591.

220. Gaines, Michael S., Christopher R. Sasso, James E. Diffendorfer, and Harald Beck. 2002. Effects of tree island size and water on the population dynamics of small mammals in the Everglades. *In* Fred H. Sklar and Arnold van der Valk (editors). *Tree Islands of the Everglades*. Kluwer Academic Publishers, Boston, Massachusetts.

221. Gallardo Mareia T., Barbara B. Martin, and Dean F. Martin. 1998. An annotated bibliography of allelopathic properties of cattails, *Typha* spp. *Florida Scientist* 61(1): 52-58.

222. Gann, George, D., Keith A. Bradley, and Steven W. Woodmansee. 2002 *Rare Plants of South Florida: their History, Conservation, and Restoration*. The Institute for Regional Conservation, Miami. Note: this publication is based on a floristic inventory database for South Florida at: *www.regionalconservation.org.*

223. Gann, Tiffany Troxler. 2001. Investigating tree island community response to increased freshwater flow in the southeastern Everglades. Masters thesis. Dept. of Biological Sciences, Florida International University, Miami, Florida.

224. Gawlik, Dale E. (editor). 2002. South Florida wading bird report, Volume 8. South Florida Water Management District, Everglades Division. West Palm Beach. Note: this publication is part of a series of annual reports. The latest of these reports prior to publication of this book is Volume 9, edited by Gaea E. Crozier and Dale E. Gawlik (2003).

225. Gawlik, Dale E., Peg Gronemeyer, and Robert A. Powell. 2002. Habitat-use patterns of avian seed dispersers in the central Everglades. *In* Fred H. Sklar and Arnold van der Valk (editors). *Tree Islands of the Everglades*. Kluwer Academic Publishers, Boston, Massachusetts.

226. Gentry, Cecil R. 1984. Hurricanes in South Florida. *In* P.J. Gleason (editor). *Environments of South Florida: Present and Past II*. Miami Geological Society, Miami, Florida.

227. Gifford, John. 1911. *The Everglades and Other Essays Relating to Southern Florida*. Everglade Land Sales Co., Kansas City, Missouri.

228. Gifford, John C. 1945. *Living by the Land*. Parker Art Printing Association, Coral Gables, Florida.

229. Gilbert, Carter R. (editor). 1992. *Volume II. Fishes. In* Ray E. Ashton, Jr. (series editor). *Rare and Endangered Biota of Florida*. University Presses of Florida, Gainesville, Florida.

230. Gilbert, Katherine M., John D. Tobe, Richard W. Cantrell, Maynard E. Sweeley, and James R. Cooper. 1995. *The Florida Wetlands Delineation Manual*. Florida Department of Environmental Protection, South Florida WMD, St. Johns River WMD, Suwannee River WMD, Southwest Florida WMD, and Northwest Florida WMD.

231. Gill, Frank B. 1995. *Ornithology*. Second edition. W.H. Freeman and Company, New York.

232. Gleason, Patrick J. 1984. Introduction: saving the wild places — a necessity for growth. *In* P.J. Gleason (editor). *Environments of South Florida: Present and Past II*. Miami Geological Society, Miami, Florida.

233. Gleason, Patrick J., H. Kelly Brooks, Arthur D. Cohen, Robert Goodrick, William G. Smith, William Spackman, and Peter Stone. 1984. The environmental significance of holocene sediments from the Everglades and saline tidal plain. *In* P.J. Gleason (editor). *Environments of South Florida: Present and Past II*. Miami Geological Society, Miami.

234. Gleason, Patrick J., Amy Ferriter, Ken Rutchey, Antonio Pernas, Bob Doren, Bob Pemberton, John Volin, and Ken Langeland. 2003. It's the end of the world as we know it – *Lygodium microphyllum* is strangling the Everglades restoration project (abstract). Joint Conference on the Science and Restoration of the Greater Everglades and Florida Bay Ecosystem, April 13-18, 2003, Palm Harbor, Florida.

235. Gleason, Patrick J. and William Spakman, Jr. 1974. Calcareous periphyton and water chemistry in the Everglades. *In* P.J. Gleason (editor). *Environments of South Florida: Present and Past*. Memoir 2: Miami Geological Society, Miami, Florida.

236. Gleason, Patrick J. and Peter A. Stone. 1975. *Prehistoric Trophic Level Status and Possible Cultural Influences on the Enrichment of Lake Okeechobee*. Central and Southern Flood Control District, West Palm Beach.

237. Gleason, Patrick J. and Peter Stone. 1994. Age, origin, and landscape evolution of the Everglades peatland. *In* Steven M. Davis and John C. Ogden (editors). *Everglades: the Ecosystem and Its Restoration*. St. Lucie Press, Delray Beach, Florida.

238. Glisson, Mark W. 1997. Management on state lands. *In* Daniel Simberloff, Don C. Schmitz, and Tom C. Brown (editors). *Strangers in Paradise: Impact and Management of Nonindigenous Species in Florida*. Island Press, Washington, D.C.

239. Godfrey, Robert K. and Jean W. Wooten. 1979 and 1981. *Aquatic and Wetland Plants of the Southeastern United States* (volumes 1 and 2). University of Georgia Press, Athens, Georgia.

240. Goebel, Ted, Michael R. Waters, and Margarita Dikova. 2003. The archaeology of Ushki Lake, Kamchatka, and the Pleistocene peopling of the Americas. *Science* (July 25) 301: 501-505.

241. Goforth, Gary, James Best Jackson, and Larry Fink. 1994. Restoring the Everglades. *Civil Engineering* 64(3): 52-55.

242. Goodrick, Robert L. 1984. The wet prairies of the northern Everglades. *In* P.J. Gleason (editor). *Environments of South Florida: Present and Past II*. Miami Geological Society, Miami, Florida.

243. Gordon, Doria R. and Kevin P. Thomas. 1997. Florida's invasion by nonindigenous plants: history, screening, and regulation. *In* Daniel Simberloff, Don C. Schmitz, and Tom C. Brown (editors). *Strangers in Paradise: Impact and Management of Nonindigenous Species in Florida*. Island Press, Washington, D.C.

244. Goreau, Thomas J. and William Zamboni de Mello. 2002. Nitrous oxide, methane, and carbon dioxide fluxes from South Florida habitats during the transition from dry to wet seasons: potential impacts of Everglades drainage and flooding on the atmosphere. *In* James W. Porter and Karen G. Porter (editors). *The Everglades, Florida Bay, and Coral Reefs of the Florida Keys: an Ecosystem Source Book*. CRC Press, Boca Raton, Florida.

245. Grant, Peter R. 1986. *Ecology and Evolution of Darwin's Finches*. Princeton University Press, Princeton, New Jersey.

246. Grant, Verne. 1963. *The Origin of Adaptations*. Columbia University Press, New York, New York.

247. Gray, P.N., R.C. Brust, and M.R. Miltner. 1993. Toward more environmentally sound agriculture in the Everglades Agricultural Area, Florida. *In* M. Moser, R.C. Prentice, and J. wan Vassem (editors). *Waterfowl and Wetland Conservation in the 1990's – a Global Perspective*. Proceeding of the IWRB Symposium, St. Petersburg Beach.

248. Griffin, John W. 2002. *Archaeology of the Everglades* (edited by Jerald T. Milanich and James J. Miller). University Press of Florida, Gainesville, Florida.

249. Guertin, Laura A., Donald F. McNeill, Barbara H. Lidz, and Kevin J. Cunningham. 1999. Chronologic model and transgressive-regressive signatures in the late neogene siliciclastic foundation (Long Key Formation) of the Florida Keys. *Journal of Sedimentary Research* 69(3): 653-666.

250. Gunderson, Lance H. 1989. Historical hydropatterns in wetland communities of Everglades National Park. *In* R.R. Sharitz and J. W. Gibbons (eds.), *Freshwater Wetlands and Wildlife*. DOE Symposium Series No. 61. United States Department of Energy, Office of Scientific and Technical Information, Oak Ridge, Tennessee.

251. Gunderson, Lance H. 1994. Vegetation of the Everglades: determinants of community composition. *In* Steven M. Davis and John C. Ogden (editors). *Everglades: the Ecosystem and Its Restoration*. St. Lucie Press, Delray Beach, Florida.

252. Gunderson, Lance H. and William F. Loftus. 1993. The Everglades. *In* William H. Martin, Stephen G. Boyce, and Arthur C. Echternacht (eds.). *Biodiversity of the Southeastern United States/Lowland Terrestrial Communities*. John Wiley & Sons, New York, New York.

253. Gunderson, L.H. and J.R. Snyder. 1994. Fire patterns in the southern Everglades. *In* Steven M. Davis and John C. Ogden (editors). *Everglades: the Ecosystem and Its Restoration*. St. Lucie Press, Delray Beach, Florida.

254. Hammer, Roger L. 2002. *Everglades Wildflowers*. The Globe Pequot Press, Guilford, Connecticut.

255. Hancock, James, and James Kushlan. 1984. *The Herons Handbook*. Harper & Row, Publishers, New York, New York.

256. Hanisch, Carola. 1998. Where is mercury deposition coming from? *Environmental Science & Technology/News* April 1, 1998, p.176-179.

257. Hanna, Alfred Jackson, and Kathryn Abbey Hanna. 1948. *Lake Okeechobee, Wellspring of the Everglades*. The American Lakes Series, Milo M. Quaife editor. The Boobs-Merrill Company Publishers, New York, New York.

258. Hanson, Kirby and George A. Maul. 1991. Florida precipitation and the Pacific El Nino, 1895-1989. *Florida Scientist* 54(3/4): 160-168.

259. Harper, R.M. 1927. Natural resources of southern Florida. 18th Annual Report, Florida Geological Survey. Tallahassee, Florida.

260. Harris, W. and W. Hurt. 1999. Introduction to soils of subtropical Florida. *In* K.R. Reddy, G.A. O'Connor, and C.L. Schelske (editors). *Phosphorus Biogeochemistry in Subtropical Ecosystems*. Lewis Publishers, Boca Raton, Florida.

261. Harshberger, J.W. 1914. The vegetation of South Florida. *Trans Wagner Free Inst. Sci. Philos.* 3: 51-89.

262. Havens, Karl E., Nicholas G. Aumen, R. Thomas James, and Val H. Smith. 1996. Rapid ecological changes in a large subtropical lake undergoing cultural eutrophication. *Ambio* 25(3): 150-155.

263. Havens, Karl E., L.A. Bull, G.L. Warren, T.L. Crisman, E.J. Phlips, and J.P. Smith. 1996. Food web structure in a subtropical lake ecosystem. *Oikos* 75: 20-32.

264. Havens, Karl E., Matthew C. Harwell, Mark A. Brady, Bruce Sharfstein, Therese L. East, Andrew J. Rodusky, Daniel Anson, and Ryan P. Maki. 2002. Large-scale mapping and predictive modeling of submerged aquatic vegetation in a shallow eutrophic lake. *The Scientific World Journal* 2: 949-965. Note: this electronic journal no longer exists. Contact the senior author at the South Florida Water Management District for information.

265. Havens, Karl E., Alan D. Steinman, and Brian Fry. 2000. Spatial variation in food web structure and function in a large, shallow subtropical lake (Lake Okeechobee, Florida, USA). *Verh. Internat. Verein. Limnol.* 27: 1-6.

266. Heald, E.J., W.E. Odum, and D.C. Tabb. 1984. Mangroves in the estuarine food chain. *In* P.J. Gleason (editor). *Environments of South Florida: Present and Past II*. Miami Geological Society, Miami.

267. Heatherington, Ann L. and Paul A. Mueller. 1997. Geochemistry and origin of Florida crustal basement terranes. *In* Anthony F. Randazzo and Douglas S. Jones (editors). *The Geology of Florida*. University Press of Florida, Gainesville, Florida.

268. Heisler, Lorraine, D. Timothy Towles, Laura A. Brandt, and Robert T. Pace. 2002. Tree island vegetation and water management in the central Everglades. *In* Fred H. Sklar and Arnold van der Valk (editors). *Tree Islands of the Everglades*. Kluwer Academic Publishers, Boston, Massachusetts.

269. Helprin, Angelo. 1887. *Explorations on the West Coast of Florida and in the Okeechobee Wilderness*. Wagner Free Institute of Science of Philadelphia.

270. Hendrix, A. Noble and William F. Loftus. 2000. Distribution and relative abundance of the crayfishes *Procambarus alleni* (Faxon) and *P. fallax* (Hagen) in southern Florida. *Wetlands* 20(1): 194-199.

271. Henry, James A., Kenneth M. Portier, and Jan Coyne. 1994. *The Climate and Weather of Florida*. Pineapple Press, Inc., Sarasota, Florida.

272. Herndon, A.K., L.H. Gunderson, and J.R. Stenberg. 1991. Sawgrass, *Cladium jamaicense*, survival in a regime of fire and flooding. *Wetlands* 11: 17-27.

273. Higer, Aaron L. and Milton C. Kolipinski. 1988. Changes in vegetation in Shark River Slough, Everglades National Park, 1940-1964. *In* Miguel Chávez Miguel (editor). *Ecología y Conservación del Delta de los ríos Usumacinta y Grijalva* (Memorias). Instituto Nacional de Investigacíon sobre Recursos Bióticos (INIREB) – División Regional Tabasco, pages 217-230. (see Chavez, 1988)

274. Hine, Albert C. 1997. Structural and paleoceanographic evolution of the margins of the Florida Platform. *In* Anthony F. Randazzo and Douglas S. Jones (editors). *The Geology of Florida*. University Press of Florida, Gainesville, Florida.

275. Hipes, Dan. Dale R. Jackson, Katy NeSmith, David Printiss, and Karla Brandt. 2001. *Field Guide to the Rare Animals of Florida*. Florida Natural Areas Inventory, Tallahassee, Florida.

276. Hobbs, H., Jr. 1942. The crayfishes of Florida. *University of Florida Biological Science Series* 3(2): v+1-179.

277. Hobbs, Horton H., Jr., and H. H. Hobbs III. 1991. An illustrated key to the crayfishes of Florida (based on first form males). *Florida Scientist* 54(1): 13-24.

278. Hoffman, Wayne, G. Thomas Bancroft, and Richard J. Sawicki. 1994. Foraging habitat of wading birds in the Water Conservation Areas of the Everglades. *In* Steven M. Davis and John C. Ogden (editors). *Everglades: the Ecosystem and Its Restoration*. St. Lucie Press, Delray Beach, Florida.

279. Hoffman, W., R.J. Sawicki, and G.T. Bancroft. (Two undated reports, A and B). Wading bird populations and distributions in the water conservation areas of the Everglades: (A) the 1988-1989 season, and (B) the 1989-1990 season. National Audubon Society Ornithological Research Unit, Tavernier, Florida.

280. Hoffmeister, John Edward. 1974. *Land from the Sea*. University of Miami Press, Coral Gables, Florida.

281. Hofstetter, Ronald H. 1984. The effect of fire on the pineland and sawgrass communities of southern Florida. *In* P.J. Gleason (editor). *Environments of South Florida: Present and Past II*. Miami Geological Society, Miami, Florida.

282. Hofstetter, Ronald H. 1988. Fire as a management tool in wetlands preservation: Florida Everglades. *In* Miguel Chávez Miguel (editor). *Ecología y Conservación del Delta de los ríos Usumacinta y Grijalva* (Memorias). Instituto Nacional de Investigacíon sobre Recursos Bióticos (INIREB) – División Regional Tabasco, pages 245-257. (see Chavez, 1988)

283. Hofstetter, Ronald H. 1991. The current status of *Melaleuca quinquenervia* in southern Florida. *In* Ted D. Center, Robert F. Doren, Ronald H. Hofstetter, Ronald L. Myers, and Louis D. Whiteaker (editors). Proceedings of the Symposium on Exotic Pest Plants. U.S. Department of the Interior, National Park Service, Washington, D.C.

284. Hogarth, Peter J. 1999. *The Biology of Mangroves*. Oxford University Press, Oxford, U.K.

285. Holling, C.S., Lance H. Gunderson, and Carl J. Walters. 1994. The structure and dynamics of the Everglades system: guidelines for ecosystem restoration. *In* Steven M. Davis and John C. Ogden (editors). *Everglades: the Ecosystem and Its Restoration*. St. Lucie Press, Delray Beach.

286. Holms, Charles W., Debra Willard, Lyn Brewster-Wingard, Lisa Wiemer, and M.E. Marot. 2002. Buttonwood embankment: the historical perspective on its role in northeastern Florida Bay sedimentary dynamics and hydrology. USGS South Florida Information Access Poster at *http://sofia.usgs.gov/publications/posters/buttonwood/print.html*.

287. Horvitz, Carol C. 1997. The impact of natural disturbances. *In* Daniel Simberloff, Don C. Schmitz, and Tom C. Brown (editors). *Strangers in Paradise: Impact and Management of Nonindigenous Species in Florida*. Island Press, Washington, D.C.

288. Horvitz, Carol C., Stephen McMann, and Andrea Freedman. 1995. Exotics and hurricane damage in three hardwood hammocks in Dade County parks, Florida. *Journal of Coastal Research* 21: 145-158.

289. Howe, William H. (editor). 1975. *The Butterflies of North America*. Doubleday & Company, Inc., Garden City, New York.

290. Hribar, Lawrence J. and Joshua J. Vlach. 2001. Mosquito (Diptera, Culicidae) and biting midge (Diptera, Ceratopogonidae) collections in the Florida Keys state parks. *Florida Scientist* 64(3): 219-223.

291. Humphrey, Stephen R. (editor). 1992. *Volume I. Mammals. In* Ray E. Ashton, Jr. (series editor). *Rare and Endangered Biota of Florida*. University Presses of Florida, Gainesville, Florida.

292. Hunt, Burton P. 1952. Food relationships between Florida spotted gar and other organisms in the Tamiami Canal, Dade County, Florida. *Transactions of the American Fisheries Society* 82: 14-33.

293. Huvane, Jacqueline K. 2002. Modern diatom distributions in Florida Bay: a preliminary analysis. *In* James W. Porter and Karen G. Porter (editors). *The Everglades, Florida Bay, and Coral Reefs of the Florida Keys: an Ecosystem Source Book*. CRC Press, Boca Raton, Florida.

294. Ives, Lieut. J.C. 1856. Military Map of the Peninsula of Florida South of Tampa Bay. U.S. War Department. *In* Steven M. Davis and John C. Ogden (editors). *Everglades: the Ecosystem and Its Restoration*. St. Lucie Press, Delray Beach, Florida.

295. Izuno, Forrest T. and A.B. Bottcher. 1994. The history of water management in South Florida. *In* A.B. Bottcher and F.T. Izuno (editors). *Everglades Agricultural Area (EAA): Water, Soil, Crop, and Environmental Management*. University of Florida Press, Gainesville, Florida.

296. James, Frances C. 1997. Nonindigenous birds. *In* Daniel Simberloff, Don C. Schmitz, and Tom C. Brown (editors). *Strangers in Paradise: Impact and Management of Nonindigenous Species in Florida*. Island Press, Washington, D.C.

297. James, R.T., V.H. Smith, and B.L. Jones. 1995. Historic trends in the Lake Okeechobee ecosystem. III. Water quality. *Archiv. Fur Hydrobiologie*, Supplement, 107: 49-69.

298. Jewell, Susan D. 1995. Exploring Wild Central Florida: a field guide to finding the natural areas and wildlife of the central peninsula. Pineapple Press, Inc., Sarasota, Florida.

299. Jewell, Susan D. 2002. Exploring Wild South Florida: a field guide to finding the natural areas and wildlife of the southern peninsula and the Florida Keys (third edition). Pineapple Press, Inc., Sarasota.

300. Johnson, Lamar. 1974. *Beyond the Fourth Generation*. The University Presses of Florida, Gainesville.

301. Johnson, F.A., F. Montalbano III, J.D. Truitt, and D.R. Eggeman. 1991. Distribution, abundance, and habitat use by mottled ducks in Florida. *Journal of Wildlife Management* 55(3): 476-482.

302. Jones, A.C., S.A Berkeley, J.A. Bohnsack, S.A. Bortone, D.K. Camp, G.H. Darcy, J.C. Davis, K.D. Haddad, M.Y. Hedgepeth, E.W. Irby, Jr., W.C. Jaap, F.S. Kennedy, Jr., W.G. Lyons, E.L. Nakamura, T.H. Perkins, J.K. Reed, K.A. Steidinger, J.T. Tilmant, and R.O. Williams. 1985. Ocean habitat and fishery resources of Florida. *In* William Seaman, Jr. (editor) *Florida Aquatic Habitat and Fishery Resources*. Florida Chapter of the American Fisheries Society, Kissimmee, Florida.

303. Jones, Douglas. 1997. The marine invertebrate fossil record of Florida. *In* Anthony F. Randazzo and Douglas S. Jones (editors). *The Geology of Florida*. University Press of Florida, Gainesville, Florida.

304. Jordan, Dennis. 1990. Mercury contamination: another threat to the Florida panther. *Endangered Species Technical Bulletin* 15(2): 1,6. U.S. Department of the Interior, Fish and Wildlife Service. Washington, D.C.

305. Joyner, B.F. 1974. Chemical and biological conditions of Lake Okeechobee, Florida, 1969-72. Technical Report. State of Florida Department of Natural Resources. Tallahasse, Floridae.

306. Kadlec, Robert H. and Susan Newman. 1992. Phosphorus removal in wetland treatment areas. Technical report prepared for the South Florida Water Management District, West Palm Beach.

307. Kale, Herbert W., II (editor). 1978. *Volume Two: Birds. In* Peter C.H. Pritchard (series editor) *Rare and Endangered Biota of Florida*. University Presses of Florida, Gainesville, Florida.

308. Kautz, Randy S. 1993. Trends in Florida wildlife habitat 1936-1987. *Florida Scientist* 56(1): 7-24.

309. Keller, Brian D. and Arthur Itkin. 2002. Shoreline nutrients and chlorophyll *a* in the Florida Keys, 1994-1997: a preliminary analysis. *In* James W. Porter and Karen G. Porter (editors). *The Everglades, Florida Bay, and Coral Reefs of the Florida Keys: an Ecosystem Source Book*. CRC Press, Boca Raton.

310. Kim, Kiho and C. Drew Harvell. 2002. Aspergillosis of sea fan corals: disease dynamics in the Florida Keys. *In* James W. Porter and Karen G. Porter (editors). *The Everglades, Florida Bay, and Coral Reefs of the Florida Keys: an Ecosystem Source Book*. CRC Press, Boca Raton, Florida.

311. Kirby, Robert R., Carl H. Hobbs, and Ashish J. Mehta. 1989. Fine sediment regime of Lake Okeechobee, Florida. A study funded by the South Florida Water Management District as part of a Lake Okeechobee phosphorus dynamics study. Coastal and Oceanographic Engineering Department, University of Florida, Gainesville, Florida.

312. Kitchens, Wiley M., Robert E. Bennetts, and Donald L. DeAngelis. 2002. Linkages between the snail kite population and wetland dynamics in a highly fragmented South Florida hydroscape. *In* James W. Porter and Karen G. Porter (editors). *The Everglades, Florida Bay, and Coral Reefs of the Florida Keys: an Ecosystem Source Book.* CRC Press, Boca Raton, Florida.

313. Klitgord, Kim D., Peter Popenoe, and Hans Schouten. 1984. Florida: a Jurassic transform plate boundary. *Journal of Geophysical Research* 89(B9): 7753-7772.

314. Kobza, Robert M. and William F. Loftus. 2003. The role of seasonal hydrology in the dynamics of fish communities inhabiting karstic refuges of the Florida Everglades (abstract). Joint Conference on the Science and Restoration of the Greater Everglades and Florida Bay Ecosystem, April 13-18, 2003, Palm Harbor, Florida.

315. Koebel, Joseph W., Jr. 1995. An historical perspective on the Kissimmee River restoration project. *Restoration Ecology* 3(3): 149-159.

316. Kreitman, Abe and Leslie A. Wedderburn. 1984. Hydrology of South Florida. *In* P.J. Gleason (editor). *Environments of South Florida: Present and Past II.* Miami Geological Society, Miami, Florida.

317. Krohne, David T. 2001. *General Ecology.* Second edition. Brooks/Cole, Pacific Grove, California.

318. Kruczynski, William L. and Fred McManus. 2002. Water quality concerns in the Florida Keys: sources, effects, and solutions. *In* James W. Porter and Karen G. Porter (editors). *The Everglades, Florida Bay, and Coral Reefs of the Florida Keys: an Ecosystem Source Book.* CRC Press, Boca Raton.

319. Kuiken, Todd and Felice Stadler. 2003. Cycle of harm: mercury's pathway from rain to fish in the environment. May 2003 report by the National Wildlife Federation, Great Lakes Natural Resource Center, Ann Arbor, Michigan.

320. Kushlan, James A. 1972. An ecological study of an alligator pond in the Big Cypress Swamp of southern Florida. Ph.D. dissertation, Department of Biology, University of Miami, Coral Gables, Florida.

321. Kushlan, James A. 1988. Impact of water management on wildlife in the Florida Everglades. *In* Miguel Chávez Miguel (editor). *Ecología y Conservación del Delta de los ríos Usumacinta y Grijalva* (Memorias). Instituto Nacional de Investigacíon sobre Recursos Bióticos (INIREB) – División Regional Tabasco, pages 231-243. (see Chavez, 1988)

322. Kushlan, James. A. 1990. Freshwater marshes. *In* Ronald L. Myers and John J. Ewel (editors). *Ecosystems of Florida.* University of Central Florida Press, Orlando, Florida.

323. Kushlan, James. A. and Thomas E. Lodge. 1974. Ecological and distributional notes on the freshwater fish of southern Florida. *Florida Scientist* 37(2): 110-128.

324. Kushlan, James. A. and Frank J. Mazzotti. 1989. Historical and present distribution of the American crocodile in Florida. *Journal of Herpetology* 23(1): 1-7.

325. Kushlan, James. A. and Deborah A. White. 1977. Nesting wading bird populations in southern Florida. *Florida Scientist* 40(1): 65-72.

326. Lakela, Olga and Robert W. Long. 1976. *Ferns of Florida.* Banyan Books, Miami.

327. Lapointe, Brian E., William R. Matzie, and Peter J. Barile. 2002. Biotic phase-shifts in Florida Bay and fore reef communities of the Florida Keys: linkages with historical freshwater flows and nitrogen loading from Everglades runoff. *In* James W. Porter and Karen G. Porter (editors). *The Everglades, Florida Bay, and Coral Reefs of the Florida Keys: an Ecosystem Source Book.* CRC Press, Boca Raton.

328. Lapointe, Brian E. and Katy Thacker. 2002. Community-based water quality and coral reef monitoring in the Negril Marine Park, Jamaica: land-based nutrient inputs and their ecological consequences. *In* James W. Porter and Karen G. Porter (editors). *The Everglades, Florida Bay, and Coral Reefs of the Florida Keys: an Ecosystem Source Book.* CRC Press, Boca Raton, Florida.

329. LaRoche, Francois B. and A.P. Ferriter. 1992. The rate of expansion of melaleuca in South Florida. *Journal of Aquatic Plant Management* 30: 62-65.

330. Layne, James N. (editor). 1978. *Volume One: Mammals. In* Peter C.H. Pritchard (series editor) *Rare and Endangered Biota of Florida.* University Presses of Florida, Gainesville, Florida.

331. Layne, James N. 1984. The land mammals of southern Florida. *In* P.J. Gleason (editor). *Environments of South Florida: Present and Past II.* Miami Geological Society, Miami, Florida.

332. Layne, James N. 1997. Nonindigenous mammals. *In* Daniel Simberloff, Don C. Schmitz, and Tom C. Brown (editors). *Strangers in Paradise: Impact and Management of Nonindigenous Species in Florida*. Island Press, Washington, D.C.

333. Lazell, James D., Jr. 1989. *Wildlife of the Florida Keys: A Natural History*. Island Press, Washington, D.C.

334. Le Conte, John. 1822. Observations on the soil and climate of East Florida. A report to the Chief of Engineers, U.S. War Department. Reprinted (1972) by the Central and Southern Florida Flood Control District, West Palm Beach.

335. Lee, Thomas N., Elizabeth Williams, Elizabeth Johns, Doug Wilson, and Ned P. Smith. 2002. Transport processes linking South Florida coastal ecosystems. *In* James W. Porter and Karen G. Porter (editors). *The Everglades, Florida Bay, and Coral Reefs of the Florida Keys: an Ecosystem Source Book*. CRC Press, Boca Raton, Florida.

336. Lefebvre, Lynn W., Miriam Marmontel, James P. Reid, Galen B. Rathbun, and Daryl P. Domning. 2001. Status and biogeography of the West Indian manatee. *In* Charles A. Woods and Florence E. Sergile (editors). *Biogeography of the West Indies: Patterns and Perspectives*. Second edition. CRC Press, Boca Raton, Florida.

337. Lehtinen, Dexter W. 1990. United States of America, plaintiff, vs. South Florida Water Management District; John R. Wodraska, Executive Director, South Florida Water Management District; Florida Department of Environmental Regulation; and Dale Twachtmann, Secretary, Florida Department of Environmental Regulation, defendants. Second Amended Complaint. U.S. District Court Southern District of Florida, Case No 88-1886-CIV-HOEVELER (signed January 4, 1990).

338. Levi, Herbert W. and Lorna R. Levi. 1968. *A Guide to Spiders and their Kin*. Golden Press, New York.

339. Levin, Ted. 2003. *Liquid Land, a Journey through the Florida Everglades*. University of Georgia Press, Athens, Georgia.

340. Lewis, R.R., III, R.G. Gilmore, Jr., D.W. Crewz, and W.E. Odum. 1985. Mangrove habitat and fishery resources of Florida. *In* William Seaman, Jr. (editor). *Florida Aquatic Habitat and Fishery Resources*. Florida Chapter of the American Fisheries Society, Kissimmee, Florida.

341. Ley, Janet A. and Carole C. McIvor. 2002. Linkages between estuarine and reef fish assemblages: enhancement by the presence of well-developed mangrove shorelines. *In* James W. Porter and Karen G. Porter (editors). *The Everglades, Florida Bay, and Coral Reefs of the Florida Keys: an Ecosystem Source Book*. CRC Press, Boca Raton, Florida.

342. Lidz, Barbara H. and Eugene A. Shinn. 1991. Paleoshorelines, reefs, and rising sea: South Florida, U.S.A. *Journal of Coastal Research* 7(1): 203-229.

343. Light, Stephen S. and J. Walter Dineen. 1994. Water control in the Everglades: a historical perspective. *In* Steven M. Davis and John C. Ogden (editors). *Everglades: the Ecosystem and Its Restoration*. St. Lucie Press, Delray Beach, Florida.

344. Lissner, Jorgen, Irving A. Mendelssohn, Bent Lorenzen, Hans Brix, Karen L. McKee, and ShiLi Miao. 2003. Interacive effects of redox intensity and phosphate availability on growth and nutrient relations of *Cladium jamaicense* (Cyperaceae). *American Journal of Botany* 90: 736-748.

345. Livingston, Robert J. 1990. Inshore marine habitats. *In* Ronald L. Myers and John J. Ewel (editors). *Ecosystems of Florida*. University of Central Florida Press, Orlando, Florida.

346. Lloyd, Jacqueline M. 1991. 1988 and 1989 Florida Petroleum Production and Exploration. Florida Geological Survey, Information Circular No. 107. Florida Department of Natural Resources.

347. Lockwood, J.L., K.H. Fenn, J.L. Curnutt, D. Rosenthal, K.L. Balent, and A.L. Mayer. 1997. Life history of the endangered Cape Sable seaside sparrow. *Wilson Bulletin* 109(4): 720-731.

348. Lodge, Thomas E. Personal observations of the author.

349. Lodge, Thomas E. 1967. A report on findings of a limnological investigation of Deep Lake (Collier County, Florida) on April 1, 1967. Unpublished report, 52 pp., Department of Biology, University of Miami, Coral Gables. (Available at the South Florida Water Management District library)

350. Lodge, T.E. 1968. The probable function of alligator holes as phosphate traps due to periodic drying of the Everglades. unpublished report, 5 pp.

351. Lodge, Thomas E., Hilburn O. Hillestad, Stephen W. Carney, and Richard B. Darling. 1995. Wetland Quality Index: a method for determining compensatory mitigation requirements for ecologically impacted wetlands. ASCE, South Florida Section annual meeting presentation, September 1995.

352. Loftin, Jan P. 1992. In the wake of Hurricane Andrew. *Florida Water* 1(2): 2-8.

353. Loftus, William F. 2000. Accumulation and fate of mercury in an Everglades aquatic food web. Ph.D. dissertation, Florida International University, Miami, Florida.

354. Loftus, William F. 2000. Inventory of fishes of Everglades National Park. *Florida Scientist* 63(1): 27-47.

355. Loftus, William F. and Oron Bass, Jr. 1992. Mercury threatens wildlife resources. *Park Science* 24(4): 18-20.

356. Loftus, William F. and Anne-Marie Eklund. 1994. Long-term dynamics of an Everglades small-fish assemblage. *In* Steven M. Davis and John C. Ogden (editors). *Everglades: the Ecosystem and Its Restoration*. St. Lucie Press, Delray Beach, Florida.

357. Loftus, William F., James D. Chapman, and Roxanne Conrow. 1990. Hydroperiod effects on Everglades marsh food webs, with relation to marsh restoration efforts. *In* Proceedings of the Fourth Triennial Conference on Science in the National Parks, Fort Collins, Colorado.

358. Loftus, William F., Robert A. Johnson, and Gordon H. Anderson. 1992. Ecological impacts of the reduction of groundwater levels in short-hydroperiod marshes of the Everglades. First International Conference on Ground Water Ecology. U.S. Environmental Protection Agency/American Water Resources Association, April 1992.

359. Loftus, William F., Robert M. Kobza, and Delissa Padilla. 2003. Fish community colonization patterns in the rocky glades wetlands of southern Florida (abstract). Joint Conference on the Science and Restoration of the Greater Everglades and Florida Bay Ecosystem, April 13-18, 2003, Palm Harbor, Florida.

360. Loftus, William F. and James A. Kushlan. 1987. Freshwater fishes of southern Florida. *Bulletin Florida State Museum, Biological Sciences* 31(4): 147-344.

361. Loftus, William F., Leo G. Nico, Jeffery Kline, Sue A. Perry, and Joel C. Trexler. Recent fish introductions into southern Florida freshwaters, with implications for the greater Everglades region (abstract). Joint Conference on the Science and Restoration of the Greater Everglades and Florida Bay Ecosystem, April 13-18, 2003, Palm Harbor, Florida.

362. Loftus, William F. and Janet W. Reid. 2000. Copepod (Crustacea) emergence from soils from Everglades marshes with different hydroperiods. *Journal of Freshwater Ecology* 15(4): 515-523.

363. Logan, Todd, Andrew C. Eller, Jr., Ross Morrell, Donna Ruffner, and Jim Sewell. 1993. Florida Panther Habitat Preservation Plan, South Florida Population. Prepared by the U.S. Fish and Wildlife Service, the Florida Game and Fresh Water Fish Commission, the Florida Department of Environmental Protection, and the National Park Service for the Florida Panther Interagency Committee (approved January 3, 1994).

364. Logan, Tom H. 1997. Florida's Endangered Species, Threatened Species and Species of Special Concern, Official Lists, August 1, 1997. Florida Fish and Wildlife Conservation Commission, Tallahassee.

365. Long, Robert W. 1984. Origin of the vascular flora of southern Florida. *In* P.J. Gleason (editor). *Environments of South Florida: Present and Past II*. Miami Geological Society, Miami, Florida.

366. Long, Robert W. and Olga Lakela. 1971. *A Flora of Tropical Florida*. University of Miami Press, Coral Gables, Florida.

367. Loope, Lloyd, Michael Duever, Alan Herndon, James Snyder, and Deborah Jansen. 1994. Hurricane impact on uplands and freshwater swamp forest. *Bioscience* 44(4): 238-246.

368. Lorenz, Jerome J., John C. Ogden, Robin D. Bjork, and George V.N. Powell. 2002. Nesting patterns of roseate spoonbills in Florida Bay 1935-1999: implications of landscape scale anthropogenic impacts. *In* James W. Porter and Karen G. Porter (editors). *The Everglades, Florida Bay, and Coral Reefs of the Florida Keys: an Ecosystem Source Book*. CRC Press, Boca Raton.

369. Loveless, Charles M. 1959. A study of the vegetation in the Florida Everglades. *Ecology* 40(1): 1-9.

370. Loveless, Charles M. 1959. The Everglades deer herd life history and management. Florida Game and Fresh Water Fish Commission, Technical Bulletin No. 6, 104pp.

371. Luer, Carlyle A. 1972. *The Native Orchids of Florida*. The New York Botanical Garden. W.S. Cowell Ltd., Ipswich, England, U.K.

372. MacArthur, R.H. and E.O. Wilson. 1967. *The Theory of Island Biogeography*. Princeton University Press, Princeton, New Jersey.

373. MacFadden, Bruce J. 1997. Fossil mammals of Florida. *In* Anthony F. Randazzo and Douglas S. Jones (editors). *The Geology of Florida*. University Press of Florida, Gainesville, Florida.

374. MacVicar, Thomas K. 1987. Rescuing the Everglades. *Civil Engineering* 57(8): 40-42.

375. MacVicar, Thomas K. and Steve S.T. Lin. 1984. Historical rainfall activity in central and southern Florida: average, return period estimates and selected extremes. *In* P.J. Gleason (editor). *Environments of South Florida: Present and Past II*. Miami Geological Society, Miami, Florida.

376. Maehr, David S. 1997. *The Florida Panther: Life and Death of a Vanishing Carnivore*. Island Press, Washington, D.C.

377. Maffei, Mark D. 1991. Melaleuca control on Arthur R. Marshall Loxahatchee National Wildlife Refuge. *In* Ted D. Center, Robert F. Doren, Ronald H. Hofstetter, Ronald L. Myers, and Louis D. Whiteaker (editors). Proceedings of the Symposium on Exotic Pest Plants. U.S. Department of the Interior, National Park Service, Washington, D.C.

378. Maffei, Mark D. 1997. Management in national wildlife refuges. *In* Daniel Simberloff, Don C. Schmitz, and Tom C. Brown (editors). *Strangers in Paradise: Impact and Management of Nonindigenous Species in Florida*. Island Press, Washington, D.C.

379. Malcom, Corey. 2002. 8,400 year-old submarine forest fire found. *Navigator* 18(2): 2-3 (publication of the Mel Fisher Maritime Heritage Society). Key West.

380. Maltby, E. and P.J. Dugan. 1994. Wetland ecosystem protection, management, and restoration: an international perspective. *In* Steven M. Davis and John C. Ogden (editors). *Everglades: the Ecosystem and Its Restoration*. St. Lucie Press, Delray Beach, Florida.

381. Manson, Daniel H. and Arnold van der Valk. 2002. Vegetation, peat elevation and peat depth on two tree islands in Water Conservation Area 3-A. *In* Fred H. Sklar and Arnold van der Valk (editors). *Tree Islands of the Everglades*. Kluwer Academic Publishers, Boston, Massachusetts.

382. Marshall, Curtis H., Roger A. Pielke Sr., and Louis T. Steyaert. 2003. Wetlands: crop freezes and land-use change in Florida. *Nature* 426(6962): 29-30.

383. Mayer, A.L. 1999. Cape Sable seaside sparrow (*Ammodramus maritimus mirabilis*) habitat and the Everglades: ecology and conservation. Ph.D. dissertation, Department of Ecology and Evolutionary Biology, University of Tennessee, Knoxville, Tennessee.

384. Mayer, A.L. and S.L. Pimm. 1998. Integrating endangered species protection and ecosystem management: the Cape Sable seaside sparrow as a case study. *In* G.M. Mace, A. Balmford, and J.R. Ginsberg (editors). *Conservation in a changing world*. Cambridge University Press, Cambridge, UK.

385. Mayr, Ernst. 1966. *Animal Species and Evolution*. The Belknap Press of Harvard University Press, Cambridge, Massachusetts.

386. Mazzotti, Frank J. 1989. Factors affecting the nesting success of the American crocodile, *Crocodylus acutus*, in Florida Bay. *Bulletin of Marine Science* 44(1): 220-228.

387. Mazzotti, Frank J. and Laura A Brandt. 1994. Ecology of the American alligator in a seasonally fluctuating environment. *In* Steven M. Davis and John C. Ogden (editors). *Everglades: the Ecosystem and Its Restoration*. St. Lucie Press, Delray Beach, Florida.

388. Mazzotti, Frank J. and M.S. Cherkiss. 2003. Status and conservation of the American crocodile in Florida: recovering an endangered species while restoring an endangered ecosystem. University of Florida, Ft. Lauderdale Research and Education Center. Technical Report. 41 pp.

389. McBride, Roy. 2002. Florida panther current verified population, distribution and highlights of field work: fall 2001 – winter 2002. Prepared for Florida Panther SubTeam of MERIT, U.S. Fish and Wildlife Service, South Florida Ecosystem Office, Vero Beach.

390. McCally, David. 1999. *The Everglades: an Environmental History*. University Presses of Florida, Gainesville, Florida.

391. McClain, Michael and Enid Y. Karr. 1989. The Florida basement: diary of a turbulent past. *The Compass* 66(2): 52-58.

392. McClellan, Guerry H. and James L. Eades. 1997. The economic and industrial minerals of Florida. *In* Anthony F. Randazzo and Douglas S. Jones (editors). *The Geology of Florida*. University Press of Florida, Gainesville, Florida.

393. McCormick, Paul V., Susan Newman, ShiLi Miao, Dale E. Gawlik, Darlene Marley, K. Ramesh Reddy, and Tom D. Fontaine. 2002. Effects of anthropogenic phosphorus inputs on the Everglades. *In* James W. Porter and Karen G. Porter (editors). *The Everglades, Florida Bay, and Coral Reefs of the Florida Keys: an Ecosystem Source Book*. CRC Press, Boca Raton, Florida.

394. McCormick, P.V. and L.J. Scinto. 1999. Influence of phosphorus loadings on wetlands periphyton assemblages: a case study from the Everglades. *In* K.R. Reddy, G.A. O'Connor, and C.L. Schelske (editors). *Phosphorus Biogeochemistry in Subtropical Ecosystems.* Lewis Publishers, Boca Raton, Florida.

395. McCrimmon, Donald A., Jr., John C. Ogden, and G. Thomas Bancroft. 2001. Great egret (*Ardea alba*). *In* A. Poole and F. Gill (editors). *The Birds of North America*, No. 570. The Birds of North America, Inc., Philadelphia, PA.

396. McDiarmid, Roy W. (editor). 1978. *Volume Three: Amphibians and Reptiles. In* Peter C.H. Pritchard (series editor) *Rare and Endangered Biota of Florida.* University Presses of Florida, Gainesville, Florida.

397. McIvor, Carole C., Janet A. Ley, and Robin D. Bjork. 1994. Changes in freshwater inflow from the Everglades to Florida Bay. *In* Steven M. Davis and John C. Ogden (editors). *Everglades: the Ecosystem and Its Restoration.* St. Lucie Press, Delray Beach, Florida.

398. McPherson, Benjamin F. 1973. Vegetation map of southern parts of subareas A and C, Big Cypress Swamp, Florida. U.S. Geological Survey Hydrologic Investigations Atlas HA-492 (supplied in the back-cover pocket of Duever et al. 1986).

399. McPherson, Benjamin F. 1984. The Big Cypress Swamp. *In* P.J. Gleason (editor). *Environments of South Florida: Present and Past II.* Miami Geological Society, Miami, Florida.

400. McPherson, Benjamin F. and Robert Halley. 1996. *The South Florida Environment – a Region Under Stress.* U.S. Geological Survey Circular 1134, U.S. Geological Survey, Denver.

401. McPherson, Benjamin F., G.Y. Hendrix, H. Klein, H.M. Tyus. 1976. The environment of South Florida, a summary report. U.S. Geological Survey Professional Paper 1011.

402. McVoy, Christopher, Winifred P. Said, Jayantha Obeysekera, and Joel VanArman. In preparation (2004). Pre-drainage Everglades Landscapes and Hydrology. South Florida Water Management District, West Palm Beach.

403. Means, D.B. and D. Simberloff. 1987. The peninsula effect: habitat-correlated species decline in Florida's herpetofauna. *Journal of Biogeography* 14: 551-568.

404. Meeder, J.F. and L.B. Meeder. 1989. Hurricanes in Florida Bay: a dominant physical process (abstract). *Bulletin of Marine Science* 44(1): 518.

405. Merritt, Richard W., Michael J. Higgins, Kennith W. Cummins, and Brigitte Vandeneeden. 1999. The Kissimmee River-Riparian marsh ecosystem, Florida: seasonal differences in invertebrate functional feeding group relationships. *In* Donald P. Batzer, Russell B. Rader, and Scott A. Wissinger (eds.). *Invertebrates in Freshwater Wetlands of North America.* John Wiley & Sons, Inc., New York.

406. Meshaka, Walter E., Jr. 1993. Hurricane Andrew and the colonization of five invading species in South Florida. *Florida Scientist* 56(4): 193-201.

407. Meshaka, Walter E., Jr. 1994. Reproductive cycle of the Indo-Pacific gecko, *Hemidactylus garnotii*, in South Florida. *Florida Scientist* 57(1,2): 6-9.

408. Meshaka, Walter. E., Jr., P.B. Butterfield, and J.B. Hauge. In press, 2003. *The Exotic Amphibians and Reptiles of Florida.* Krieger Publ. Co., Melbourne, Florida.

409. Meshaka, Walter E., Jr., William F. Loftus, and Todd Steiner. 2000. The herpetofauna of Everglades National Park. *Florida Scientist* 63(2): 84-103.

410. Meshaka, Walter E., Jr., Ray Snow, Oron L. Bass, Jr., and William B Robertson, Jr. 2002. Occurrence of wildlife on tree islands in the southern Everglades. *In* Fred H. Sklar and Arnold van der Valk (editors). *Tree Islands of the Everglades.* Kluwer Academic Publishers, Boston, Massachusetts.

411. Meyer, Frederick W. 1989. Hydrology, ground-water movement, and subsurface storage in the Floridan aquifer system in southern Florida. U.S. Geological Survey Professional Paper 1403-G.

412. Miao, S.L. and W.F. DeBusk. 1999. Effects of phosphorus enrichment on structure and function of sawgrass and cattail communities in the Everglades. *In* K.R. Reddy, G.A. O'Connor, and C.L. Schelske (editors). *Phosphorus Biogeochemistry in Subtropical Ecosystems.* Lewis Publishers, Boca Raton, Florida.

413. Milanich, Jerald T. 1994. *Archaeology of Precolumbian Florida.* University Presses of Florida, Gainesville, Florida.

414. Milanich, Jerald T., and Charles H. Fairbanks. 1980. *Florida Archaeology.* Academic Press, New York, New York.

415. Milius, Susan. 2003. After West Nile Virus: what will it do to birds and beasts of North America. *Science News* 163(13): 203-205.

416. Miller, James A. 1997. Hydrogeology of Florida. *In* Anthony F. Randazzo and Douglas S. Jones (editors). *The Geology of Florida*. University Press of Florida, Gainesville, Florida.

417. Missimer, Thomas M. 1984. The geology of South Florida: a summary. *In* P.J. Gleason (editor). *Environments of South Florida: Present and Past II*. Miami Geological Society, Miami.

418. Mitsch, William J. and James G. Gosselink. 1993. *Wetlands*, second edition. Van Nostrand Reinhold, New York, New York.

419. Moler, Paul E. (editor). 1992. *Volume III. Amphibians and Reptiles*. *In* Ray E. Ashton, Jr. (series editor). *Rare and Endangered Biota of Florida*. University Presses of Florida, Gainesville, Florida.

420. Molnar, George, Ronald H. Hofstetter, Robert F. Doren, Louis D. Whiteaker, and Michael T. Brennan. 1991. Management of *Melaleuca quinquenervia* within East Everglades wetlands. *In* Ted D. Center, Robert F. Doren, Ronald H. Hofstetter, Ronald L. Myers, and Louis D. Whiteaker (editors). Proceedings of the Symposium on Exotic Pest Plants. U.S. Department of the Interior, National Park Service, Washington, D.C.

421. Mooij, Wolf M., Robert E. Bennetts, Wiley M. Kitchens, and Donald L. DeAngelis. 2002. Exploring the effect of drought extent and interval on the Florida snail kite: interplay between spatial and temporal scales. *Ecological Modelling* 149: 25-39.

422. Mora, C. and P. F. Sale. 2002. Are populations of coral reef fish open or closed? *Trends in Ecology and Evolution* 17(9): 422-428.

423. Morris, Allen. 1995. *Florida Place Names*. Pineapple Press, Inc., Sarasota.

424. Morgan, Gary S. 2001. Patterns of extinction in West Indian bats. *In* Charles A. Woods and Florence E. Sergile (editors). *Biogeography of the West Indies: Patterns and Perspectives*. Second edition. CRC Press, Boca Raton.

425. Moustafa, M.Z., S. Newman, T.D. Fontaine, M.J. Chimney, and T.C. Kosier. 1999. Phosphorus retention by the Everglades nutrient removal: an Everglades stormwater treatment area. *In* K.R. Reddy, G.A. O'Connor, and C.L. Schelske (editors). *Phosphorus Biogeochemistry in Subtropical Ecosystems*. Lewis Publishers, Boca Raton.

426. Murphy, J. Brendan and R. Damiam Nance. 1992. Mountain belts and the supercontinent cycle. *Scientific American* 266:84-91.

427. National Geographic Society. 2002. *Field Guide to the Birds of North America* (fourth edition). National Geographic Society, Washington, D.C.

428. National Park Service. 2002/2000. Big Cypress National Preserve and Everglades National Park, Florida (maps/descriptions for visitor use). U.S. Department of the Interior. GPO: 2002 – 491-282/40191 and GPO: 2000 – 460-976/00344 respectively.

429. National Research Council. 2001. *Compensating for Wetland Losses under the Clean Water Act*. National Academy Press, Washington, D.C.

430. National Research Council. 2001. *Aquifer Storage and Recovery in the Comprehensive Everglades Restoration Plan: a Critique of the Pilot Projects and Related Plans for ASR in the Lake Okeechobee and Western Hillsboro Areas*. National Academy Press, Washington, D.C.

431. National Research Council. 2002. *Regional Issues in Aquifer Storage and Recovery for Everglades Restoration*. National Academies Press, Washington, D.C.

432. National Research Council. 2002. *Florida Bay Research Programs and their Relation to the Comprehensive Everglades Restoration Plan*. National Academies Press, Washington, D.C.

433. National Research Council. 2003. *Adaptive Monitoring & Assessment for the Comprehensive Everglades Restoration Plan*. National Academies Press, Washington, D.C.

434. National Research Council. 2003. *Does Water Flow Influence Everglades Landscape Patterns?* The National Academies Press, Washington, D.C.

435. National Research Council. 2003. *Science and the Greater Everglades Ecosystem Restoration*. The National Academy Press, Washington, D.C.

436. Neill, Wilfred T. 1969. *The Geography of Life*. Columbia University Press, New York, New York.

437. Nellis, David W. 1994. *Seashore Plants of South Florida and the Caribbean*. Pineapple Press, Inc., Sarasota, Florida.

438. Nelson, Gil. 2000. *The Ferns of Florida*. Pineapple Press, Inc., Sarasota, Florida.

439. Nelson, Terry A., Ginger Garte, Charles Featherstone, Harold R. Wanless, John H. Trefry, Woo-Jun Kang, Simone Metz, Carlos Alvarez-Zarikian, Terri Hood, Peter Swart, Geoffrey Ellis, Pat Blackwelder, Leonore Tedesco, Catherine Slouch, Joseph F. Pachut, and Mike O'Neal. 2002. Linkages between the South Florida peninsula and coastal zone: a sediment-based history of natural and anthropogenic influences. *In* James W. Porter and Karen G. Porter (editors). *The Everglades, Florida Bay, and Coral Reefs of the Florida Keys: an Ecosystem Source Book*. CRC Press, Boca Raton, Florida.

440. Newcott, William R. 1993. Lightning, nature's high-voltage spectacle. *National Geographic* 184(1): 83-103.

441. Niering, William A. 1990. Vegetation dynamics in relation to wetland creation. *In* Jon A. Kusler and Mary E. Kentula (editors). *Wetland Creation and Restoration: the Status of the Science*. Island Press, Washington, D.C.

442. NOAA. 1980. Eastern United States Coastal and Ocean Zones Data Atlas. Prepared by G. Carleton Ray, M. Geraldine McCormick-Ray, James A. Dobbin. Charles N. Ehler, and Daniel J. Basta for the Council on Environmental Quality, Executive Office of the President, and the Office of Coastal Zone Management, National Oceanic and Atmospheric Administration, U.S. Department of Commerce, Washington, D.C.

443. Noel, Jill M., Andy Maxwell, William J. Platt, and Linda Pace. 1995. Effects of Hurricane Andrew on cypress (*Taxodium distichum* var. *nutans*) in South Florida. *Journal of Coastal Research* 21: 184-196.

444. Nott, M.P., O.L. Bass Jr., D.M. Fleming, S.E. Killeffer, N. Frahley, L. Manne, J.L. Curnutt, T.M. Broods, R. Powell, and S.L. Pimm. 1998. Water levels, rapid vegetation changes, and the endangered Cape Sable seaside sparrow. *Animal Conservation* 1(1): 23-32.

445. Odum, Howard T., Elisabeth C. Odum, and Mark T. Brown. 1998. *Environment and Society in Florida*. Lewis Publishers, Boca Raton, Florida.

446. Odum, William E. 1970. Pathways of energy flow in a South Florida estuary. Ph.D. dissertation, University of Miami, Coral Gables, Florida.

447. Odum, William E. and Carole C. McIvor. 1990. Mangroves. *In* Ronald L. Myers and John J. Ewel (editors). *Ecosystems of Florida*. University of Central Florida Press, Orlando, Florida.

448. Ogden, John C. 1985. The wood stork. *Audubon Wildlife Report 1985*. The National Audubon Society, New York, New York.

449. Ogden, John C. 1994. A comparison of wading bird nesting colony dynamics (1931-1946 and 1974-1989) as an indication of ecosystem conditions in the southern Everglades. *In* Steven M. Davis and John C. Ogden (editors). *Everglades: the Ecosystem and Its Restoration*. St. Lucie Press, Delray Beach, Florida.

450. Ogden, John C., J.A. Kushlan, and J.T. Tilmant. 1978. The food habits and nesting success of wood storks in Everglades National Park, 1974. U.S. Department of the Interior, National Park Service. Natural Resources Report Number 16.

451. Ogden, John C. and Betsy Trent Thomas. 1985. A colonial wading bird survey in the central Llanos of Venezuela. *Colonial Waterbirds* 8(1): 23-31.

452. Olmsted, Ingrid, and Lloyd L. Loope. 1984. Plant communities of Everglades National Park. *In* P.J. Gleason (editor). *Environments of South Florida: Present and Past II*. Miami Geological Society, Miami.

453. Orem, William H., Debra A. Willard, Harry E. Lerch, Anne L. Bates, Ann Boylan, and Margo Comm. 2002. Nutrient geochemistry of sediments from two tree islands in Water Conservation Area 3B, the Everglades, Florida. *In* Fred H. Sklar and Arnold van der Valk (editors). *Tree Islands of the Everglades*. Kluwer Academic Publishers, Boston, Massachusetts.

454. Orians, Gordon H., Michael Bean, Russell Lande, Kent Loftin, Stuart Pimm, R. Eugene Turner, and Milton Weller. 1992. *Report of the Advisory Panel on the Everglades and Endangered Species*. Audubon Conservation Report No. 8. National Audubon Society, New York, New York.

455. Palmer, Jay W. 1998. Florida and global warming. *Florida Scientist* 61(2): 96-105.

456. Parker, Garald. 1984. Hydrology of the pre-drainage system of the Everglades in Southern Florida. *In* P.J. Gleason (editor). *Environments of South Florida: Present and Past II*. Miami Geological Society, Miami, Florida.

457. Parker, Garald G., G.E. Ferguson, S.K. Love, and others. 1955. Water resources of southeastern Florida, with special reference to the geology and groundwater of the Miami area. U.S. Geological Survey Water-Supply Paper 1255.

458. Paulson, D.R. 1966. The dragonflies (Odonata: Anisoptera) of southern Florida. Ph.D. dissertation, University of Miami, Coral Gables, Florida.

459. Pennak, Robert W. 1978. *Fresh-water Invertebrates of the United States* (second edition). John Wiley & Sons, New York, New York.

460. Penny, Malcolm. 1991. *Alligators and Crocodiles*. Crescent Books, New York, New York.

461. Perfit, Michael R., and Ernest E. Williams. 1989. Geological contraints [sic] and biological retrodictions in the evolution of the Caribbean Sea and its islands. *In* C.A. Woods (editor). *Biogeography of the West Indies, Past, Present, and Future*. Sandhill Crane Press, Gainesville, Florida.

462. Pesnell, G.L., and R.T. Brown, III. 1977. The major plant communities of Lake Okeechobee, Florida, and their associated inundation characteristics as determined by gradient analysis. South Florida Water Management District Technical Publication No. 77-1.

463. Peterson, Roger Tory and Virginia Marie Peterson. 2002. *Field Guide to the Birds of Eastern and Central North America*. Fifth Edition. Houghton Mifflin Company, New York.

464. Phillips, Walter S. 1940. A tropical hammock on the Miami (Florida) limestone. *Ecology* 21(2): 166-175.

465. Pielou, E.C. 1991. *After the Ice Age*. University of Chicago Press, Chicago, Illinois.

466. Pimm, Stuart L., Gary E. Davis, Lloyd Loope, Charles TD. Roman, Thomas J. Smith III, and James T. Tilmant. 1994. Hurricane Andrew. *Bioscience* 44(4): 224-229.

467. Pimm, Stuart L., Julie L. Lockwood, Clinton N. Jenkins, John L. Curnutt, M. Philip Nott, Robert D. Powell, and Oron L. Bass, Jr. 2002. Sparrow in the Grass: a report on the first ten years of research on the Cape Sable Seaside Sparrow (*Ammodramus maritimus mirabilis*). Report under Cooperative Agreements CA5280-7-9016 (to the Univ. of Tennessee, Principal Investigator Stuart L. Pimm) and CA5280-0-0010 (to the Univ. of California at Santa Cruz, Principal Investigator Julie L. Lockwood).

468. Pirkle, E.C., W.H. Yoho, and C.W. Hendry, Jr. 1970. Ancient sea level stands in Florida. Florida Department of Natural Resources, Bureau of Geology, Geological Bulletin No. 52, Tallahassee, Florida.

469. Porter, Clyde L., Jr. 1967. Composition and productivity of a subtropical prairie. *Ecology* 48(6): 937-942.

470. Porter, James W., Bladimir Kosmynin, Kathryn L. Patterson, Karen G. Porter, Walter C. Japp, Jennifer L. Wheaton, Keith Hackett, Matt Lybolt, Chris P. Tsokos, George Yanev, Douglas M. Marcinek, John Dotten, David Eaken, Matt Patterson, Ouida W. Meier, Mike Brill, and Phillip Dustan. 2002. Detection of coral reef change by the Florida Keys coral reef monitoring project. *In* James W. Porter and Karen G. Porter (editors). *The Everglades, Florida Bay, and Coral Reefs of the Florida Keys: an Ecosystem Source Book*. CRC Press, Boca Raton, Florida.

471. Post, William and Jon S. Greenlaw. 2000. The present and future of the Cape Sable seaside sparrow. *Florida Field Naturalist* 28(3): 93-160.

472. Powell, Allyn B., Donald E. Hoss, William F. Hettler, David S. Peters, and Stephanie Wagner. 1989. Abundance and distribution of ichthyoplankton in Florida Bay and adjacent waters. *Bulletin of Marine Science* 44(1): 35-48.

473. Powell, George V.N., Robin D. Bjork, John C. Ogden, Richard T. Paul, A. Harriett Powell, and William B. Robertson, Jr. 1989. Population trends in some Florida Bay wading birds. *Wilson Bulletin* 101(3): 436-457.

474. Pratt, Harry D. and Kent S. Littig. 1971. Mosquitos of Public Health Importance and their Control. U.S. Department of Health, Education, and Welfare. Public Health Service. Health Services and Mental Health Administration, Bureau of Community Environmental Management. Atlanta, Georgia.

475. Puri, H.S. and R.O. Vernon. 1964. Summary of the geology of Florida and a guidebook to the classic exposures. Florida Geological Survey Special Publication 5 (revised).

476. Quirolo, DeeVon. 2002. The role of a nonprofit organization, Reef Relief, in protecting coral reefs. *In* James W. Porter and Karen G. Porter (editors). *The Everglades, Florida Bay, and Coral Reefs of the Florida Keys: an Ecosystem Source Book*. CRC Press, Boca Raton, Florida.

477. Rader, Russell B. 1994. Macroinvertebrates of the northern Everglades: species composition and trophic structure. *Florida Scientist* 57(1,2): 22-33.

478. Rader, Russell B. 1999. The Florida Everglades: natural variability, invertebrate diversity, and foodweb stability. *In* Donald P. Batzer, Russell B. Rader, and Scott A. Wissinger (eds.). *Invertebrates in Freshwater Wetlands of North America*. John Wiley & Sons, Inc., New York.

479. Rader, Russell B. and Curtis J. Richardson. 1992. The effects of nutrient enrichment on algae and macroinvertebrates in the Everglades: a review. *Wetlands* 12(2): 121-135.

480. Rader, Russell B. and Curtis J. Richardson. 1994. Response of macroinvertebrates and small fish to nutrient enrichment in the northern Everglades. *Wetlands* 14(2): 134-146.

481. Radice, Paul and Bill Loftus. 1995. The Miami cave crayfish: a colorful new crustacean for the aquarium. *Tropical Fish Hobbyist*, July 1995: 112-116.

482. Raloff, Janet. 2003. Why the mercury falls: heavy-metal rains may trace to oxidants, including smog. *Science News* 163(5): 72-74.

483. Randall, John M., Roy R. Lewis III, and Deborah B. Jensen. 1997. Ecological restoration. *In* Daniel Simberloff, Don C. Schmitz, and Tom C. Brown (editors). *Strangers in Paradise: Impact and Management of Nonindigenous Species in Florida*. Island Press, Washington, D.C.

484. Randazzo, Anthony F. 1997. The sedimentary platform of Florida: Mesozoic to Cenozoic. *In* Anthony F. Randazzo and Douglas S. Jones (editors). *The Geology of Florida*. University Press of Florida, Gainesville, Florida.

485. Randazzo, Anthony F. and Robert B. Halley. 1997. Geology of the Florida Keys. *In* Anthony F. Randazzo and Douglas S. Jones (editors). *The Geology of Florida*. University Press of Florida, Gainesville, Florida.

486. Randazzo, Anthony F. and Douglas S. Jones (editors). 1997. *The Geology of Florida*. University Press of Florida, Gainesville, Florida.

487. Reddy, K.R., E. Lowe, and T. Fontaine. 1999. Phosphorus in Florida's ecosystems: analysis of current issues. *In* K.R. Reddy, G.A. O'Connor, and C.L. Schelske (editors). *Phosphorus Biogeochemistry in Subtropical Ecosystems*. Lewis Publishers, Boca Raton, Florida.

488. Reeder, Pamela B. and Steven M. Davis. 1983. Decomposition, nutrient uptake and microbial colonization of sawgrass and cattail leaves in Water Conservation Area 2A. Technical Publication #83-4, South Florida Water Management District, West Palm Beach.

489. Reese, Ronald S. 2002. Inventory and review of aquifer storage and recovery in southern Florida. U.S. Geological Survey Water-Resources Investigations Report 02-4036. Tallahassee, Florida.

490. Reese, Ronald S. and Kevin J. Cunningham. 2000. Hydrogeology of the gray limestone aquifer in southern Florida. U.S. Geological Survey Water-Resources Investigations Report 99-4213 (prepared in cooperation with the South Florida Water Management District) Tallahassee.

491. Reich, Christopher D., Eugene A. Shinn, Todd D. Hickey, and Ann B. Tihansky. 2002. Tidal and meteorological influences on shallow marine groundwater flow in the upper Florida Keys. *In* James W. Porter and Karen G. Porter (editors). *The Everglades, Florida Bay, and Coral Reefs of the Florida Keys: an Ecosystem Source Book*. CRC Press, Boca Raton.

492. Reimold, Robert J. and William H. Queen (editors). 1974. *Ecology of Halophytes*. Academic Press, Inc., New York.

493. Reiskind, Jonathan. 2001. The contribution of the Caribbean to the spider fauna of Florida. *In* Charles A. Woods and Florence E. Sergile (editors). *Biogeography of the West Indies: Patterns and Perspectives*. Second edition. CRC Press, Boca Raton, Florida.

494. Rhoads, Peter B. 1970. An interim report on a study of the life history and ecology of the Everglades crayfish, *Procambarus alleni* (Faxon). Unpublished paper, Department of Biology, University of Miami (source: Everglades National Park reference library, Pam. box #52, Pam. #43-R).

495. Rich, Earl. 1990. Observations on feeding by *Pomacea paludosa*. *Florida Scientist* 53(suppl.): 13.

496. Richardson, C.J. 1999. The role of wetlands in storage, release, and cycling of phosphorus on the landscape: a 25-year retrospective. *In* K.R. Reddy, G.A. O'Connor, and C.L. Schelske (editors). *Phosphorus Biogeochemistry in Subtropical Ecosystems*. Lewis Publishers, Boca Raton, Florida.

497. Richardson, Curtis J. 2003. Predicting trophic level responses to phosphorus concentrations in the Everglades (abstract). Joint Conference on the Science and Restoration of the Greater Everglades and Florida Bay Ecosystem, April 13-18, 2003, Palm Harbor, Florida.

498. Richardson, John R., Wade L. Bryant, Wiley M. Kitchens, Jennifer E. Mattson, and Kevin R. Pope. 1990. An Evaluation of Refuge Habitats and Relationship to Water Quality, Quantity and Hydroperiod, a Synthesis Report. Prepared for the Arthur R. Marshall Loxahatchee National Wildlife Refuge, Boynton Beach, Florida by the Florida Cooperative Fish and Wildlife Research Unit, University of Florida, Gainesville, Florida.

499. Richardson, John. R. and Tim T. Harris. 1995. Vegetation mapping and change detection in the Lake Okeechobee marsh ecosystem. *In* N.G. Aumen and R.G. Wetzel (editors). Ecological studies on the littoral and pelagic systems of Lake Okeechobee, Florida (USA). *Archiv für Hydrobiologie*, special issue: Advances in Limnology, 45:1-356.

500. Richardson, Laurie L. and Paul V. Zimba. 2002. Spatial and temporal patterns of phytoplankton in Florida Bay: utility of algal accessory pigments and remote sensing to assess bloom dynamics. *In* James W. Porter and Karen G. Porter (editors). *The Everglades, Florida Bay, and Coral Reefs of the Florida Keys: an Ecosystem Source Book*. CRC Press, Boca Raton, Florida.

501. Risi, J.A., H.R. Wanless, L.P. Tedesco, and S. Gelsanliter. 1995. Catastrophic sedimentation from Hurricane Andrew along the southwest Florida coast. *J. Coastal Res.* 21: 82-102.

502. Robblee, M.B., T.R. Barber, P.R. Carlson, Jr., M.J. Durako, J.W. Fourqurean, L.K. Muehlstein, D. Porter, L.A. Yarbro, R.T. Zieman, J.C. Zieman. 1991. Mass mortality of the tropical seagrass *Thalassia testudinum* in Florida Bay (USA). *Marine Ecology Progress Series* 71: 297-299.

503. Robbin, Daniel M. 1984. A new Holocene sea level curve for the upper Florida Keys and Florida reef tract. *In* P.J. Gleason (editor). *Environments of South Florida: Present and Past II*. Miami Geological Society, Miami, Florida.

504. Robertson, William B., Jr. 1954. Everglades fire — past, present and future. *Everglades Natural History* 2(1): 10-16.

505. Robertson, William B., Jr. 1989. *Everglades: the Park Story* (revised edition). Florida National Parks and Monuments Association, Inc., Homestead, Florida.

506. Robertson, William B., Jr. and Peter C. Frederick. 1994. The faunal chapters: contexts, synthesis, and departures. *In* Steven M. Davis and John C. Ogden (editors). *Everglades: the Ecosystem and Its Restoration*. St. Lucie Press, Delray Beach, Florida.

507. Robertson, William B., Jr. and James A Kushlan. 1984. The southern Florida avifauna. *In* P.J. Gleason (editor). *Environments of South Florida: Present and Past II*. Miami Geological Society, Miami, Florida.

508. Robins, C. Richard, Reeve M. Bailey, Carl E. Bond, James R. Brooker, Ernest A. Lachner, Robert N. Lea, and W.B. Scott. 1991. *Common and Scientific Names of Fishes from the United States and Canada* (fifth edition). American Fisheries Society, Special Publication 20, Bethesda, Maryland.

509. Robins, C. Richard and G. Carleton Ray. 1986. *A Field Guide to Atlantic Coast Fishes of North America*. Houghton Mifflin, Boston, Massachusetts.

510. Rodgers, James A., Jr., Herbert W. Kale II, and Henry T. Smith. 1996. *Volume V. Birds. In* Ray E. Ashton, Jr. (series editor). *Rare and Endangered Biota of Florida*. University Presses of Florida, Gainesville, Florida.

511. Roman, Charles T., Nicholas G. Aumen, Joel C. Trexler, Robert J. Fennema, William F. Loftus, and Michael A. Soukup. 1994. Hurricane Andrew's impact on freshwater resources. *Bioscience* 44(4): 247-255.

512. Rosendahl, P.C., and P.W. Rose. 1982. Freshwater flow rates and distribution within the Everglades marsh. *In* R.D. Cross and D.L. Williams (editors). *Proceedings of the National Symposium on Freshwater Inflow to Estuaries, Coastal Ecosystems Project*. U.S. Fish and Wildlife Service, Washington, D.C.

513. Ross, Charles A. (editor). 1989. *Crocodiles and Alligators*. Facts on File, Inc., New York, New York.

514. Ross, Michael S., Evelyn E. Gaiser, John F. Meeder, and Matthew T. Lewin. 2002. Multi-taxon analysis of the "white zone," a common ecotonal feature of South Florida coastal wetlands. *In* James W. Porter and Karen G. Porter (editors). *The Everglades, Florida Bay, and Coral Reefs of the Florida Keys: an Ecosystem Source Book*. CRC Press, Boca Raton, Florida.

515. Rothra, Elizabeth Ogren. 1995. *Florida's Pioneer Naturalist: the Life of Charles Torrey Simpson*. University Press of Florida, Gainesville, Florida.

516. Runde, D.E. 1991. Trends in wading bird nesting populations in Florida: 1976-1978 vs. 1986-1989 (final performance report). Florida Game and Fresh Water Fish Commission Nongame Wildlife Program, 90p.

517. Rutherford, Edward S., James T. Tilmant, Edith B. Thue, and Thomas W. Schmidt. 1989. Fishery harvest and population dynamics of spotted seatrout, *Cynoscion nebulosus*, in Florida bay and adjacent waters. *Bulletin of Marine Science* 44(1): 108-125.

518. Sammarco, P. W. and M. L. Heron. 1994. The bio-physics of marine larval dispersal. Washington, D.C., American Geophysical Union.

519. Schardt, Jeffrey D. 1997. Maintenance control. *In* Daniel Simberloff, Don C. Schmitz, and Tom C. Brown (editors). *Strangers in Paradise: Impact and Management of Nonindigenous Species in Florida*. Island Press, Washington, D.C.

520. Schelske, C.L., F.J. Aldridge, and W.F. Kenney. 1999. Assessing nutrient limitation and trophic state in Florida lakes. *In* K.R. Reddy, G.A. O'Connor, and C.L. Schelske (editors). *Phosphorus Biogeochemistry in Subtropical Ecosystems*. Lewis Publishers, Boca Raton, Florida.

521. Schmitz, Don C., Daniel Simberloff, Ronald H. Hofstetter, William Haller, and David Sutton. 1997. The ecological impact of nonindigenous plants. *In* Daniel Simberloff, Don C. Schmitz, and Tom C. Brown (editors). *Strangers in Paradise: Impact and Management of Nonindigenous Species in Florida*. Island Press, Washington, D.C.

522. Schemnitz, S.D. and J.L. Schortemeyer. 1973. The impact of half tracks and airboats on the Florida Everglades environment. *In* Proceedings of the 1973 Snowmobile and Off-the-Road-Vehicle Research Symposium. Technical Report No. 9. Michigan State University, East Lansing, Michigan.

523. Schmidt, Walter. 1997. Geomorphology and physiography of Florida. *In* Anthony F. Randazzo and Douglas S. Jones (editors). *The Geology of Florida*. University Press of Florida, Gainesville, Florida.

524. Schneider, Douglas L. 1973. The limnology of a meromictic, sinkhole lake in Southern Florida. Ph.D. dissertation, Department of Biology, University of Miami, Coral Gable, Floridas.

525. Scott, Gerald P., Michael R. Dewey, Larry J. Hansen, Ralph E. Owen, and Edward S. Rutherford. 1989. How many mullet are there in Florida Bay? *Bulletin of Marine Science* 44(1): 89-107.

526. Scott, Thomas M. 1997. Miocene to Holocene history of Florida. *In* Anthony F. Randazzo and Douglas S. Jones (editors). *The Geology of Florida*. University Press of Florida, Gainesville, Florida.

527. Scurlock, J. Paul. 1987. *Native Trees and Shrubs of the Florida Keys*. Laurel & Herbert, Inc., Lower Sugerloaf Key, Florida.

528. Sears, William H. 1974. Archeological perspectives on prehistoric environment in the Okeechobee basin savannah. *In* P.J. Gleason (editor). *Environments of South Florida: Present and Past*. Memoir 2: Miami Geological Society, Miami, Florida.

529. Sepulveda, M.S. 1997. Mercury contamination of great egret (*Ardea albus*) nestlings from southern Florida and its effects on health and survival. Masters thesis, University of Florida.

530. Sepulveda, M.S., P.C. Frederick, M.G. Spalding, and G.E. Jr. Williams. 1999. Mercury contamination in free-ranging great egret (*Ardea albus*) nestlings from southern Florida. *Environmental Contamination and Toxicology* 18(5): 985-992.

531. Sepulveda, M.S., G.E. Williams, P.C. Frederick, and M.S. Spalding. 1999. Effects of mercury on health and first year survival of free-ranging great egrets (*Ardea albus*) from southern Florida. *Archives of Environmental Contamination and Toxicology* 37: 369-376.

532. Sibley, David Allen. National Audubon Society The Sibley Guide to Birds. Alfred A. Knopf, New York.

533. Simberloff, Daniel. 1997. The biology of invasions. *In* Daniel Simberloff, Don C. Schmitz, and Tom C. Brown (editors). *Strangers in Paradise: Impact and Management of Nonindigenous Species in Florida*. Island Press, Washington, D.C.

534. Simberloff, Daniel. 1997. Eradication. *In* Daniel Simberloff, Don C. Schmitz, and Tom C. Brown (editors). *Strangers in Paradise: Impact and Management of Nonindigenous Species in Florida*. Island Press, Washington, D.C.

535. Simberloff, Daniel, Don. C. Schmitz, and Tom C. Brown. 1997. Why we should care. *In* Daniel Simberloff, Don C. Schmitz, and Tom C. Brown (editors). *Strangers in Paradise: Impact and Management of Nonindigenous Species in Florida*. Island Press, Washington, D.C.

536. Simberloff, Daniel, Don C. Schmitz, and Tom C. Brown (editors). 1997. *Strangers in Paradise: Impact and Management of Nonindigenous Species in Florida*. Island Press, Washington, D.C.

537. Simmons, Glen and Laura Ogden. 1998. *Gladesmen*. University Press of Florida, Gainesville.

538. Sklar, Fred H. and Arnold van der Valk (editors). 2002. *Tree Islands of the Everglades*. Kluwer Academic Publishers, Boston, Massachusetts.

539. Sklar, Fred H. and Arnold van der Valk. 2002. Tree islands of the Everglades: an overview. *In* Fred H. Sklar and Arnold van der Valk (editors). *Tree Islands of the Everglades*. Kluwer Academic Publishers, Boston, Massachusetts.

540. Sklar, Fred, Christopher McVoy, Randy VanZee, Dale E. Gawlik, Ken Tarbonton, David Rudnick, ShiLi Miao, and Tom Armentano. 2002. The effects of altered hydrology on the ecology of the Everglades. *In* James W. Porter and Karen G. Porter (editors). *The Everglades, Florida Bay, and Coral Reefs of the Florida Keys: an Ecosystem Source Book*. CRC Press, Boca Raton, Florida.

541. Slater, Harold H., William J. Platt, David B. Baker, and Hester A. Johnson. 1995. Effects of Hurricane Andrew on damage and mortality of trees in subtropical hardwood hammocks of Long Pine Key, Everglades National Park, Florida, USA. *Journal of Coastal Research* 21: 197-207.

542. Small, John. K. 1914. Exploration in the Everglades and on the Florida Keys. *Journal of the New York Botanical Garden* 15: 69-79.

543. Small, John. K. 1916. Royal Palm Hammock. *Journal of the New York Botanical Garden* 17: 165-172.

544. Small, John. K. 1916. A cruise to the Cape Sable region of Florida. *Journal of the New York Botanical Garden* 17: 189-202.

545. Small, John. K. 1918. A winter collecting trip in Florida. *Journal of the New York Botanical Garden* 19: 69-77.

546. Small, John. K. 1918. Botanical exploration in Florida in 1917. *Journal of the New York Botanical Garden* 19: 279-290.

547. Small, John. K. 1919. Coastwise dunes and lagoons: a record of botanical exploration in Florida in the spring of 1918. *Journal of the New York Botanical Garden* 20: 191-207.

548. Small, John. K. 1920. Of grottoes and ancient dunes: a record of exploration in Florida in December, 1918. *Journal of the New York Botanical Garden* 21: 45-54.

549. Small, John. K. 1921. Old trails and new discoveries: a record of exploration in Florida in the spring of 1919. *Journal of the New York Botanical Garden* 22: 25-40 and 49-64.

550. Small, John. K. 1921. Historic Trails, by land and water: a record of exploration in Florida in December 1919. *Journal of the New York Botanical Garden* 22: 193-222.

551. Small, John. K. 1924. The land where spring meets autumn. *Journal of the New York Botanical Garden* 25: 53-94.

552. Small, John. K. 1928. The Royal Palm — Roystonea regia. *Journal of the New York Botanical Garden* 29: 1-9.

553. Small, John. K. 1933. *Manual of the Southeastern Flora*. Published by the author, Press of the Science Press Printing Company. Lancaster, Pennsylvania.

554. Small, John. K. 1933. An Everglade Cypress Swamp. *Journal of the New York Botanical Garden* 34: 261-267.

555. Smith, Douglas L. 1982. Review of the tectonic history of the Florida basement. *Tectonophysics* 88: 1-22.

556. Smith, Douglas L. and Kenneth M. Lord. 1997. Tectonic evolution and geophysics of the Florida basement. *In* Anthony F. Randazzo and Douglas S. Jones (editors). *The Geology of Florida*. University Press of Florida, Gainesville, Florida.

557. Smith, J.P. and M.W. Collopy. 1995. Colony turnover, nest success and productivity, and causes of nest failure among wading birds (Ciconiiformes) at lake Okeechobee, Florida (1989-1992). *In* N.G. Aumen and R.G. Wetzel (editors). Ecological studies on the littoral and pelagic systems of lake Okeechobee, Florida (USA). *Archiv für Hydrobiologie*, special issue: Advances in Limnology, 45: 1-356.

558. Smith, J.P., J.R. Richardson, and M.W. Collopy. 1995. Foraging habitat selection among wading birds (Ciconiiformes) at Lake Okeechobee, Florida, in relation to hydrology and vegetative cover. *In* N.G. Aumen and R.G. Wetzel (editors). Ecological studies on the littoral and pelagic systems of lake Okeechobee, Florida (USA). *Archiv für Hydrobiologie*, special issue: Advances in Limnology, 45: 1-356.

559. Smith, Ned P. and Patrick A. Pitts. 2002. Regional-scale and long-term transport patterns in the Florida Keys. *In* James W. Porter and Karen G. Porter (editors). *The Everglades, Florida Bay, and Coral Reefs of the Florida Keys: an Ecosystem Source Book*. CRC Press, Boca Raton, Florida.

560. Smith, Thomas J., III, J. Harold Hudson, Michael B. Robblee, George V. N. Powell, and Peter J. Isdale. 1989. Freshwater flow from the Everglades to Florida Bay: a historical reconstruction based on fluorescent banding in the coral *Solenastrea bournoni*. *Bulletin of Marine Science* 44(1): 274-282.

561. Smith, Thomas J., III, Michael B. Robblee, Harold R. Wanless, and Thomas W. Doyle. 1994. Mangroves, hurricanes, and lightning strikes. *Bioscience* 44(4): 256-262.

562. Smith, Tommy R. and Oron L. Bass, Jr. 1994. Landscape, white-tailed deer, and the distribution of Florida panthers in the Everglades. *In* Steven M. Davis and John C. Ogden (editors). *Everglades: the Ecosystem and Its Restoration*. St. Lucie Press, Delray Beach, Florida.

563. Snyder, George H. 1994. Soils of the EAA. *In* A.B. Bottcher and F.T. Izuno (editors). *Everglades Agricultural Area (EAA): Water, Soil, Crop, and Environmental Management*. University of Florida Press, Gainesville.

564. Snyder, G.H. and J.M. Davidson. 1994. Everglades agriculture: past, present, and future. *In* Steven M. Davis and John C. Ogden (editors). *Everglades: the Ecosystem and Its Restoration*. St. Lucie Press, Delray Beach.

565. Snyder, James R., Alan Herndon, and William B. Robertson, Jr. 1990. South Florida rockland. *In* Ronald L. Myers and John J. Ewel (editors). *Ecosystems of Florida*. University of Central Florida Press, Orlando.

566. Sogard, Susan M., George V.N. Powell, and Jeff G. Holmquist. 1989. Spatial distribution and trends in abundance of fishes residing in seagrass meadows on Florida Bay mudbanks. *Bulletin of Marine Science* 44(1): 179-199.

567. South Florida Water Management District. 1997. Surface Water Improvement and Management (SWIM) Plan — Update to Lake Okeechobee. South Florida Water Management District, West Palm Beach.

568. South Florida Water Management District and Florida Department of Environmental Protection. 2004. *Everglades Consolidated Report*. South Florida Water Management District, West Palm Beach (note: the compact disk of this report also contains the reports for 2000, 2001, 2002 and 2003, and the Everglades Forever Act).

569. Spalding, M.G., R.D. Bjork, G. V.N. Powell, and S.F. Sundlof. 1994. Mercury and cause of death in great white herons. *Journal of Wildlife Management* 58(4): 735-739.

570. Spalding, M.G., P.C. Frederick, H.C. McGill, S.N. Bouton, and L.R. McDowell. 2000. Methylmercury accumulation in tissues and its effects on growth and appetite in captive great egrets. *Journal of Wildlife Diseases* 36: 411-422.

571. Spalding, M.G., P.C. Frederick, H.C. McGill, S.N. Bouton, L.J. Richey, I.M. Schumacher, S.G.M. Blackmore, and J. Harrison. 2000. Histologic, neurologic, and immunologic effects of methylmercury in captive great egrets. *Journal of Wildlife Diseases* 36: 423-435.

572. Safford, W.E. 1919. Natural history of Paradise Key and the nearby Everglades of Florida. In The Annual Report of the Board of Regents of the Institution for the Year Ending June 30, 1917. Smithsonian Institution, Washington, D.C.

573. Steinman, A.D., K.E. Havens, N.G. Aumen, R.T. James, Kang-Ren Jin, J. Zang, and B.H. Rosen. 1999. Phosphorus in Lake Okeechobee: sources, links, and strategies. *In* K.R. Reddy, G.A. O'Connor, and C.L. Schelske (editors). *Phosphorus Biogeochemistry in Subtropical Ecosystems*. Lewis Publishers, Boca Raton, Florida.

574. Steinman, Alan D., Karl E. Havens, Hunter J. Carrick, and Randy VanZee. 2002. The past, present, and future hydrology and ecology of Lake Okeechobee and its watersheds. *In* James W. Porter and Karen G. Porter (editors). *The Everglades, Florida Bay, and Coral Reefs of the Florida Keys: an Ecosystem Source Book*. CRC Press, Boca Raton, Florida.

575. Stephens, John C. 1984. Subsidence of organic soils in the Florida Everglades—a review and update. *In* P.J. Gleason (editor). *Environments of South Florida: Present and Past II*. Miami Geological Society, Miami, Florida.

576. Stevens, Amanda J., Zachariah C. Welch, Philip C. Darby, and H. Franklin Percival. 2002. Temperature effects on Florida applesnail activity: implications for snail kite foraging success and distribution. *Wildlife Society Bulletin* 30(1): 75-81.

577. Stevenson, George. B. 1992. *Trees of Everglades National Park and the Florida Keys*. Florida National Parks & Monuments Association, Inc., Homestead, Florida.

578. Stevenson, Henry M and Bruce H. Anderson. 1994. *The Birdlife of Florida*. University Press of Florida, Gainesville, Florida.

579. Steward, Kerry K. 1984. Physiological, edaphic and environmental characteristics of Everglades sawgrass communities. *In* P.J. Gleason (editor). *Environments of South Florida: Present and Past II*. Miami Geological Society, Miami, Florida.

580. Stober, Q.J., K. Thornton, R. Jones, J. Richards, C. Ivey, R. Welch, M Madden, J. Trexler, E. Gaiser, D. Scheidt, and S. Rathbun. 2001. South Florida Ecosystem Assessment: Phase I/II – (summary) Everglades Stressor Interactions: Hydropatterns, Eutrophication, Habitat alteration, and Mercury Contamination. Publication EPA 904-R-01-002, U.S. Environmental Protection Agency, Atlanta.

581. Stone, Peter A., Patrick J. Gleason, and Gail L. Chmura. 2002. Bayhead tree islands on deep peats of the northeastern Everglades. *In* Fred H. Sklar and Arnold van der Valk (editors). *Tree Islands of the Everglades*. Kluwer Academic Publishers, Boston, Massachusetts.

582. Strong, Allan M. and G. Thomas Bancroft. 1994. Patterns of deforestation and fragmentation of mangrove and deciduous seasonal forests in the upper Florida Keys. *Bulletin of Marine Science* 54(3): 795-804.

583. Sturrock, David. 1959. *Fruits for Southern Florida*. Southern Printing Co., Inc., Stuart, Florida.

584. Swift, David R. 1984. Periphyton and water quality relationships in the Everglades water conservation areas. *In* P.J. Gleason (editor). *Environments of South Florida: Present and Past II*. Miami Geological Society, Miami, Florida.

585. Sykes, Paul W., Jr., and Gloria S. Hunter. 1978. Bird use of flooded agricultural fields during summer and early fall and some recommendations for management. *Florida Field Naturalist*, 6(2): 36-43.

586. Tabb, Durbin C., Taylor R. Alexander, Terence M. Thomas, and Nancy Maynard. 1967. The physical, biological and geological character of the area south of C-111 canal in extreme southeastern Everglades National Park, Florida. Report to the U.S. National Park Service, ML 67103.

587. Tabb, Durbin C. and Martin A. Roessler. 1989. History of studies on juvenile fishes of coastal waters of Everglades National Park. *Bulletin of Marine Science* 44(1): 23-34.

588. Tauvers, Peter R. and William R. Muehlberger. 1987. Is the Brunswick magnetic anomaly really the alleghanian suture? *Tectonics* 6(3): 331-342.

589. Taylor, William Randolph. 1960. *Marine Algae of the Eastern Tropical and Subtropical Coasts of the Americas*. The University of Michigan Press, Ann Arbor, Michigan.

590. Tebeau, Charlton W. 1968. *Man in the Everglades*. University of Miami Press, Coral Gables, Florida.

591. Tebeau, Charlton W. 1984. Exploration and early descriptions of the Everglades, Lake Okeechobee, and the Kissimmee River. *In* P.J. Gleason (editor). *Environments of South Florida: Present and Past II*. Miami Geological Society, Miami, Florida.

592. Tedesco, L.P., H.R. Wanless, L.A. Scusa, J.A. Risi, and S. Gelsanliter. 1995. Impact of Hurricane Andrew on South Florida's sandy coastlines. *Journal of Coastal Research* 21: 59-82.

593. Thompson, Fred G. 1999. An identification manual for the freshwater snails of Florida. *Walkerana* 10(23): 1-96.

594. Tilmant, James T. 1989. A history and an overview of recent trends in fisheries of Florida Bay. *Bulletin of Marine Science* 44(1): 3-22.

595. Tilmant, James T., Richard W. Curry, Ronald Jones, Alina Szmant, Joseph C. Zieman, Mark Flora, Michael B. Robblee, Dewitt Smith, R.W. Snow, and Harold Wanless. 1994. Hurricane Andrew's effects on marine resources. *Bioscience* 44(4): 230-237.

596. Tilmant, James T., Edward S. Rutherford, and Edith B. Thue. 1989. Fishery harvest and population dynamics of red drum (*Sciaenops ocellatus*) from Florida Bay and adjacent waters. *Bulletin of Marine Science* 44(1): 126-138.

597. Timmer, C. Elroy and Stanley S. Teague. 1991. Melaleuca eradication program: assessment of methodology and efficacy. *In* Ted D. Center, Robert F. Doren, Ronald H. Hofstetter, Ronald L. Myers, and Louis D. Whiteaker (editors). Proceedings of the Symposium on Exotic Pest Plants. U.S. Department of the Interior, National Park Service, Washington, D.C.

598. Tomlinson, P.B. 1986. *The Botany of Mangroves*. Cambridge University Press, London.

599. Tomlinson, P.B. 2001. *The Biology of Trees Native to Tropical Florida,* second edition. Harvard Printing and Publications Services, Allston, Massachusetts.

600. Toops, Connie M. 1979. *The Alligator: Monarch of the Everglades*. The Everglades Natural History Association, Inc., Homestead, Florida.

601. Toth, Louis A. 1987. Effects of hydrologic regimes on lifetime production and nutrient dynamics of sawgrass. Technical Publication #87-6, South Florida Water Management District, West Palm Beach.

602. Toth, Louis A. 1993. The ecological basis of the Kissimmee River restoration plan. *Florida Scientist* 56(1): 25-51.

603. Tracy, B.A. and W.A. Brandon. 1995. A combined wave and surge hindcast for the coast of Florida during Hurricane Andrew. *Journal of Coastal Research* 21: 49-58.

604. Trexler, Joel C., William F. Loftus, Frank Jordan, John H. Chick, Karen L. Kandl, Thomas C. McElroy, and Oron L. Bass, Jr. 2002. Ecological scale and its implications for freshwater fishes in the Florida Everglades. *In* James W. Porter and Karen G. Porter (editors). *The Everglades, Florida Bay, and Coral Reefs of the Florida Keys: an Ecosystem Source Book*. CRC Press, Boca Raton, Florida.

605. Trexler, Joel C., William F. Loftus, Frank Jordan, Jerome J. Lorenz, John H. Chick and Robert M. Kobza. 2000. Emperical assessment of fish introductions in a subtropical wetland: an evaluation of contrasting views. *Biological Invasions* 2: 265-277.

606. Turgeon, Donna D., Arthur E. Bogan, Eugene V. Coan, William K. Emerson, William G. Lyon, William L. Pratt, Clyde F.E. Roper, Amelie Scheltema, Fred G. Thompson, and James D. Williams. 1988. *Common and Scientific Names of Aquatic Invertebrates from the United States and Canada: Mollusks*. American Fisheries Society Special Publication 16, Bethesda, Maryland.

607. Turnbull, R.E., F.A. Johnson, and D.H. Brakhage. 1989. Status, distribution, and foods of fulvous whistling-ducks in South Florida. *Journal of Wildlife Management* 53(4): 1046-1051.

608. Turner, Richard L. 1998. Effects of submergence on embryonic survival and development rate of the Florida applesnail, *Pomacea paludosa*: implications for egg predation and marsh management. *Florida Scientist* 61(2): 118-129.

609. Ugarte, Christina A. and Kenneth G. Rice. 2003. Abundance and diet of *Rana grylio* across South Florida wetlands (abstract). Joint Conference on the Science and Restoration of the Greater Everglades and Florida Bay Ecosystem, April 13-18, 2003, Palm Harbor, Florida.

610. Upchurch, Sam B. and Anthony F. Randazzo. 1997. Environmental geology of Florida. *In* Anthony F. Randazzo and Douglas S. Jones (editors). *The Geology of Florida*. University Press of Florida, Gainesville, Florida.

611. U.S. Army Corps of Engineers. c. 1981. An Environmental Assessment Related to the Construction, Use and Maintenance of the Okeechobee Waterway, Florida. Technical report, Jacksonville District, Contract No. DACW17-80-0048.

612. U.S. Army Corps of Engineers and South Florida Water Management District. 1999. Central and Southern Florida Project Comprehensive Review Study Final Integrated Feasibility Report and programmatic Environmental Impact Statement. Final report, appendices, and ATLSS.

613. U.S. Environmental Protection Agency. 2000. Total maximum daily load (TMDL) development for total phorphorus, Lake Okeechobee, Florida. USEPA, Region 4.

614. U.S. Fish and Wildlife Service, Southeast Region. 1999. *South Florida Multi-Species Recovery Plan*.

615. U.S. Fish and Wildlife Service. 2002. Landscape Conservation Strategy for the Florida Panther in South Florida. Vero Beach, Florida. 191 pp.

616. U.S. Fish and Wildlife Service. 2003. Wood Stork Report: a newsletter dedicated to sharing information about the wood stork 2(1): 1-15.

617. U.S. Fish and Wildlife Service, South Florida Ecosystem Restoration Office and U.S. Department of the Interior, National Park Service, Everglades National Park. 1997. Balancing on the Brink: the Everglades and the Cape Sable seaside sparrow.

618. U.S. Senate. 1911. Everglades of Florida: acts, reports, and other papers, state and national, relating to the Everglades of the State of Florida and their reclamation. 62nd Congress, 1st session, Senate Document No. 89. Government Printing Office, Washington, D.C.

619. VanArman, Joel. 1984. South Florida's estuaries. *In* P.J. Gleason (editor). *Environments of South Florida: Present and Past II*. Miami Geological Society, Miami, Florida.

620. VanArman, Peggy G. 2003. Biology and ecology of epigean crayfish that inhabit Everglades environments, *Procambarus alleni* (Faxon) and *Procambarus fallax* (Hagen). Ph.D. dissertation, Nova Southeastern University Oceanographic Center, Dania Beach, Florida.

621. van der Valk, Arnold, and Fred Sklar. 2002. What we know and should know about tree islands. *In* Fred H. Sklar and Arnold van der Valk (editors). *Tree Islands of the Everglades*. Kluwer Academic Publishers, Boston, Massachusetts.

622. Vignoles, Charles. 1823. *Observations upon the Floridas.* (Citation in text is from Will, 1977)

623. Wade, Dale, John Ewel, and Ronald H\ofstetter. 1980. *Fire in South Florida Ecosystems.* U.S. Dept of Agriculture, Forest Service Technical Report SE-17. Southeastern Forest Experiment Station, Asheville, North Carolina.

624. Walker, W.W., Jr. 1999. Long-term water quality trends in the Everglades. *In* K.R. Reddy, G.A. O'Connor, and C.L. Schelske (editors). *Phosphorus Biogeochemistry in Subtropical Ecosystems.* Lewis Publishers, Boca Raton, Florida.

625. Waller, Bradley G. 1988. Hydrologic effects of drainage and water management on the wetland ecosystem of South Florida (U.S.A.). *In* Miguel Chávez Miguel (editor). *Ecología y Conservación del Delta de los ríos Usumacinta y Grijalva* (Memorias). Instituto Nacional de Investigacíon sobre Recursos Bióticos (INIREB) – División Regional Tabasco, pages 201-215. (see Chavez, 1988)

626. Walsh, Gerald E. 1974. Mangroves: a review. *In* Robert J. Reimold and William H Queen (editors). *Ecology of Holophytes.* Academic Press, Inc., New York, New York.

627. Walters, Carl J. and Lance H. Gunderson. 1994. A screening of water policy alternatives for ecological restoration in the Everglades. *In* Steven M. Davis and John C. Ogden (editors). *Everglades: the Ecosystem and Its Restoration.* St. Lucie Press, Delray Beach, Florida.

628. Walters, Carl, Lance Gunderson, and C.S. Holling. 1992. Experimental polices for water management in the Everglades. *Ecological Applications* 2(2): 189-202.

629. Walters, Jeffrey R. (chair), Steven R. Beissinger, John W. Fitzpatrick, Russell Greenberg, James D. Nichols, H. Ronald Pulliam, and David W. Winkler. 1999. Cape Sable seaside sparrow panel review: final report. American Orithologists' Union panel report to the Science Coordination Team, South Florida Ecosystem Restoration Task Force.

630. Wanless, Harold R. 1984. Mangrove sedimentation in geologic perspective. *In* P.J. Gleason (editor). *Environments of South Florida: Present and Past II.* Miami Geological Society, Miami, Florida.

631. Wanless, Harold R. 1989. The inundation of our coastlines (past, present and future with a focus on South Florida). *Sea Frontiers* 35(5): 264-271.

632. Wanless, Harold R. Personal communications, 2003. Department of Geological Sciences, Univ. of Miami, Coral Gables, Florida.

633. Wanless, Harold R., P. Oleck, L.P. Tedesco, B.E. Hall. 2000. Next 100 years of evolution of the greater Everglades ecosystem in response to anticipated sea level rise: nature, extent and causes. Greater Everglades Ecosystem Restoration Science Conference, Naples, Fl., December, 2000, pp. 174-176.

634. Wanless, Harold R. and Matthew G. Tagett. 1989. Origin, growth and evolution of carbonate mudbanks in Florida Bay. *Bulletin of Marine Science* 44(1): 490-514.

635. Wanless, Harold R., Randall W. Parkinson, and Lenore P. Tedesco. 1994. Sea level control on stability of Everglades wetlands. *In* Steven M. Davis and John C. Ogden (editors). *Everglades: the Ecosystem and Its Restoration.* St. Lucie Press, Delray Beach, Florida.

636. Wanless, H.R., L.P. Tedesco, J.A. Risi, B.G. Bischof, and S. Gelsanliter. 1995. The Role of Storm Processes on the Growth and Evolution of Coastal and Shallow Marine Sedimentary Environments in South Florida, Field Trip Guide, The 1st SEPM Congress on Sedimentary Geology, St. Petersburg, Florida, 179p.

637. Ward, Daniel B. (editor). 1979. *Volume Five: Plants. In* Peter C.H. Pritchard (series editor) *Rare and Endangered Biota of Florida.* University Presses of Florida, Gainesville, Florida.

638. Ward, Daniel B. 2003. Introduction to the question: native or not? Number 1 in the series, native or not: studies of problematic species. *The Palmetto.* Quarterly Magazine of the Florida Native Plant Society 22(2): 7-9.

639. Ware, Forrest J., Homer Royals, and Ted Lange. 1990. Mercury contamination in Florida largemouth bass. *Proceedings of the Annual Conference of the Southeastern Association of Fish and Wildlife Agencies* 44: 5-12.

640. Warren, Gary L. 1997. Nonindigenous freshwater invertebrates. *In* Daniel Simberloff, Don C. Schmitz, and Tom C. Brown (editors). *Strangers in Paradise: Impact and Management of Nonindigenous Species in Florida.* Island Press, Washington, D.C.

641. Warren, G.L., M.J. Vogel, and D.D. Fox. 1995. Trophic and distributional dynamics of Lake Okeechobee sublittoral benthic invertebrate communities. *In* N.G. Aumen and R.G. Wetzel (editors). Ecological studies on the littoral and pelagic systems of lake Okeechobee, Florida (USA). *Archiv für Hydrobiologie,* special issue: Advances in Limnology, 45: 1-356.

642. Webb, S. David. 1990. Historical biogeography. *In* Ronald L. Myers and John J. Ewel (editors). *Ecosystems of Florida*. University of Central Florida Press, Orlando, Florida.

643. Weisskoff, Richard. 2000. Missing pieces in ecosystem restoration: the case from the Florida Everglades. *Economic Systems Research* 12(3): 271-303.

644. Welch, Roy, Marguerite Madden, and Robert Doren. 2002. Maps and GIS databases for environmental studies of the Everglades. *In* James W. Porter and Karen G. Porter (editors). *The Everglades, Florida Bay, and Coral Reefs of the Florida Keys: an Ecosystem Source Book*. CRC Press, Boca Raton.

645. Wetzel, Paul R. 2002. Tree island ecosystems of the world. *In* Fred H. Sklar and Arnold van der Valk (editors). *Tree Islands of the Everglades*. Kluwer Academic Publishers, Boston, Massachusetts.

646. Wetzel, Paul R. 2002. Analysis of tree island communities. *In* Fred H. Sklar and Arnold van der Valk (editors). *Tree Islands of the Everglades*. Kluwer Academic Publishers, Boston, Massachusetts.

647. Whitaker, John O., Jr. 1996. *The Audubon Society Field Guide to North American Mammals*. Second edition. Alfred A. Knopf, New York.

648. White, Peter S. 1994. Synthesis: Vegetation pattern and process in the Everglades ecosystem. *In* Steven M. Davis and John C. Ogden (editors). *Everglades: the Ecosystem and Its Restoration*. St. Lucie Press, Delray Beach, Florida.

649. Wilkes, Sherwood C. (principal investigator). 1989. A brief overview of the South New River Swamp. Report prepared for the Environmental Coalition of Florida by the Discovery Center Museum, Ft. Lauderdale, Florida. For more information, contact the Biological Resources Division, Department of Planning and Environmental Protection, Broward County Government.

650. Will, L. E. 1977. *A Cracker History of Lake Okeechobee*. The Glades Historical Society, Belle Glade, Florida. (Note: originally published in 1964 by The Great Outdoors Publishing Co., the 1977 republication contains important appendices that enhance the credibility of this important historical reference)

651. Willard, D. A. 2003. Tree islands of the Everglades – a disappearing resource. U.S. Geological Survey Open File Report 03-26.

652. Willard, Debra A., Charles W. Holmes, Michael S. Korvela, Daniel Manson, James B. Murray, William H. Orem, and D. Timothy Towles. 2002. Paleoecological insights on fixed tree island development in the Florida Everglades: I. environmental controls. *In* Fred H. Sklar and Arnold van der Valk (editors). *Tree Islands of the Everglades*. Kluwer Academic Publishers, Boston, Massachusetts.

653. Williams, Austin B., Lawrence G. Abele, Darryl L. Felder, Horton H. Hobbs, Jr., Raymond B. Manning, Patsy A. McLaughlin, and Isabel Pérez Farfante. 1989. *Common and Scientific Names of Aquatic Invertebrates from the United States and Canada: Decapod Crustaceans*. American Fisheries Society Special Publication 16, Bethesda, Maryland.

654. Williams, John M. and Iver W. Duedall. 2002. *Florida Hurricanes and Tropical Storms, 1871-2001*. University Press of Florida, Gainesville, Florida.

655. Williams, Kimberlyn, Katherine C. Ewel, Richard. P. Stumpf, Francis E. Putz, and Thomas W. Workman. 1999. Sea-level rise and coastal forest retreat on the west coast of Florida, USA. *Ecology* 80(6): 2045-2063.

656. Willoughby, Hugh L. 1898. *Across the Everglades, a Canoe Journey of Exploration*. Fifth edition. Florida Classics Library Edition (1992), Port Salerno, Florida.

657. Winsberg, Morton D. 1990. *Florida Weather*. University of Central Florida Press, Orlando.

658. Wolf, Paul R. and Charles D. Ghilani. 2001. *Elementary Surveying: An Introduction to Geomatics*, 10th edition. Prentice Hall.

659. Wood, Ferguson and Nancy G. Maynard. 1974. Ecology of the micro-algae of the Florida Everglades. *In* P.J. Gleason (editor). *Environments of South Florida: Present and Past*. Memoir 2: Miami Geological Society, Miami.

660. Wright, Myron H., Jr., Ray O. Green, Jr., and Norman D. Reed. 1972. The swallow-tailed kite: graceful aerialist of the Everglades. *National Geographic* 142(4): 496-505.

661. Wunderlin, Richard P. 1998. *Guide to the Vascular Plants of Florida*. University Press of Florida, Gainesville. Note: this book is being updated. Current information can be found at: *www.plantatlas.usf.edu*

662. Wyss, Alexander James. 1996. Nesting ecology of Fulvous whistling-ducks in the Everglades Agricultural Area of southern Florida. Masters thesis, Auburn University, Auburn, Alabama.

663. Yates, S.A. 1974. An autecological study of sawgrass, *Cladium jamaicense*, in southern Florida. Masters thesis, Department of Biology, University of Miami, Coral Gables, Florida.

664. Yegang, Wu, Ken Rutchey, Weihe Guan, Les Vilchek, and Fred H. Sklar. 2002. Spatial simulations of tree islands for Everglades restoration. *In* Fred H. Sklar and Arnold van der Valk (editors). *Tree Islands of the Everglades*. Kluwer Academic Publishers, Boston, Massachusetts.

665. Yeung, C., M. M. Criales, and T. N. Lee. 2000. Unusual larval abundance of *Scyllarides nodifer* and *Albunea* sp. in the Florida Keys during the intrusion of low salinity Mississippi flood water in September, 1993. *Journal of Geophysical Research* 105(C12): 28741-28758.

666. Yeung, C., D. L. Jones, M. M. Criales, T. L. Jackson, and W. J. Richards. 2001. Countercurrent flow and the influx of spiny lobster *Panulirus argus* postlarvae into Florida Bay: influence of eddy transport. *Mar. Freshwater Res.* 52: 1217-1232.

667. Yeung, C. and T. N. Lee. 2002. Larval transport and retention of the spiny lobster, *Panulirus argus*, in the coastal zone of the Florida Keys, U.S.A. *Fisheries Oceanography* 11(5): 286-309.

668. Zolczynski, Stephen J., Jr. and William D. Davis. 1976. Growth characteristics of the northern and Florida subspecies of largemouth bass and their hybrid, and a comparison of catchability between the subspecies. *Transactions of the American Fisheries Society* 105(2): 240-243.

669. Zona, Scott. 1997. The genera of Palmae (Araceae) of the southeastern United States. *Harvard Papers in Botany*, No. 11: 71-107.

670. Zuckerman, Bertram. 1993. *The Kampong, The Fairchild's Tropical Paradise*. National Tropical Botanical Garden and Fairchild Tropical Garden.

Index

A